WRITING WORLDS

The purpose of this book is to explore issues of geographical description from a post-structuralist sensibility. Focusing on landscape representation, the authors organise their discussion around the three themes of discourse, text and metaphor. Each theme is used as a potential entry point into understanding the shape and substance of particular kinds of geographical writings: the discourse of economics, geopolitics and urban planning, travellers' descriptions, propaganda maps, cartography and geometry, poetry and painting.

The book argues that representations of landscape – the city, the countryside or wilderness – are not mimetic, but rather a product of the nature of the discourse in which they are written. Though the landscape representations explored by the authors varies considerably – travellers' accounts of Niagara Falls to Turner's painting of Leeds – each is a written world within a discrete discourse. These essays all participate in the ongoing project of deconstructing geographical discourse to explore the dynamics of power in the representation of landscape.

Trevor J. Barnes is Associate Professor of Geography at the University of British Columbia. His main research interests are in economic geography, and he is the author with Eric Sheppard of *The Capitalist Space Economy* (Unwin Hyman 1990). **James S. Duncan** is Professor of Geography at Syracuse University. He is a cultural geographer and is the co-editor with John Agnew of *The Power of Place* (Unwin Hyman 1989) and author of *The City as Text* (Cambridge University Press 1990).

WRITING WORLDS

Discourse, text and metaphor in the representation of landscape

Edited by
Trevor J. Barnes
and
James S. Duncan

London and New York

First published 1992
by Routledge
11 New Fetter Lane, London EC4P 4EE

Simultaneously published in the USA and Canada
by Routledge
a division of Routledge, Chapman and Hall, Inc.
29 West 35th Street, New York, NY 10001

Typeset in Scantext September by
Leaper & Gard Ltd, Bristol
Printed and bound in Great Britain by
Biddles Ltd, Guildford and King's Lynn

British Library Cataloguing in Publication Data
Writing worlds: Discourse, Text and Metaphor in the
Representation of Landscape.
I. Geography
I. Barnes, Trevor J. II. Duncan, James S.
910

ISBN 0-415-05499-0
ISBN 0-415-06983-1 (pbk)

Library of Congress Cataloging in Publication Data
Writing Worlds: Discourse, Texts, and Metaphors in the Representation
of Landscape/ Edited by Trevor J. Barnes and James S. Duncan
Includes bibliographical references and index
ISBN 0-415-05499-0
1. Geography–Methodology 2. Discourse analysis
3. Deconstruction (Literary analysis) I. Barnes, Trevor J.
II. Duncan, James S.
G70.W75 1992
910′.014–dc20

To Claire and Michael
and to Jimmy

CONTENTS

ILLUSTRATIONS

Plates

ILLUSTRATIONS

Figures

CONTRIBUTORS

Trevor J. Barnes is Associate Professor in Geography at the University of British Columbia

Michael R. Curry is Assistant Professor in Geography at the University of California, Los Angeles

Stephen Daniels is Lecturer in Geography at the University of Nottingham

James S. Duncan is Professor of Geography at Syracuse University

Nancy G. Duncan is an Adjunct Professor of Geography at Syracuse University

J.B. Harley is Professor in Geography at the University of Wisconsin at Milwaukee

Leslie W. Hepple is Senior Lecturer in Geography at the University of Bristol

Judith Kenny is an Assistant Professor in Geography at the University of Wisconsin at Milwaukee

Patrick McGreevy is Assistant Professor in Geography and Earth Sciences at Clarion University at Pennsylvania

Gunnar Olsson is Professor in Planning at Nordplan, Stockholm

John Pickles is Associate Professor in Geography at the University of Kentucky at Lexington

Jonathan Smith is Assistant Professor in Geography at Texas A&M University

Gearóid Ó Tuathail (Gerard Toal) is Lecturer in Geography at the University of Liverpool

PREFACE AND ACKNOWLEDGEMENTS

In a late sixteenth-century reprint of Munster's *Cosmographia* there appeared a rather curious map of Europe in the shape of a queen (plate 1). What are we to make of such a map? We could simply say that the map is wrong, that it is a misrepresentation of the shape of the coastline of Europe. But surely this would be to miss the point. For the purpose of maps, in spite of the rhetoric of many positivistic cartographers, is not mimetic (to mirror in summary form an objective world beyond the map), but to communicate ideas within a cultural and political context. Therefore, the questions that we should ask of such a map are: under what cultural and political conditions did such a representation seem useful, and what rhetorical devices did it use to communicate? Such questions escape the fetishism of mimesis that prevails in cartography, and descriptive geography more generally, by problematizing representation.

It is precisely the problem of representation as it pertains to the geographical world that concerns us in this volume. In particular, the book addresses the issue of writing worlds; that is, examining what it is that we do as geographers when we represent landscapes, real or imagined, through our writings. Our use of the term 'writing', it should be noted, is extremely broad, and cuts across the range of human inscriptions – from writing words to drawing maps. But the one thing that we believe geographical writing is not is a faithful duplication of an external reality. Much more is going on than mechanical reproduction. To describe what geographers are doing when they write is thus the common purpose of the papers that comprise this book. Furthermore, in carrying out that task, all the essays, to varying degrees, draw upon one or more of three concepts: discourse, text and metaphor. These terms are then used to organize the essays within the collection.

To appreciate directly the usefulness of discourse, text and metaphor, we can return to Munster's queen-shaped map of Europe, and use this triad of terms to make sense of this seemingly peculiarly written world. First, it is quite clear that the map is metaphorical and that the coastline of Europe has been drawn in the shape of a queen. But how might we read the metaphor? Is Europe to be seen as a queen who reigns over the other continents, Africa and Asia arranged like courtiers at the margins of the map? Or is she a queen reigning over the fractious individual states of Europe who are like her subjects? Alternatively, is she to be seen as a woman caring for and nourishing the states who are her children? Yet again, perhaps she is Europa, the ancient Greek personification of European culture stolen from the coast of Tyre by Zeus himself? No matter which metaphor we choose, the presumption is that the map is a text that is to be read. Furthermore it presumes intertextuality. For each of these metaphors assumes a prior knowledge of cultural and political texts that speak of the troubled relationship between Europe and Africa and Asia: of the states of Europe in relation to each other and to the emerging 'idea' of Europe itself; of the idea of sovereignty and the reciprocal duties of ruler and ruled; of ideas of femininity and its relation to the family; and of a knowledge of Greek mythology. As the above suggests, this map is not simply metaphorical, nor simply a text (although there is nothing simple about either of these). Rather this map is part of the discursive field of European politics in the late sixteenth century, within which states were increasingly thinking of themselves as European (something they did not do in the Middle Ages). This discourse of Europe was not an amorphous idea that was merely 'in the air'; it was actualized in political practice within different textual spheres: diplomacy, of course, but also plays, and maps such as the one we are considering here. The purpose of the discourse, and of this particular map, was to encourage peace among the states of Europe in the name of the collectivity to which, it was argued, they all belonged. To raise the issue of mimesis is irrelevant to this particular map, for it shifts the discursive field from that of politics to that of positivist science.

Before concluding we would like to acknowledge the discursive field from which this volume emerged. It arose from two special sessions on 'Text and landscape' that were presented in Portland, Oregon, at the 1987 annual meeting of the Association of American Geographers. We would like to thank all the participants in those twin sessions – presenters, discussants and questioners. In addition, Trevor Barnes would like to thank the Department of Geography, Bristol University, for the material support and intellectual companionship that allowed him to complete his portion of the book there during his sabbatical year 1989–90. Finally, two of the essays in the volume were published previously, and we would like to thank the University of Toronto Press for giving us permission to reprint Brian Harley's paper 'Deconstructing the map' and for the journal *Turner Studies* for allowing us to republish Stephen Daniels's paper 'The implications of industry: Turner and Leeds'.

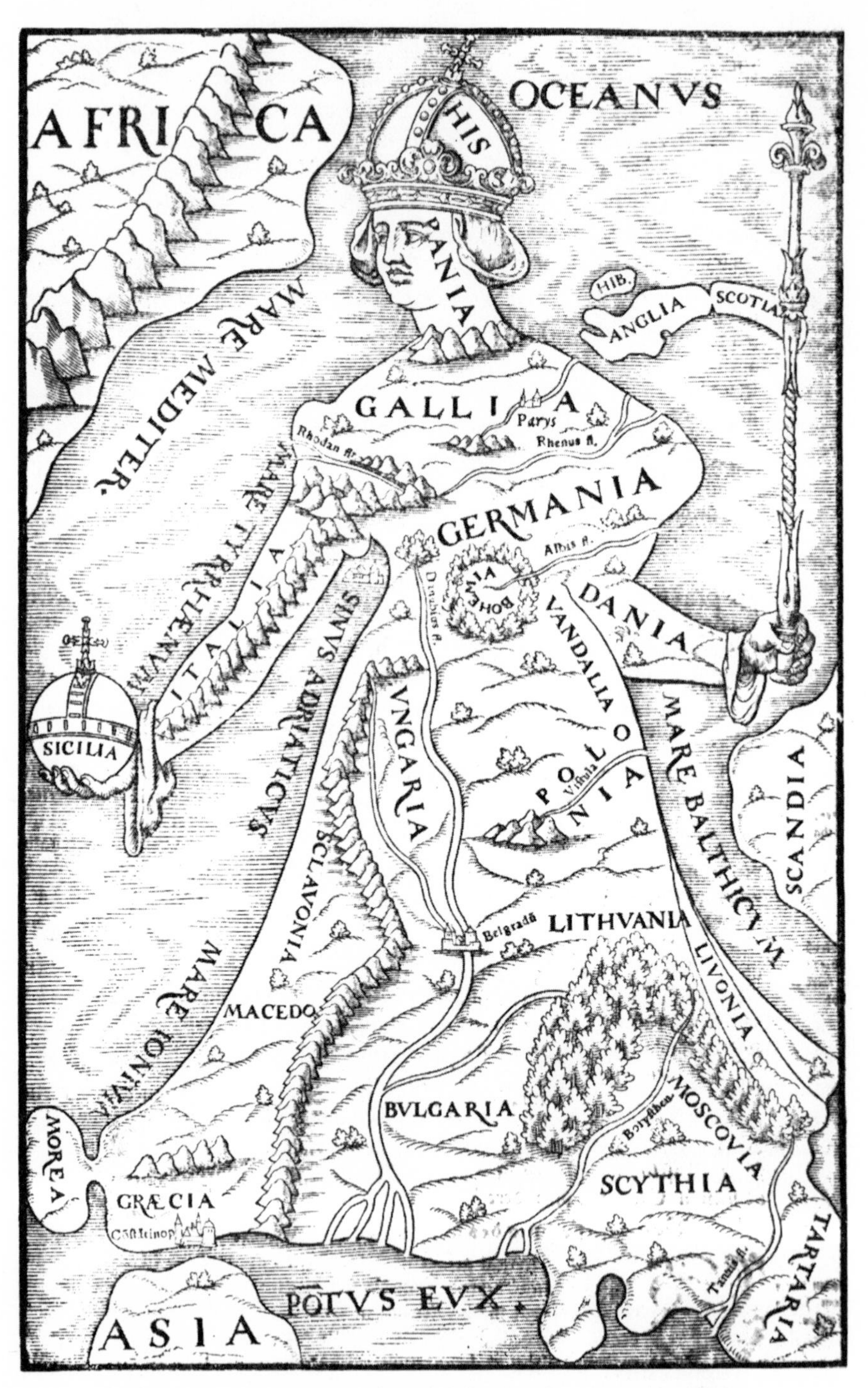

Plate 1 Europe as a queen, from S. Münster, *Cosmographia*, 1588

1

INTRODUCTION

Writing worlds

Trevor J. Barnes and James S. Duncan

'How about writing a composition for me for English? I'll be up the creek if I don't get the goddam thing in by Monday. The reason I ask. How 'bout it?' ...

'What on?' I said.

'Anything. Anything descriptive. A room. Or a house. Or something you once lived in or something – you know. Just as long as it's descriptive as hell.' ... 'Just don't do it too good, is all,' he said. 'That sonuvabitch Hartzell thinks you're a hot shot in English, and he knows you're my roommate. So I mean don't stick all the commas and stuff in the right place.'

'That's something else that gives me a royal pain. I mean if you're good at writing compositions and somebody starts talking about commas. Stradlater was always doing that. He wanted you to think that the only reason he was lousy at writing compositions was because he stuck all the commas in the wrong place.... God, how I hate that stuff.'

(J. D. Salinger, *The Catcher in the Rye*, 1951, pp. 32–3)

INTRODUCTION

Very little attention is paid to writing in human geography. This is ironic, given that the very root meaning of the word 'geography' is literally 'earth writing' (from the Greek *geo*, meaning 'earth', and *graphien*, meaning 'to write'). It is also ironic in another sense, because the one thing that links all geographers of whatever stripe is that they write. Clearly they write about different things, but whatever their speciality they all face the same problem of facing a blank page; that is, the difficulty of convincingly representing to their audience in written form the things that they claim to have done (archival research, literature reviews, fieldwork, abstract theorizing, and so on).

In spite of the fact that human geographers write for a living, until recently the actual process of writing was considered unproblematic. Even now some think that human geography is easier than physical geography because the former only involves writing (sticking the commas in the right place), whereas the latter involves the perplexities of mathematics and the like. Lying behind this flippancy towards writing is a particular view of language use, one that we will call 'naive realism' or objectivism. Following Eagleton (1983: 134), 'in the ideology of [naive] realism ... words are felt to link up with their thoughts or objects in essentially right and incontrovertible ways: the word becomes the only proper way of viewing this object or experiencing this thought'. The result is that the task of writing is the mechanical one of bolting words together in the right order so that the final construction represents the thought or object modelled. In this sense 'earth writing', rather than 'writing about the earth', was a good description of what geographers thought they were doing. Earth came with its own labels, and provided that they were in the correct sequence one's written account was always a mirror representation. Anyone could do it as long as there were a dictionary and style manual at hand.

In the last decade or so this view of writing has been challenged. Rather than unproblematic, writing is now seen as utterly problematic. This should not surprise us. From our own experience we know that writing is very hard; that words often do not come out the way that we mean them. But there is more to this than individuals struggling to find *le mot juste*. In particular, a number of researchers argue that there is now a general 'crisis of representation' in the human sciences (the phrase has been recently popularized by Marcus and Fischer 1986). Pieces of the world, it is suggested, do not come with their own labels, and thus representing 'out there' to an audience must involve more than just lining up pieces of language in the right order. Instead, it is humans that decide how to represent things, and not the things themselves. As Gregory and Walford (1989c: 2) write, 'our texts are not mirrors which we hold up to the world, reflecting its shapes and structures immediately and without distortion. They are, instead, creatures of our own making, though their making is not entirely of our own choosing.' At least three consequences follow.

First, once we sever the supposed one-to-one link between language and brute reality, the notion that writing mirrors the world is untenable. For there is no pre-interpreted reality that writing reflects. We should note that such a view now seems a general one (at least outside human geography). As the anthropologist Clifford Geertz (1988: 137) writes:

> 'telling it like it is' is hardly more adequate a slogan for ethnography than for philosophy since Wittgenstein (or Gadamer), history since Collingwood (or Ricoeur), literature since Auerbach (or Barthes), painting since Gombrich (or Goodman), politics since Foucault (or Skinner), or physics since Kuhn (or Hesse).

But if our writing does not reflect some bedrock reality, then what does it reflect? The different scholars cited above each has his/her own particular answer, but most agree that it must involve yet prior interpretations. That is, our texts draw upon other texts, that themselves are based on yet different texts, and so on. In the vocabulary of literary theory there is only intertextuality, defined as 'the process whereby meaning is produced from text to text rather than, as it were, between text and world' (Rylance 1987: 113). The consequence is that writing is constitutive, not simply reflective; new worlds are made out of old texts, and old worlds are the basis of new texts. In this world of one text careening off another, we cannot appeal to any epistemological bedrocks in privileging one text over another. For what is true is made inside texts, not outside them. As the anthropologist James Clifford (1986: 22) writes:

> A conceptual shift, 'tectonic' in its implications, has taken place. We ground things, now, on a moving earth. There is no longer any place of overview (mountaintop) from which to map human ways of life, no Archimedian point from which to represent the world.

Second, writing about worlds reveals as much about ourselves as it does about the worlds represented. Given the point made by Clifford, that when we write we do so from a necessarily local setting (there is no mountain top), the worlds we represent are inevitably stamped with our own particular set of local interests, views, standards, and so on. To understand critically our own representations, and also those of others, we must therefore know the kinds of factors bearing upon an author that makes an account come out the way it does. Once again, Clifford (1986: 6) is useful in providing a partial checklist: the social context in which the piece was written, the institutional setting (audience, intellectual tradition, school of thought), the genre of which it is a part (textbook, scholarly article, newspaper piece), the political position that sustains the authority of the author (colonial administrator, Third World academic, Western journalist) and, finally, the historical context that makes all the above factors contingent on particular times and places. We can readily appreciate Clifford's concern, as an anthropologist focusing on the cultural 'other', with detailing the various 'local' interests that are implicated in the construction of ethnographic representations (see, for example, his discussion of the factors influencing Malinowski's ethnography; Clifford 1988). But other, 'harder' disciplines, such as economics or even physics, are no less immune. For example, Mirowski (1988) deftly unpacks the influences (biographical, political, sociological, institutional) that led to the acceptance and success of a dominant physical metaphor in neoclassical economics. While Bloor (1982) argues that primarily contemporary political and religious factors were behind the constructions of, and controversies over, early theories of light in physics. The broader point is that when we 'tell it like it is' we are also 'telling it like we are'.

Finally, in writing about worlds, we must pay attention to our rhetoric, as well as the rhetoric of others. Under the rubric of objectivism, rhetorical devices such as metaphors, irony, similes and the like are useless, if not nonsensical, aspects of language; at best they obfuscate the truth, and at worst they pervert it. At every opportunity they should be extirpated. With the recent assault on objectivism, there is growing recognition that, far from being merely decorative, rhetorical devices are central to conveying meaning. They are the means by which we persuade our audience that we really did the things that we say we did. In this light, objectivism itself is just one of many different strategies to convince; like Hemingway's no style, it is a style in itself. Moreover, as Hayden White (1978) argues, it is impossible to get away from such writing strategies or tropes, because they constitute the very object of study. As White (1978: 2) writes, a trope 'is the shadow from which all realistic discourse tries to free itself. This flight, however, is futile; for tropics is the process by which all discourse constitutes the object which it pretends only to describe realistically and to analyze objectively.' Recognition of the centrality of rhetoric is now increasingly found in a number of disciplines: intellectual history (White 1978; LaCapra 1983), anthropology (Clifford and Marcus 1986; Clifford 1988; Geertz 1988), economics (McCloskey, 1985) and even mathematics (Kline 1980). In human geography there has been some discussion by Gregory (1989a). In particular, Gregory criticizes the traditional trope of narrative found in historical geography because of the sense of order and closure that it imparts. Instead, he searches for other writing tropes that celebrate the complexity and openness of 'day-to-day lives of particular people in particular places' (Gregory 1989a: 89). That said, there are very few examples of human geographers exploring alternative tropes.

The purpose of this book is to show, by focusing on landscape broadly conceived, the importance of these arguments about writing and representation for human geography. Clearly landscape is an ambiguous term, and we will not provide any definitive definition of it here. Following Daniels and Cosgrove (1988: 1), landscapes:

> may be represented in a variety of materials and on many surfaces – in paint on canvas, in writing on paper, in earth, stone, water, and vegetation on the ground. A landscape park is more palpable but no more real, nor less imaginary, than a landscape painting or poem.

In our collection, landscape is represented in a diverse set of forms, including theoretical models of the space economy, propaganda maps, travellers' accounts of Niagara Falls, Turner's pictures of northern English cities, and more besides. Although the means and purposes are different, linking these different subjects is the attempt at geographical representation: to represent respectively the economic world of profit and loss, the political world of ideology, the physical world of natural wonder, and the visual world of industrialization.

In each of these four examples the three points made about representation above are very relevant. First, the representations are not a mirror copy of some external reality. In some cases this is plainly so, as with propaganda maps, but it is also true of travellers' accounts of Niagara Falls. For, as Patrick McGreevy's chapter confirms, such accounts are based on previous representations; that is, travellers' texts are about other texts and not some pristine falls itself. Second, to understand each of the representations fully we must know something about the context of its authors and audience. Again this even includes such unlikely cases as mathematical modelling in economic geography. As Barnes argues, to comprehend the success of the abstract rational landscapes portrayed by neoclassical economic geographers it is essential to know that it is underpinned by a physical metaphor borrowed from nineteenth-century physics. The very prestige of science in which the metaphor originated endowed authority on its authors and ensured a favourable reception. Finally, in each of these representations examined one needs to explore the tropes employed – that is, the styles used to persuade the reader. The graphical devices used in the drawing of propaganda maps are perhaps the most obvious case, but they also occur in mathematical models of economic geography (here metaphor is dominant – the mathematical equation convinces because we are ready to accept that it is like some aspect of human behaviour over space).

Now clearly we are using 'writing' in a very broad sense to include such activities as drawing maps, making plans, and even painting. Our justification is that each of these activities, as investigated here at least, involves representing the landscape. With the task of representation the central one (whether accomplished through the medium of words, drawings or paintings), the three general problems discussed above necessarily pertain.

The next section of this introductory essay carries forward the arguments made here by suggesting that three key concepts that emerge from the new literature on writing and representation are text, discourse and metaphor. Specifically, each of the essays in the book draws upon at least one of the three concepts, often more than one, in discussing some substantive case of landscape writing and representation. The final section of the introduction provides a synopsis of each of the essays, showing how each author draws upon one or more of the triad – text, discourse, metaphor – in presenting his or her argument.

TEXTS, DISCOURSES AND METAPHORS

Texts

The notion of text used in this volume is not the traditional one of a printed page or a volume sitting on a shelf in the library. Rather, following Roland Barthes and other contemporary literary theorists and cultural anthropologists, we use an expanded concept of the text: one that includes other cultural productions such as paintings, maps and landscapes, as well as social, economic and political institutions. These should all be seen as signifying practices that are read, not passively, but, as it were, rewritten as they are read. This expanded notion of texts originates from a broadly post-modern view, one that sees them as constitutive of reality rather than mimicking it – in other words, as cultural practices of signification rather than as referential duplications. For, just as written texts are not simply mirrors of a reality outside themselves, so cultural productions, such as landscapes, are not 'about' something more real than themselves. But although not referential, such practices of signification are intertextual in that they embody other cultural texts and, as a consequence, are communicative and productive of meaning. Such meaning, however, is by no means fixed; rather, it is culturally and historically, and sometimes even individually and momentarily, variable.

It is, however, one thing to argue that the world is like a text, and another to demonstrate it convincingly. Ricoeur's 'The model of the text' (1971) attempts to do just that, though. Ricoeur poses two questions: first, is the model of the text a good paradigm of social science? And, second, is the method of textual interpretation relevant? He argues that for four reasons both questions should be answered positively.

First, he suggests that the principal characteristics of written discourse also describe social life more generally. In particular, meaning in written discourse is concretized when it is inscribed or textualized. Similarly, in social life, institutional objectification, recurrent patterns of behaviour and monuments in the built environment take on the same type of fixity.

Second, within written works an author's intentions and the meaning of the text often cease to coincide; in other words, the text escapes its author (Barthes 1987g). In the same way, institutionalized patterns of action are frequently detached from their collective agents. Deeds have consequences that are often unintended, thereby making it impossible to identify the authors of complex events.

Third, written texts frequently have an importance beyond the initial context in which they were composed. They are interpreted and reinterpreted in the light of changing circumstances. Similarly, social events and institutions are subject to continual reinterpretation.

Finally, the meaning of a text is unstable, dependent upon the wide range of interpretations brought to bear upon it by various different readers. Similarly, social productions and institutions also address a wide range of possible interpreters. But these interpreters are not free to make of the text what they like, but are subject to discursive practices of specific textual communities (Stock 1990). As we will argue below, this implied notion of subjugation need not be as deterministic as in some interpretations of Foucault's work on discursive practices. Rather, we contend simply that to speak, write or read one must do so within the conceptual framework of specific discourses.

The social-life-as-text metaphor is easily applicable to landscape because it too is a social and cultural production. Thus a landscape possesses a similar objective fixity to that of a written text. It also becomes detached from the intentions of its original authors, and in terms of social and psychological impact and material consequences the various readings of landscapes matter more than any authorial intentions. In addition, the landscape has an importance beyond the initial situation for which it was constructed, addressing a potentially wide range of readers. In short, landscapes are characterized by all those features that Ricoeur identifies as definitive of a text.

Reflecting on the text metaphor, the intellectual historian Dominick LaCapra (1983: 19) suggests that, although it involves a certain 'linguistic inflation', its value is in allowing us to understand the problems involved in taking '"reality" or "context" as unproblematic ground or a gold standard'. Following his reasoning, we suggest that 'text' is also an appropriate trope to use in analysing landscapes because it conveys the inherent instability of meaning, fragmentation or absence of integrity, lack of authorial control, polyvocality and irresolvable social contradictions that often characterize them; characteristics that are demonstrated in the papers that follow.

It is probably in anthropology that we find the most recent use of the text metaphor. In particular, Clifford Geertz (1973: 1988) has been a tireless and enthusiastic proponent of culture as text. For him, culture is something that is 'read' by an ethnographer as one might read written material. Furthermore, he argues that this is not simply an academic pursuit, but one that everyone practises.

In a similar vein, Marcus and Fischer (1986: 26) argue that the principal benefit of the textual analogy for ethnographers is that it focuses their attention on the relationship between the ethnographer's interpretation and that of his/her informants. It makes the researcher reflect critically on the practice of ethnography that for too long, they argue, masqueraded under positivistic social science as an unproblematic description of reality, narrated by an author whose presence is masked by the rhetoric of absence.

More generally, the (postmodern) view of text championed by such anthropologists problematizes the very notion of 'representation' that has become so entrenched in ethnography and, we should add, some foreign area studies in geography. The question of representation that is raised is both epistemological and moral. On the one hand, it poses the question of the translatability of cultural difference – how do we know that our point of view is also that of the native whom we seek to represent? On the other hand, it also raises the ethical question of the morality of speaking for others. This is especially pertinent, as Clifford (1988) makes clear, in anthropology. Furthermore, the concept of text that is invoked in such work also has a critical function by encouraging the interrogation of textual strategies, especially those that presume objectivity.

Associated with the use of the concept 'text' is intertextuality, that is, the textual context of a literary work (Eagleton 1983). Originating in literary criticism, the term has sparked increasing interest among historians (LaCapra 1983; Stock 1990). Likewise, anthropologists (Clifford and Marcus 1986; Tyler 1987b; Geertz 1973) have explored the notion in order to problematize ethnographic descriptions. Tyler (1987b), for example, argues that although most ethnographic accounts are portrayed as objectivist description based on field research, they are better described as intertextual works, highly mediated by a traditional corpus of anthropological monographs and theories.

Furthermore, it is not simply our accounts of the world that are intertextual; the world itself is intertextual. Places are intertextual sites because various texts and discursive practices based on previous texts are deeply inscribed in their landscapes and institutions. We construct both the world and our actions towards it from texts that speak of who we are or wish to be. Such 'texts in the world' then recursively act back on the previous texts that shaped them. This perspective is explicit in the work of some geographers (Cosgrove and Daniels 1988; Duncan 1990; Duncan and Duncan 1988) and in a number of the essays in this volume.

Discourses

Texts, in the broad sense that we construe them, are constitutive of larger, even more open-ended, structures termed discourses. The latter are frameworks that embrace particular combinations of narratives, concepts, ideologies and signifying practices, each relevant to a particular realm of social action. Between discourses words may have different connotations, causing people who ostensibly speak the same language to talk past one another, often without realizing it. This is because words or other signifiers within discourses have no natural connection with their signifieds (concepts) or their referents. Rather, the relation is socially constructed and therefore variable.

Under this view, discourses are practices of signification, thereby providing a framework for understanding the world. As such, discourses are both enabling as well as constraining: they determine answers to questions, as well as the questions that can be asked. More generally, a discourse constitutes the limits within which ideas and practices are considered to be natural; that is, they set the bounds on what questions are considered relevant or even intelligible. These limits are by no means fixed, however. This is because discourses are not unified, but are subject to negotiation, challenge and transformation. For power relations within a social formation are communicated, and sometimes resisted, precisely through the medium of particular discourses.

In addition, the production and reproduction of discourses are also linked to institutions. Within this context, discourses shape the positioning of individuals in an institution, and the discourses so adopted, in turn, depend upon an individual's position there. Thus discourses constitute standpoints that are defined largely by their relationship to other discourses.

The concept of discourse is a key term in poststructuralism and in postmodernism more generally. It represents a clear break with earlier ahistorical categories of humanism and structuralism such as human nature, timeless meaning or universal rationality. Although structuralism successfully decentres the individual and, in this sense, is clearly a break with humanism (modernism), it is not fully a postmodern project in that it posits transhistorical structures underlying discourses (best seen in Lévi-Strauss's work). Poststructural discourse theory, however, sees discourses as conventional and historical. It assumes that discourses, and the 'truths' that they construct, vary among cultural groups and among classes, races, gender-based or other groups whose interests may clash. In the most anarchic versions even the existence of different interest groups within society is questioned, along with the very notion of culture (dismissed as a liberal humanist 'totalizing' or 'essentializing' concept; Cottom 1989: 50).

Although competing discourses may evolve among opposing interest groups, there may be a relatively stable discursive formation in which these competing discourses coexist. All the politically engaged classes or other interest groups in a society may support, albeit not uncritically, the hegemonic discourses. Alternatively, there may be open clashes between groups whose presuppositions are based in antagonistic discourses. Roland Barthes (1987d: 200) says that 'discourse (discursivity) moves in its historical impetus by clashes. A new discourse can only emerge as the paradox which goes against ... the surrounding or preceding doxa.'

Much of the work in discourse theory derives from Foucault's studies of the relations between knowledge, discourses, representations and power. Because of its appeal to 'common sense' or its scientific status, knowledge in the form of representations is in itself a power rather than simply a reflection of power relations in the 'real' world 'beyond' the academy, the media or government task force. In this sense, discourses have a naturalizing power which is largely unseen. In his genealogies of discourses Foucault (1967; 1978) studies the construction of discourses and their institutionalization. He argues that it is their association with institutions that legitimates the 'truths' that they produce. The power of discourses derives not so much from the abstract ideas they represent as from their material basis in the institutions and practices that make up the micro-political realm which Foucault sees as the source of much of the power in a society.

Within postmodern ethnography (Clifford 1988; Clifford and Marcus 1986; Marcus and Fischer 1986) there is a related, but somewhat different, notion of discourse that also calls attention to the crisis of representation in social science. Discourse in this literature often refers to a dialogue between researchers and those they study. This dialogue replaces the monologue of an author, a self-authorized 'authority' who represents 'others'. It calls for polyvocality in ethnographic texts and for rhetorical devices that call attention to the artifice in writing ethnographic 'description'. It also calls for the deconstruction of the internal contradictions within dominant discourses about cultural 'opposites'. This view of discourse adopts the same scepticism about unmediated access to reality as other postmodern or poststructural discourse theory, and shares in celebrating difference and opposition.

Metaphors

One form in which discourses can be presented, shaped and can gain authority is as metaphors. Debate and controversy over metaphors, however, has persisted since the ancient Greeks. On the one hand, many have viewed them as, at best, frivolous and ornamental or, at worst, obfuscatory and logically perverted. For example, Hobbes (1962: 34) wrote in *Leviathan* that 'when we use words metaphorically; that is in other senses than that they were ordained for; ... [we] thereby deceive others'. This same sentiment was taken up by the British empiricists and the Viennese positivists in the late nineteenth and early twentieth centuries, finally reaching human geography in the 1960s during the heyday of the 'quantitative revolution'. Then David Harvey (1967: 551) wrote that 'the form in which ... metaphors are cast seems to hinder objective judgment'.

On the other hand, there has always been a strong undercurrent celebrating the centrality of metaphor. It is found in Vico's work, and Nietzsche's, but in the twentieth century it was first eloquently and forcefully stated by I. A. Richards in his *Philosophy of Rhetoric* (1936). There he wrote that, rather than being 'a sort of happy extra trick with words' (p. 90), metaphor is 'the omnipresent principle of language' (p. 92). In the last fifteen years, interest in metaphor has strengthened, and its role has been explored in disciplines as diverse as physics (Hesse 1980a), anthropology (Gudeman 1986), economics (McCloskey 1985; Mirowski 1988) and human geography (Buttimer 1982).

This increased interest in metaphor partly stems from the critique mounted on objectivism, and discussed in the first section. For objectivism was almost exclusively concerned with validation, that is, developing procedures to check the correspondence between theory and the real world. However, it became increasingly clear in science, at least, that one could not express the real world in its own terms, but only in theoretical ones. Thus it made no sense to check theory against some neutral outside world, because the outside world itself was apprehended only through theory. Interest, therefore, switched from concern about validation to concern about how theories themselves were formulated and developed. Science was then seen 'less as an inert body of positive knowledge and more as an on-going activity' (Cameron 1983: 263). To understand the development and formulation of theory, a number of people turned to metaphor.

Although there is much controversy over the meaning of metaphor, most agree that it asserts a similarity between two or more different things, for example: culture is a text. Here one thing, culture, is 'metaphorically redescribed' (the phrase is Hesse's 1980b) in terms of a frame of reference with which it is not usually associated, a written text. In general, that clash of different frames of reference can produce all manner of effects: incredulity, a smirk or, after sufficient time, a Nobel Prize-winning novel or theory of physics. Of course, in most cases the metaphor is simply forgotten, but in a few the bringing together of two hitherto unrelated things is the creative spark for something much more. As Buttimer (1982: 90) writes, 'metaphor ... touches a deep level of understanding ..., for it points to the process of learning and discovery – to those analogical leaps from the familiar to the unfamiliar which rally imagination and emotion as well as intellect'. In this sense, metaphor provides a bridge for understanding the development and formulation of theory.

There has been considerable discussion as to how metaphors produce the creative spark. The most common approach is called the interaction view, first proposed by Max Black (1962). Here the idea is that the meaning of words is not fixed, but includes all kinds of associations and kindred concepts. This slackness in meaning enables ideas associated with one thing to be transferred easily to another thing. For example, because the meaning of culture is not definitively set, but includes a penumbra of ideas and concepts, it can be seen to share features with something very different, written texts. This view of metaphor has recently been challenged by Davidson (1979) and Rorty (1989), who claim that metaphors have no meaning other than their literal one, which is nonsensical. Davidson and Rorty are not dismissing metaphor, however. They just do not think one needs to talk about 'transference of meaning', which makes metaphor seem slightly mystical. Rather, precisely because metaphors are literally nonsensical, they are the jolt, the *frisson*, that makes us see the world in a different way; a way that could not be imagined before the metaphor was used (Rorty 1989: 16).

However metaphors create new angles on the world, once they are 'savored rather than spat out' (Rorty 1989: 18) they gradually acquire a habitual use. At that point they are dead metaphors. The significance for the 'on-going activity of science' is that dead metaphors become equivalent to the literal, and are then the basis on which new metaphors are coined. To put it in Kuhn's terms, when we use dead metaphors we are engaged in 'normal' science, not the revolutionary one that comes about only through producing new metaphors. That said, we should always remember that, although dead metaphors are equivalent to the literal, they are not equivalent to some outside real world. They are socially and culturally constructed entities that emerged in an earlier process of metaphorical redescription. As such, metaphors relate to other metaphors and, as in the case of texts, not to some pre-linguistic brute reality.

Turning to our immediate concern of writing, it is helpful when considering metaphor to make a distinction between 'big' and 'small' metaphor use. Big metaphors are those that lie behind general research methods, and schools of thought, while small metaphors are those that pepper individual pieces of writing. Now, clearly, there is some relationship between the two, although there may well be cases where they are inconsistent. In either case, though, the purpose of using metaphor is rhetorical, to persuade the reader that the writer's view is correct. In fact Kenneth Burke (1950), the influential American literary critic, makes metaphor one of his four master tropes, along with irony, synecdoche and metonymy.[1] That trope works precisely by appealing to our desire to reduce the unfamiliar to the familiar; in other words, metaphors persuade by saying that things that we thought were outside our ken (and thereby disconcerting) are really a lot like other things that we know very well.

White (1978) provides examples of the rhetorical power of both big and small metaphors. With regard to small metaphor use, he argues that the persuasiveness of A. J. P. Taylor's book *The Course of German History* is a consequence of its frequent appeal to dead metaphors that, in turn, evoke in the reader's mind an assurance of objectivity. As White (1978: 114) writes: 'it is seldom noted how the effect of "objectivity" can be attained by the use of nonpoetic language, that is to say, by language in which dead metaphors rather than vivid ones provide the substance of the discourse'. With regard to big metaphor use, White points to the very method of historical narrative itself. A narrative's power, he argues, rests on its ability to evoke the familiar, and thereby to persuade:

> properly understood, histories ought never to be read as unambiguous signs of the events they report, but rather as symbolic structures, extended metaphors, that 'liken' the events reported in them to some form with which we have already become familiar in our literary culture.
>
> (White 1978: 91)

From White's comments it should be clear that metaphors are implicated in the very fabric of society and social processes; if they are to work they must resonate against an existing set of social and cultural representations. But in the resonance there is often conflict, intellectual and sometimes physical. Accepting new metaphors and their power to do new things frequently entails jettisoning old metaphors: Blakes's old metaphors got in the way of Yeats's new ones, just as David Harvey's old physical metaphors for achieving spatial prediction got in the way of his new ones for achieving urban social justice. As Arib and Hesse (1986: 156) write, 'metaphor is potentially revolutionary'. To establish whether the revolution succeeds, and for how long it succeeds, entails examining both internal issues (e.g. the metaphor's logical consistency), and external ones (e.g. social relationships of power and vested interest). More generally, one should not see metaphor as right or wrong, or static in any sense. Metaphors are tasteful or tasteless (Davidson 1979), appropriate or inappropriate (Arib and Hesse 1986), useful or a hindrance (Rorty 1989). Furthermore, what is tasteful or appropriate or useful emerges only out of the broader social context in which such metaphors are embedded, and that context is dynamic. As a result, the nature of society, including the academy, is measured in part by the kind of 'metaphors it induces or allows, and the [kind] ... of judges of metaphor that it educates or rewards' (Booth 1979: 62). By our metaphors you shall know us.

In summary, the notions of text, discourse and metaphor have emerged as powerful concepts over the last twenty years. Originally defined in terms of literary criticism, they now have a much wider importance, as we have tried to show here. We conclude this chapter with a brief precis of each of the following twelve essays, seeing their discussion of texts, discourses or metaphors as pivotal concepts in their broader claims about the writing and representation of landscape.

DISCUSSION OF THE PAPERS

In spite of all their differences, there are a number of commonalities in the twelve papers that constitute the volume. The most basic is that landscapes, social action, paintings, maps, language and, of course, written documents are all held to be susceptible to textual interpretation. Although the authors of these chapters have differences of opinion over the nature of that textual interpretation, they are in broad agreement that mimetic representation is a pipe dream that should be abandoned. Another related area of broad agreement is that the objects of enquiry, whether they be landscapes, maps or government documents, must be approached intertextually; showing the way that texts from other conceptual realms cross-cut, transform and, in turn, are transformed by the texts in question. The authors, however, differ in their opinions about how unstable this renders the world. Finally, for all the authors, the concept of power is central, not only to the constitution of the objects of study, but to their representation. Particular attention is devoted to the latter, to how power is inscribed in subjects and objects, and how power must be opened up to critique, no matter where it is found. Note that this unwillingness to privilege and exempt from critique any locus of power is one hallmark of the postmodern attitude, although a number of the essayists would be wary of that appellation.

The first essay, by James and Nancy Duncan, investigates the landscape interpretation of the semiotician and literary theorist Roland Barthes, who was one of the founders of both structuralism and poststructuralism. The focus of the Duncans' concern is primarily theoretical and methodological; that is, they are concerned less with what Barthes thought about landscapes than with how he thought about them. Choosing five of Barthes's books and articles written between the mid-1950s and his death in 1980, they trace his intellectual journey through these works from a structural, semiotic type of landscape interpretation to a poststructural approach. The implications of this shift are simultaneously traced within the realm of political discourse as Barthes moves from his early demystification of the landscapes of the bourgeoisie towards a postmodern critique of power itself, wherever it takes root in the landscape, and the realization that the search for any 'true' text underlying landscape history is a modernist fantasy.

Following is Stephen Daniels's essay, which focuses upon a very different text, a water-colour of Leeds painted by Turner in the early nineteenth century. Daniels reveals the textual quality of this painting by performing a virtuoso reading of it, showing how it 'speaks' both of facets of Leeds's industrialization as well as of the intellectual and political space occupied by the artist. Daniels accomplishes this reading by proceeding intertextually, tacking back and forth between the painting of Leeds and a series of other paintings by Turner and other artists, poetry, prose, city histories of Leeds, biography and contemporary political documents. In this manner Daniels illuminates not only the painting of Leeds but also Turner and his time. While the author does not dwell upon the methodological implications of his work, this chapter serves as a fine illustration of the intertextual approach to interpretation.

Rather than interpreting the human-made landscape of industrialization, Patrick McGreevy turns his attention to a naturally forming one, Niagara Falls. But even the interpretation of nature must be through a textual account. Moreover, such texts vary according to the concerns of both authors and audiences. In particular, using written accounts of Niagara Falls over the last hundred years – letters, newspaper clippings, diaries, poems and novels – McGreevy neatly brings together issues of text and metaphor. For he argues that such written texts have been informed by the metaphor of death; visitors do not see a pristine falls but one, as it were, that was already written upon by the death metaphor. Within this general interpretation, McGreevy undertakes two tasks. First, to show that the metaphor of death has crept out of the written texts and found its way into the landscape that humans have created around Niagara. This refers not only to the inanimate human-made landscape, but also to the actions of humans themselves (the suicides are the most obvious example). Second, to show that the metaphor of death informing the falls is not some universal sentiment, but arose in a particular social and cultural context, the nineteenth century and its associated Victorian values. Furthermore, having emerged within that context, the metaphor was subsequently used by contemporary writers to construct an internal geography of the falls itself.

Jonathan Smith's essay that follows is also concerned with linking metaphors and texts, in this case a 'big' metaphor that is adopted by many of the authors in this volume, landscape as text. In particular, he is concerned with unpacking the metaphor to understand its wider implications, including those of power relations. Focusing on the issue of scenic beauty, Smith argues that the metaphor of landscape as text implies committed readers. In the modern world, however, the idea of sustained reading is disparaged, among non-academics at least. In contemporary society the landscape is now read in terms of phrases from glossy brochures and the flickering images of television sets. For this reason many individuals are now barred from the kind of reading that is necessary to appreciate the quality of the physical environment. As a consequence, we cannot expect landscape beauty to serve as text that will elicit serious commentaries from those outside the academy. As a result, the very meaning of the metaphor landscape as text implies a certain elitism by those who use it.

Gunnar Olsson is also concerned with the implicit power relationships that subsist in metaphor use, but in this case they are the metaphor of the sign. In particular, drawing upon the work of Derrida, Foucault, Girard and Lacan, Olsson deconstructs the word 'is', which he argues is a key concept in the vocabulary of power. He organizes his argument around the deconstruction of three signs (sets of lines) showing how each presupposes a different mode of thought and power relations. The first is the equals sign (=), which belongs to logic and draws its power from metonymy. It promises something that it can never deliver; an equality between two things that are not the same. The second is the slash (/). It belongs to dialectics and receives its power from metaphor. Finally, the third is the Saussurean bar (—) which splits and unites the two functions of the sign. All three are used in attempts to control meaning in different ways. All are lines of power.

Michael Curry shares many of Olsson's concerns about power but he couches them in historical and sociological terms. In particular, he examines the development and implications of two features of twentieth-century academic discourse in human geography: first, an architectonic impulse that desires to create an ordered, hierarchical system; and, second, a reconceptualization of the concrete away from simple description, towards the development of a systematic ontology. Curry exemplifies these two features by focusing on the changing meanings of just one idea, 'forms of life', which has been an important concept within at least four different areas of geography over the last century: within the French school, beginning with Vidal de la Blache; within urban sociology, beginning with Louis Wirth; within some areas of quantitative geography, in particular Q analysis; and, finally, within the emerging set of writings on a post-modern geography. In all cases, Curry finds the move towards both an architectonic approach and a systematic ontology regrettable; it is a move, he argues, that says much about the power and authority of scientism within our society, and the way in which such power and authority infect all levels of the academy. Rejecting any version of 'forms of life' that is tainted by scientism, he returns to Wittgenstein's original use of the term. For Wittgenstein's view, argues Curry, is non-architectonic, and celebrates the diverse concrete experiences of the everyday without reducing them to an ontological theory. Furthermore, under a Wittgensteinian interpretation, 'forms of life' is an inherently reflexive notion; it behoves human geographers who use the term to apply it to themselves, thereby enabling them to understand and criticize the social relations, including those of the power of scientism, in which they are enmeshed.

Trevor Barnes's paper that follows is also concerned with the power of science, but in only one sub-area of the discipline, economic geography. Combining notions of text and metaphor, Barnes argues that the economic landscapes constructed by neoclassical and Marxist economic geographers are underlain by respectively a physical and a biological metaphor. As a consequence, such landscapes do not represent some given external reality, but reflect only the internal logic of the metaphor employed. Barnes then asks whether the internal logic of these two metaphors is appropriate to the respective tasks set them. In particular, drawing upon the work of Mary Hesse, he outlines three criteria of appropriate metaphor use. Applying such criteria in turn, Barnes concludes that both the physical and the biological metaphors used in economic geography are inappropriate. In general, he suggests that both types of metaphor suffer from their narrowness; that they are unable to cope with the diversity of space and place, and therefore either ignore it or reduce it to something less than it is.

The paper by Leslie Hepple is also concerned with metaphor use, in this case their political consequences on the ground. He argues that much recent geopolitical discourse in the southern cone countries of Latin America is informed by an organistic metaphor of the state; one that has been appropriated by the military and used to justify their own vested power interests. Initially the organistic metaphor was used to justify external disputes over national territory with neighbouring countries. More recently, however, it has been employed to direct and legitimate internal military intervention. For once groups within a country opposed to the military interest were labelled cancerous, and thereby harmful to the organism of the state, justification existed for eradication. The physical brutality of the internal security forces was directly predicated upon the language of metaphor.

Gearóid Ó Tuathail, in the chapter that follows, is also concerned with the geopolitical. In particular, he explores the manner in which South Africa is represented in foreign policy discourse in the United States. Using a broad range of texts by government officials and foreign policy 'experts', he focuses upon what he terms the two major 'scripts' that constitute the representation of South Africa: the first of a morally repugnant place and the second of a strategically important place. Drawing upon the Baudrillard's notion of hyperreality, defined as the simulations of the real which ultimately become more real than the real itself, Tuathail launches a postmodern critique of American foreign policy. He argues that the two central strategic claims of loss of resources and communist threat are both hyperreal. He then draws links between a racist South Africa and racist America, suggesting that this affinity in part accounts for why Americans script a morally repugnant South Africa as a 'tragedy'. The author uses the notion of script and hyperreality not to undermine the existence of 'the real'; after all, the notion of hyperreality presupposes and depends upon a notion of the real. Rather he uses these concepts to reveal the irony and paradox in 'official conceptions of the real'.

Judith Kenny also chooses to interpret a politically charged text, albeit at the municipal level. She studies the Portland, Oregon, Comprehensive Plan, proceeding intertextually by drawing the interconnections between the plan, a broader set of political writings and the specific texts swirling around a particularly controversial land use case in the city. She organizes her interpretation of the relationship between the plan and the land use case around the concept of a discursive field, the range of competing discourses relevant to a particular realm of social practice. She skilfully demonstrates how competing discourses within an American liberal ideology and American planning practice allow groups with different perceived interests (merchant groups, neighbourhood groups, city planners, city politicians) to produce different readings of the plan. As such she argues for a politics of reading that ultimately becomes translated into the material fabric of the city.

John Pickles examines yet a different type of politically embedded text, the propaganda map. More broadly, he conceives of all maps as a form of discourse, which calls into question any correspondence theory of map representation. A more satisfactory way to approach maps, he argues, is through hermeneutics. Drawing upon the methodological insights of Ricoeur and others, he adopts an intertextual approach to maps in general and propaganda maps in particular. Such an intertextual reading of maps explores both the texts in which they are embedded and the contexts into which they are projected. The author's choice of propaganda maps is particularly apposite for questioning correspondence theory, because it challenges the alleged difference between the 'good' and the 'bad' map, and between the 'true, scientific' map and the 'false, political' map. Pickles convincingly argues that cartography faces a crisis of representation, and shows one way in which the cartographic discourse might be reconstructed.

Enlarging upon Pickles's argument, Brian Harley argues that geography needs an ethnography of modern cartography. He urges geographers to stop seeing maps as part of an objective scientific system and begin to see them as part of a cultural system. This suggests that the link between cartographic representation and reality can be better comprehended through social theory than through positivism. To this end the author using Derridean deconstruction theory and Foucault's notion of discourse attempts to sketch the outline of such a theoretical approach.Arguing that maps are texts which are susceptible to deconstruction, he reveals not only the rhetorical devices used in cartographic communication, but also the inherent intertextuality of maps. He claims that cartography as at present constituted normalizes and disciplines the world and as such is a part of the power/knowledge matrix. Bringing together the Derridean and Foucauldian strands of his argument, he contends that cartography's position within this matrix is made more powerful because of its 'sly rhetoric of neutrality'.

Finally, the Afterword is less a conclusion than a critical reflection on the volume. Believing that some readers will see this collection as a contribution to a postmodern human geography, we reflect critically on the use of that latter term. Ironically our subsequent doubts about postmodernism are themselves born out of a postmodern attitude: one where self-reflection and self-criticism are paramount. Like Peer Gynt, the Great Boyg blocks our way; we can only 'Go round about, go round about.'

NOTE

1 White (1978) argues that, because synecdoche and metonymy are themselves forms of metaphor, Burke in effect provides only two master tropes.

2

IDEOLOGY AND BLISS

Roland Barthes and the secret histories of landscape

James S. Duncan and Nancy G. Duncan

INTRODUCTION

Roland Barthes was a remarkable essayist; as a philosopher and semiologist he was a keen observer of society and an incisive critic of its cultural texts. Texts, for Barthes, are not simply the written word; they are composed of signs found in all cultural productions. 'The world,' he wrote in 1964, 'is full of signs, but these signs do not all have the fine simplicity of the letters of the alphabet, of road signs, or of military uniforms: they are infinitely more complex.' In his early work as a semiologist he sought to decipher society's signs and to reveal the complexity and instability beneath the apparent simplicity of the everyday cultural landscape.[1]

Barthes was fascinated by the mythologies of daily life, whereby objects which are in actuality historical, cultural, conventional and therefore contingent seem the most innocent, the most natural and inevitable. Through Barthes's eyes one sees the world exposed and demystified; one's 'natural attitude' towards the environment is shattered as the apparent innocence of landscapes is shown to have profound ideological implications. In this case ideology refers to the use of signifiers to express and surreptitiously justify the dominant values of an historical period.

The theoretical underpinnings of Barthes's work lie within semiology, the science of signs. Building on the work of Ferdinand de Saussure (1966), Barthes greatly expanded the notion of a text and the analogies between spoken or written languages and many other communication systems based on such diverse cultural phenomena as fashion, etiquette and urban morphology.[2] As a science of signs, semiology is a type of structuralism whereby signs as objects in the environment (signifiers) and the concepts they recall (signifieds) are elements of a 'language'. The use of the term 'language' here denotes a purely relational system in that there is an arbitrary and conventional relationship between the signifier, which is the word or object used to signify, the concept signified, and the referent. In his later work, however, Barthes began more and more to seek out and celebrate the instability and inherent emptiness of signs, to the point where his earlier attempts at a science of signs, having evolved into a poetics of signs, eventually became what he referred to as an 'erotics' of signs. This later work was clearly poststructural in character.

Although to our knowledge Barthes never used the term 'secret histories' to describe his work, we feel that the phrase captures his lifelong goal. At various points during his career he sought these histories in different places. At first, using Marx and Saussure as guides, he sought them beneath ideology, by uncovering the mythologies of the bourgeoisie. Later, following Freud and Lacan, he found them buried in the unconscious 'memory of the body'.

In this chapter we will examine his search for these secret histories. We will trace his shift from structuralism to poststructuralism by examining changes in the way he interprets landscapes. As such our goals are both to provide a discussion of a little examined facet of Barthes's work as well as to offer insights into the nature of structural and poststructural landscape interpretation. We will thus focus our attention on five of Barthes's texts that deal most directly with landscape interpretation. These texts, which span the last two decades of his life, show how he progressively emptied the sign of all cognitive meaning until towards the end of his life he argued that one reads a landscape 'according to the memory of the body' (1977a: 20).

Barthes held that social reality is composed of many signifying systems, of which the landscape is one. Taking linguistics as a model for analysing all signifying systems, he builds upon Saussure's distinctions, such as those between *langue* and *parole* and between signifier, signified and sign. All such signifying systems, he argues, are products of historical and cultural conventions. He further argues that the signs of which these systems are composed are 'healthier', that is, less ideologically deceptive, when they draw attention to their arbitrariness and do not appear as innocent, unmotivated, eternal or natural. His early semiological work, although scientific, even scientistic, in its formality and analytical classificatory schemes, was nevertheless highly original and uniquely perceptive about the mythologies which help to sustain everyday life.

Myth occurs when objects suitable for communication, landscapes included, become 'appropriated by society', in Barthes's words. They become signifiers and thus are emptied of content. It is the sign, defined as the combination of the signifier (the object) and the signified (the concept with which the signified is contingently associated), that is full of meaning. However, with real material objects as opposed to words there is always a facticity which the myth must 'deform'. In 'Myth today' (1986a: 122) Barthes says that in language 'the signifier cannot distort anything at all because the signifier, being empty, arbitrary, offers no resistance to it'. But in the case of other semiological systems he argues that the objects have a full history from which they become alienated. Thus emptied, myth is a mere form, a mythical signifier, ready to acquire a new fullness, a new contingent existence that appeals to a particular group of readers. As such it acquires a second-order meaning.

It might be noted that Barthes's use of the term 'history', while not self-contradictory, differs in emphasis depending upon whether he contrasts it with nature or with myth. In the first case he emphasizes the contingency and changing nature of history and in the second he emphasizes its facticity. In both cases, however, he uses it to refer to an empirical reality. This is not surprising, given his use of words such as 'ideology', 'myth', 'deform' or 'uncover', 'reveal' and 'demystify', as all imply a realist ontology that he never entirely gives up, but with which he becomes increasingly uncomfortable. Barthes uses history to mean historical reality as opposed to myth when he says that, if a signifier becomes empty and thus ready to be appropriated into a mythological system, it becomes impoverished and 'history evaporates' (1986a: 117). Throughout much of his writing, however, he uses the term 'history', or 'culture', to contrast with nature in order to point out the common mistake of naturalization in which that which is in fact contingent is seen as either universal or more widespread and longer-lasting than in fact it is.

'THE BLUE GUIDE'

One of Barthes's most famous works is *Mythologies* (1986b), a collection of essays written between 1954 and 1956. Among them is 'The *Blue Guide*' (*Guide Bleu*) (1986c), about the widely used Hachette World Travel Guides. 'The *Blue Guide*' provides an early example of Barthes's landscape interpretation in which he analyses both the mythology surrounding travel and the claim of the travel guide to be a primary tool of landscape appreciation and an essential bourgeois educational aid to vision and cultural awareness. Instead, Barthes argues, the travel guide functions as 'an agent of blindness' that focuses the traveller's attention on a limited range of landscape features, thereby 'overpowering' or even 'masking' the 'real' spectacle of human life and history and simultaneously providing an illusion of cultural stability and continuity.

Barthes analyses the historical and sociological background of the *Blue Guide*'s equation of the scenic with the picturesque. He sees it as a dominant, but not entirely unchallenged, contemporary bourgeois ideology originating in a nineteenth-century 'Helvetico–Protestant morality' that promotes the aesthetic appreciation of uneven ground, mountains, gorges, torrents and defiles. This ideology is described as a 'hybrid compound of the cult of nature and of puritanism' which espouses 'regeneration through clean air, moral ideas at the sight of mountain tops, summit climbing as civic virtue, etc.'.

He also sees the *Blue Guide* as implicated in the promotion of an individualistic ideology which identifies morality with 'effort and solitude'. He says that in the nineteenth century, with the introduction of sightseeing in mountain terrain from the comfort of a railway carriage, travelling came to be seen as a way of 'buying effort'. In this way the bourgeois could attain the morally uplifting benefits of mountain climbing through a kind of passive travel. Of effort, he says, one could 'keep its image and essence without feeling any of its ill-effects' (1986c: 74).

The *Blue Guide* is shown to be a mystification of sociological and political realities in its nearly exclusive concern with monuments. Using the *Blue Guide* to Spain as his example, Barthes points out that, besides picturesque scenery, only monuments are extolled, these being primarily religious. According to the *Blue Guide* mythology, the history of art is Christian and 'one travels only to visit churches'. He further claims that the seemingly benign observations in the *Guide* on the prosperity of the country are a statistical 'myth-alibi' that masks the real, lived social conditions and class relations of the present and of the past, thus making the landscapes virtually indecipherable.

According to Barthes the *Guide* systematically obscures the barbaric side of Christianity that might otherwise be seen in the landscape of southern Spain. The text masks the fact that a violent, nationalist Christianity has defaced the earlier achievements of Muslim civilization: the mosque at Cordoba, for example, 'whose wonderful forest of columns is at every turn obstructed by massive blocks of altars, or a colossal Virgin (set up by Franco) denaturing the site which it aggressively dominates' (1986c: 75). Such a scene, he argues, 'should help the French bourgeois to glimpse at least once in his life that historically there is also a reverse side to Christianity'. But such a glimpse is denied the tourist by a guide that can see only official Christianity.

The *Blue Guide* also effectively depopulates the landscape of real flesh-and-blood Spaniards, putting in their place only representatives of ideal types. According to the *Guide* 'the Basque is an adventurous sailor, the Levantine a light-hearted gardener, the Catalan a clever tradesman, and the Cantabrian a sentimental highlander' (1986c: 75). Barthes identifies this orientation as the 'disease of thinking in essences, which is at the bottom of every bourgeois mythology of man'. (It is ironic to note here Barthes's own essentialist phrasing.) Such stereotyped categories become mere tropes, synecdoches, in which the part stands for the whole. For example, one commentator noted that the lightheartedness of the Levantine gardener stands for a general 'lack of emotional complexity' and by implication 'primitivism and cultural inferiority' (Silverman 1983: 29).

Barthes is not interested in the landscape *per se*. He is not, at least at this point in his career, a lover of landscapes. He simply sees landscape as one among many systems of communication, as a system of objects open to appropriation by society. While any substance can be endowed with significance, mythical systems tend to utilize materials that have already been made suitable for communication. These are materials whose signifying function can be assumed, so that they can be analysed formally apart from their substantive qualities.

To produce a semiology of landscape, then, one must investigate the ideological process by which landscape signs become second-order signifiers. Thus when a signifier – that is, an element in a signifying system such as a landscape – is joined with a signified or the concept to be communicated it becomes a sign. Next this sign can become a signifier itself when it is associated with a new concept (signified), and thus a second-order sign is formed. This happens when the meaning of a sign which is historically and culturally variable, and thus contingent, dissipates and becomes open to new meanings. The form which may have had a rich, full history can become an empty signifier, to be filled again with a new history, another contingent significance, to become a new sign. Barthes says that the new signified concept deforms and alienates the historical discourse, changing it into an element in a metalanguage.

Barthes argues that by feigning naturalness or innocence myths depoliticize the world. More correctly, myth appears to depoliticize, for in actuality it operates in the service of conservatism (of either the political left or the right). Therefore, if one is to demystify or repoliticize the study of myth systems, one must start with an examination of the signifieds which have been robbed or cleansed of their history. Barthes describes how this process operates:

> The Spain of the Blue Guide has been made for the tourist, and 'primitives' have prepared their dances with a view to an exotic festivity ... This miraculous evaporation of history is another form of a concept common to most bourgeois myths: the irresponsibility of man.
>
> (1986a: 151)

It should be noted that at no point did Barthes invoke a claim to truth or metalinguistic method which would set semiology above the signifying practices he analyses. He fully recognized that the critical study of mythology, like the study of ideology, is made highly problematic by the lack of an accessible reality against which to compare it. Even at the relatively early scientific, structuralist stage in his career when Barthes wrote *Mythologies* he understood the epistemological difficulties of empiricism. Nevertheless he believed it necessary to attempt to contrast myth with reality, however inaccessible. For example, he states:

> wine is objectively good, and *at the same time*, the goodness of wine is a myth: here is the aporia. The mythologist gets out of this the best he can: he deals with the goodness of wine, not with the wine itself ...
>
> (1986a: 158)

'THE EIFFEL TOWER'

'The Eiffel Tower' (1984), another of Barthes's mythologies, is a brilliant essay written for a literate French audience familiar with the then popular theory of structuralism. Here Barthes is playful in his use of structuralist concepts. He presents the Eiffel Tower as the example *par excellence* of the pure signifier, the empty signifier to which a multitude of meanings can be attached. He contrasts it with objects more clearly defined by the uses to which they are put. The functional obviousness of churches or sports arenas tends to obscure their inherent instability of meaning.

The Eiffel Tower is presented as a clear example of the ambiguity of landscape features and more generally of the instability of the relation between signifier and signified in the semiology of landscapes. In his decoding of the Eiffel Tower and its relationship to Paris, Barthes demonstrates the marvellous polyvocality of this most inexhaustible signifier. In contrast to his earlier scientific pretensions and fascination with systematic analysis, Barthes demonstrates in this essay that landscape interpretation can be not only compelling and provocative but also a highly creative, even idiosyncratic, exercise. His analysis of the Eiffel Tower is anything but formulaic. Although typically structuralist in its preoccupation with dualities, it is peculiarly Barthean in its endless, playful fascination with the opposition of emptiness and fullness.

As a 'universal symbol of Paris', Barthes says, 'there is no journey to France which isn't made, somehow, in the Tower's name, . . . it belongs to the universal language of travel' (1984: 3–4). Seeing towers in general as mythical symbols, cosmic axes, lines connecting heaven and earth, he says that the Eiffel Tower also belongs to the universal language of religion. But the tower means much more. In fact, Barthes argues, it means everything and anything because of its emptiness (1984: 4). For contemporary Parisians it has simply always been there; like a feature of nature, it is taken for granted. Thus it becomes an empty sign, ready, waiting to receive meaning. 'It attracts meaning the way a lightning rod attracts thunderbolts' (1984: 5).

The tower has a dual character, for it is an object seen by all and yet, if one is up in it, it becomes an object for seeing. The tower, therefore, is a singular monument which, as he puts it, 'transgresses this habitual divorce of seeing and being seen' (1984: 5). Here we witness the kind of bipolar analysis, the reversals, by which Barthes the structuralist was enthralled. Unfolding this structuralist line of argument, he says that the tower gives one a splendid view of Paris, but that belvederes usually look out upon nature. This touristic perspective of the fine view, therefore, he claims, implies a naturalist mythology (1984: 8). The tower transforms the city into a kind of nature. He says that it gives a romantic dimension to what is frequently grim (1984: 8).

The tower provides a bird's eye view that allows the visitor to read the text of the city – in other words, allows the structure of the city to become visible. The city becomes an intelligible object from the tower, without, Barthes argues, 'losing anything of its materiality' (1984: 9). The tower, he says, provides us with a new category – that of 'concrete abstraction' (1984: 9). The view from the tower allows the visitor to link in his or her own mind all the familiar urban landmarks into a system. Thus 'every visitor to the tower makes structuralism without knowing it' (1984: 9). Barthes sees this as an intellectual exercise of deciphering the geographical, historical and social spaces of Paris which inevitably interrupts the pleasure of the panoramic view. This decipherment affords the visitor a power over Paris, a single point from which to appropriate the great city.

The tower is a pure, virtually empty sign because it means so many things. It is the 'zero degree of monument' (1984: 7), says Barthes, alluding to his earlier attempts within the realm of literary criticism to achieve completely contentless writing (1987a). Barthes says that we go to the tower 'to participate in a dream' which is stimulated by the very emptiness of the object (1984: 7). It is its uselessness that affords it its powerful oneiric function. He says it is more a crystallizer of meaning than a meaningful object in and of itself (1984: 7).

POSTSTRUCTURAL MOVES

The period around 1970 when Barthes wrote one of his best known books, *S/Z* (1987b), as well as two of the texts that we are considering here, 'Semiology and the urban' (1986d) and *Empire of Signs* (1987c), is one of transition from his more scientific, structural analyses to his more elusive poststructural mode of writing.[3] In these and later texts Barthes abandons his quest for formal structural analysis, developing instead an anarchic, enigmatic, fragmented and 'scriptable' (intentionally difficult to read) style of writing. Despite its unsettling call to the reader to be an active participant in the overturning of taken-for-granted assumptions, the exclusionary effect of his difficult style could be read as revealing a French intellectual's post-1968 scepticism about the political efficacy of writing.

This scepticism made him turn away from his earlier attempts at demystification towards poststructuralism. And yet this move was itself a highly charged political act. In an interview given to *Le Magazine Littéraire* in 1975 he argued that when he wrote his mythologies in the late 1950s 'arrogant discourse came only from the right' (1986h: 219). But by the mid-1970s such arrogance was also to be found, he argued, on the left. Myths, he realized, follow the majority, and by the mid-1970s approximately half the French population was voting for the left. But, as he could not bring himself to expose the mythologies of the left, he chose to abandon the work on mythologies altogether (1986i: 270–1).

But to abandon the mythologies was not to abandon the political. Rather it was to engage in a more relentless critique of the political. The problem, he says, is power itself:

> Power is everywhere ... it is perpetual. It never gets tired, it goes on and on, like a calendar. Power is plural. I thus have the feeling that my private war is not with power but with powers, wherever they are.
>
> (1986i: 269)

For 'fascism is the constant temptation of power, its natural element, what comes in through the back door after it has been tossed out the front' (1986i: 271). Nihilism, he concludes, is the only possible philosophy for our current situation (1986j: 155).

His poststructuralism represents a two-pronged attack on power. The first sally is against language and the second against the core of the Enlightenment project, rationalism, or mind itself. The critique of mythologies, he now believes, does not go far enough. The only fundamental attack on bourgeois culture is an attack on its language, for 'bourgeois culture is within us: in our syntax, in the way we speak'. This is more effective, he argues, than the most terroristic act, for the latter can and will be recuperated by bourgeois culture (1986e: 162). The ultimate terroristic act is to criticize bourgeois language under the very reign of the bourgeoisie (1986l: 197). Such a critique, when effective, rocks the very foundations of power.

The second form of resistance that Barthes argues for is based upon pleasure and the body. He wants to 'unrepress' pleasure, to free it from the censure that societies, whether their morality be Christian, rationalistic or Marxist, have imposed upon it (1986h: 205). To seek pleasure, therefore, is to revolt against society:

> Pleasure . . . does not depend on a logic of understanding and on sensation; it is a drift, something both revolutionary and asocial, and it cannot be taken over by any collectivity, any mentality, any ideolect.
>
> (1987d: 23)

He couples this with a rejection of what Nietzsche terms 'gregarity' and seeks to marginalize himself from the oppressive conformism that characterizes society. He argues for a new individualism which constitutes:

> the only protest that no power can tolerate: protest through withdrawal. Power can be affronted through attack or defense, but withdrawal is what society can assimilate the least.
>
> (1986m: 364)

A unifying thread connecting the mythologies with his poststructural work is the political. However, his conception of the political changed from the politics of ideology to the politics of pleasure (Fiske 1989: 63). Below we explore in greater depth this second, in some ways more radical, politics and examine Barthes's landscapes of pleasure. It might be noted that hyperindividualism may be radical in France, which has a highly centralized state, but in the United States it plays into the hands of the conservative right wing.

In such writings as *The Pleasure of the Text* (1987d) Barthes develops what he refers to as an erotics of reading in which the text is seen as an object of erotic pleasure. Barthes thus produces a materialistic theory in which the body becomes, as he puts it, his 'mana word'. Barthes says that 'the pleasure of the text is that moment when my body pursues its own ideas – for my body does not have the same ideas I do' (1987d: 17). In this work he appears to be in search of the ineffable and in fact succeeds in producing a text which can give immense pleasure of the kind he describes. Through voluptuous metaphors, captivating turns of phrase, extravagant neologisms which glisten with their newness, and discomfiting critiques of scientific pretension, Barthes conveys, albeit momentarily and elusively, the pleasure of the text and the tedium of stereotyped language. However, in our opinion, some of the most compelling moments in the text recall his earlier writings, especially his *Mythologies*.

Perhaps the most important distinction he makes is between *plaisir* and *jouissance*. These terms are also among the most difficult to translate. Richard Miller, his translator, substitutes 'pleasure' and 'bliss', but 'bliss' may convey a more metaphysical, mystical form than the bodily sense of *jouissance* that is intended. The distinction is also somewhat analogous to that between his *lisible* (readable, comfortable) and his *scriptable* (writable, difficult, requiring the active participation of the reader as writer) in that *plaisir* is the more comfortable form of pleasure and *jouissance* the more demanding and tension-producing.[4] The text of pleasure is the text that 'grants euphoria'. And the text of bliss disorients. The text that gives pleasure (*plaisir*) is the text that:

> is linked to a comfortable practice of relatively passive reading, whereas the text of *jouissance* is the text that imposes a state of loss, the text that discomfits, unsettles the reader's historical, cultural, psychological assumptions, the consistency of his tastes, values, and memories, brings to a crisis his relation with language.
>
> (1987d: 14)

Here Barthes relates *jouissance* to disruption and the ecstasy of demystification, to what he calls 'disfiguration'.

Jonathan Culler (1983: 95–7) claims that, despite Barthes's continuing war on the taking for granted as natural that which is historical and contingent, his invocation of the body fails to escape from the contemporary French and specifically Lacanian mythology of Desire which, Culler claims, simply functions as a new name for Nature. Thus Barthes's use of bodily metaphors, his search for a materialistic solution to the problem of subjectivity in a culturally and socially conditioned world, is not an explanatory move; there seems to be no social scientific impulse left here. Any grounding or degree of certainty is radically individual and of the body, not of the mind.

The 'erotics' of signs he now proposes has been progressively emptied of social or conventional significance, and thus of its reproducibility. Thus, despite its materialism, Barthes's later work, both in literature and in landscape interpretation, moves away from cultural interpretation and historical analysis. Instead it seems to emanate from a desire to transcend historicity by grounding itself in the transhistorical body.

'SEMIOLOGY AND THE URBAN'

In his article 'Semiology and the urban', first published in French in 1971, Barthes begins to play with the notion of sociality as eroticism and the materiality and stability of signifiers as independent from the instability of signifieds. He argues that the urban landscape is a text in which signifiers become signifieds in an endless chain of metaphors. Therefore one can never achieve a satisfactory, much less definitive, interpretation of a landscape; all participants in the urban drama write landscape poetry as they wend their own particular paths through the city streets. He quotes Victor Hugo's comment that those who move through the city are readers (1986d: 90) and adds his own observation that they can also be thought of as writers who, like the reader/writer of the 100,000 million poems of Queneau, can create a different poem by changing only a single line (1986d: 95).

Barthes sees all social interaction in the city as discourse – erotic discourse. By 'erotic' here he refers to social intercourse with the Other. The Other is a concept that appeals to the structuralist in Barthes, as it emphasizes the oppositional, dual nature of the self and of society. A city such as Paris is stimulating, challenging, ludic, because of its heterogeneity. He contrasts the centre of Paris with its periphery: 'Paris as a centre was always experienced semantically by the periphery as the place where we play the other' (1986d: 96). He sees a rhythm of signification in the city as the 'opposition, alternation and juxtaposition of marked and unmarked elements'. Here we see Barthes, the structuralist, again finding juxtaposition and opposition more interesting than content. The problem of semiotics as Barthes sees it is to analyse the landscape as if it were literally, not metaphorically, composed of the same types of elements as a language or text. But, he wonders, how can we pass beyond the metaphorical use of the concept of language? To appreciate fully Barthes's notion of landscape as a text, it is necessary to be aware of his definition of a text, which can be found in his 1971 article 'From work to text' (1987e). Although in this particular essay he does not explicitly analyse landscapes as texts, it is clear from his other writings, such as 'Semiology and the urban' and *Empire of Signs*, which will be discussed below, that he considers them to be such. He does say in 'From work to text', however, that it was a landscape that triggered in him the 'vivid idea' of a text. He describes it thus:

> on the side of the valley, a wadi flowing down below ['wadi' is there to bear witness to a certain feeling of unfamiliarity]; what he perceives is multiple, irreducible, coming from a disconnected heterogeneous variety of substances and perspectives: lights, colors, vegetation, heat, air, slender explosions of noises, scant cries of birds, children's voices from over the other side, passages, gestures, clothes of inhabitants near or far away. All these incidents are half identifiable: they come from codes which are known but their combination is unique, founds the stoll in a difference repeatable only as difference.
>
> (1987e: 159)

In this essay Barthes defines text in a broad fashion as a space in which there is a weaving together of symbols to create an irreducible plurality of meaning. It is a signifying practice that abolishes the distinction between writing and reading, production and consumption. A text does not occupy a space as does a work on a library shelf, but is a field within which there is an activity of production, of signification. A text has no closure, is an endless process of communication in which authors and contexts of origin are not privileged. Texts are not subject to Marxist, psychoanalytic or other hermeneutic exercises. He states that the text is 'plural'; 'it answers not to an interpretation, even a liberal one, but to an explosion, a dissemination'. Obviously Barthes has moved away from his earlier semiological ambition to discover and pin down specific codes. We can also see here that the term 'text' is appropriate to describing a landscape as a 'field' within which an open-ended process of signification takes place. In fact the spatial metaphors used by Barthes and poststructuralists in general, especially when writing about texts, are quite striking.

EMPIRE OF SIGNS

Empire of Signs (1987c) is a quintessentially poststructural work in that it undermines the normally taken-for-granted relationship between signifiers (in this case Japanese cultural objects, including, importantly, Japanese landscapes) and their signifieds (the concepts associated by the Japanese or by Westerners with these objects). He accomplishes this by writing an entire book about Japanese landscapes, Japanese artefacts and Japanese cultural values, about a whole Japanese interrelated symbolic system that he claims is not actually *about* Japan. His claim, made on the first page of the book and reiterated in various places throughout the book, purposely creates a tension in the reader, who finds it difficult to bear in mind that this book which appears to be all about Zen, Japanese food, haiku, the urban morphology of Tokyo and its railway stations and department stores, which is even illustrated with photographs of Japan, is not really about Japan; he claims that it is not in any sense a representation of Japanese reality. This tension is presumably meant to unsettle and impress the reader with the idea that there is no necessary relation between signifiers and referents and to raise the question of the highly problematic notion of representation. Here we can see he has all but abandoned the realist ontologies of his mythologies period.

Empire of Signs is, he says, merely about the possibility of difference, about an opposite, an Other, viewed in relation to France and more generally the Occident. It is, he claims, about the possibility of 'an unheard-of symbolic system, one altogether detached from our own' (1987c: 3). He says that he could have equally created a fictive nation and a fictive set of cultural features. However, he thinks it more effective to take Japan not as a reality to represent, as Westerners are wont to do, but as a 'reserve of features whose manipulation – whose invented interplay' (1987c: 3) affords him an opportunity to write about irreconcilable differences. With this system, which he sees as utterly detached from our own, Barthes intends to make us question our own cultural systems in a radical way. Here his wish to use Japanese cultural features as a contrasting system in order to denaturalize the Western taken-for-granted is reminiscent of the earlier Barthes of the mythologies.

Before we analyse this work as an example of Barthes's poststructural landscape analysis, let us briefly comment on it in the light of his earlier perspective in *Mythologies* on the Western tendency to empty signifiers belonging to a cultural Other until they become totally alienated from their very real historical and social contexts. The example that he provides in this earlier work is the way that the French press, through the use of a photograph (what could be more natural?), masks the history of European racism and oppression in its African colonies:

> I am at the barber's and a copy of *Paris-Match* is offered to me. On the cover, a young negro in a French uniform is saluting, with eyes uplifted, probably fixed on a fold of the tricolour ... whether naively or not, I see very well what it signifies to me: that France is a great Empire, that all her sons, without any colour discrimination, faithfully serve under her flag, and that there is no better answer to the detractors of an alleged colonialism than the zeal shown by this negro in serving his so-called oppressors. I am therefore again faced with a great semiological system: there is a signifier, itself already formed with a previous system (a black soldier is giving the French salute); there is the signified (it is here a purposeful mixture of Frenchness and militariness); finally, there is the presence of the signified through the signifier.
>
> (1986a: 116)

Barthes would have agreed that such appropriation of a cultural Other on the part of the French journalist is a clear example of the type of orientalism Edward Said (1979) so successfully describes and exposes as a dangerous product of Western imperialism. While Barthes in his *Mythologies* is clearly alive to the dangers of such an appropriation, in *Empire of Signs* he succumbs to that long-standing European temptation, which Baudet (1988) traces back to classical Greece, to appropriate the Other for European purposes.

This is not to reveal in Barthes any inconsistency that he would not claim for himself; he never wished to remain in any intellectual location for long, having referred to his Chair at the Collège de France as a 'wheelchair', always rolling away from earlier positions (1982: 474). However, we would argue that Barthes's critiques from his *Mythologies* period are particularly compelling for the geographer and other social scientists, and thus we feel justified in using his own mode of analysis from this earlier period to question the validity and the usefulness of his emptying of Japanese signifiers. In other words, he alienates cultural features considered by most Japanese to be highly significant from the historical reality of Japan. While Barthes's exercise can be viewed as the undermining of the assumed relationship between signifiers and signifieds, this is nevertheless a poststructural project, a thoroughly French purpose that is being served by the appropriation of the Japanese system of objects and values, wrenched from their real historical context.

Although Barthes makes problematic the relation between Japan as a real historical entity and the subject of his essay, he does admit to some form of description. And this is, of course, a highly selective description:

> What I was describing was definitely not the technological, capitalistic Japan but a phantasmatic Japan.
>
> (1986e: 158)

He elaborates on his view of Japan:

> The Japan I wrote about . . . finds itself completely plunged into modernity and yet so close to the feudal period that it can maintain a kind of semantic luxury which has not yet been flattened out, tamed by mass civilization, by the consumer society.
>
> (1986e: 158)

Here we can see that Barthes, like other orientalists before him, erases the modern in the Orient, which is associated with the West, and privileges the past, the traditional, that which is Europe's Other. The emptiness of Japan which Barthes so craves is an emptiness which he claims cannot be found in the West. The signs of Japan, he writes:

> are not written in books but traced on the silk of life . . . the sign systems with their extraordinary virtuosity, their subtlety, their strength and elegance are, in the end, empty. They are empty because they do not refer to an ultimate signified, as our own signs do, hypostatized in the name of God, science, reason, law, and so forth. . . . In the West there comes a point when . . . everything in the world comes to a halt with God, who is the keystone of the arch, since God can only be a signified, never a signifier: how could God ever mean anything besides himself? While in Japan, as I read things, there is no supreme signified to anchor the chain of signs, there is no keystone, which permits signs to flourish with great subtlety and freedom.
>
> (1986f: 98–9)

Although Barthes rejects the orientalist claim to represent true knowledge of the Orient, and in fact questions the very possibility of achieving such an aim, this does not mean that Barthes successfully escapes the charge of orientalism. Like that of so many Europeans who have engaged in the study of the Other, Barthes's work reveals much more about European symbolic systems and values than it does about those of the Other. The difference – and it is an important one – is that Barthes's stated purpose in the book is to do precisely that, while for other writers this outcome has been inadvertent.

Barthes revels in the fact that he cannot understand the Japanese language. Whereas the orientalist scholar believes that knowledge of the language of the Other is an essential requirement for sound scholarship, Barthes wishes to avoid the inevitable 'recuperation' associated with translation. Recuperation results from the assumption that there is a one-to-one correspondence between the elements of two different systems and entails the denial of irreducible difference. Barthes thus attacks one of the fundamental values of the orientalist – that of language training. The European's ability to translate a foreign language, or to translate a whole set of cultural symbols, for the benefit, use, enjoyment, intellectual stimulation and, ultimately, for the power of the society of the translator, has generally been seen as an unquestioned good.

Barthes says that Japan (the real country) has 'starred him with any number of flashes'; or better still 'afforded him with a situation of writing'. We can see the benefit to European intellectuals that Barthes believes he can extract from Japan in the quotation below. Here he explains the advantage he feels he has as a complete stranger with a very superficial knowledge of the Japanese cultural landscape and no knowledge of the Japanese language. Thus in reference to the language he says:

> The dream: to know a foreign (alien) language and yet not to understand it: to perceive the difference in it without that difference ever being recuperated by the superficial sociality of discourse, communication or vulgarity; . . . to undo our own 'reality' under the effect of other formulations . . . to descend into the untranslatable, to experience its shock without ever muffling it, until everything occidental in us totters and the rights of the father tongue vacillate.
>
> (1987c: 6)

It might also be noted that this is a very brave and ambitious dream, and that is interesting, especially for the many geographers and others who invest their careers in area studies, in that it is so opposed to their normal scholarly practice of translating what is alien into familiar terms.

We can see here again that Barthes's poststructural project as represented in *Empire of Signs* has some obvious similarities with his earlier work on mythologies. His purpose here, as in *Mythologies*, is to undermine the tendency to see one's own culture as natural.

Without written or spoken language Barthes skims across the Japanese landscape, searching out, and collecting into a system, signifiers of all kinds. He claims to find that:

> in this country [Japan] the empire of signifiers is so immense, so in excess of speech, that the exchange of signs remains of a fascinating richness, mobility, and subtlety, despite the opacity of language, sometimes even as a consequence of that opacity.
>
> (1987c: 9–10)

In the chapter entitled 'Centre-city, empty centre' Barthes argues that the Western city must have a city centre if it is not to cause profound uneasiness. The centre is 'full'. He says:

> it is here that the values of civilization are gathered and condensed: spirituality (churches), power (offices), money (banks), merchandise (department stores), language (agoras: cafes and promenades); to go downtown

> or to the centre-city is to encounter the social 'truth', to participate in the proud plenitude of 'reality'.
>
> (1987c: 30)

By contrast, he says (Tokyo) has an empty centre. Here he puts the word Tokyo in brackets undoubtedly to renew in the reader the tension caused by the ambiguity over whether he is or is not representing the real empirical city of Tokyo. He says:

> The entire city turns around a site both forbidden and indifferent, a residence concealed beneath foliage, protected by moats, inhabited by an emperor who is never seen, which is to say, literally, by no one knows who.
>
> (1987c: 30)

Barthes is intrigued with the ambiguity or, as he says, paradox of Tokyo's centre because it is empty but forbidden, hidden, and a site that can command only indifference, a 'sacred nothing'. Tokyo is a system of routes and spaces spread out in circular detours and returns which avoid the empty centre. It is a centre that does not radiate power, does not fire the imagination with its plenitude, but which supports the urban morphology with its emptiness. It is this same emptiness which Barthes sees embodied in Japanese culture, in its art and in Zen Buddhism. Zen, the religion which strives for emptiness, is seen as a force which 'writes' the landscape.

It is not at all surprising to those who know Barthes's writing that he should be attracted to the Zen experience, in which emptiness plays such an important part. Much of Barthes's writings have been based on the duality empty/full and have involved a search for emptiness and the absence of meaning (1986k). The concept of white writing which he outlines in *Writing Degree Zero* (1987a) is an example. Fullness represents the core of Western society and for him it is entirely negative:

> There is nothing more difficult to 'admit', for a Western mind, than this emptiness (which everything within us longs to fill up, through that obsession with the phallus, the father, the 'master-word' ... fullness is remembrance (the past, the Father), while its neurotic form is repetition, and its social form is stereotype.
>
> (1986g: 117–18)

We see again Barthes's French orientalism in his celebration of irrationality in Japanese culture. To him Tokyo serves as a reminder that the rational is merely one system among many. This view of rationality is similar to Foucault's. Like Foucault, Barthes also employs the exotic as a conceptual Other against which to destabilize one's own taken-for-granted, to impress oneself with the arbitrariness of the cultural.

Barthes compares Tokyo's unnumbered urban toponymy with the highly organized system of street addresses found in most other large urban centres.

Whereas in Western cities the efficiency of named and numbered street addresses is thoroughly ingrained in the popular imagination, the Japanese system of postal addresses is a purely bureaucratic affair. He states flatly that Tokyo is a city of streets without names. The postal service has a master plan of districts and blocks which means nothing to the average resident. What replaces the convenience of Western cities and of cities such as Kyoto (which Barthes identifies as essentially a Chinese city) is Tokyo's charm and 'writability'. The visitor's sense of discovery is heightened by the intense visual and corporeal experience of having to be guided by many strangers through the city. According to Barthes, 'to visit a place [in Tokyo] for the first time is to begin to write it' (1987c: 36). He says:

> This city can be known only by an activity of an ethnographic kind: you must orient yourself in it not by book, by address, but by walking, by sight, by habit, by experience; here every discovery is intense and fragile, it can be repeated or recovered only by memory of the trance it has left in you: to visit a place for the first time is thereby to begin to write the address not being written, it must establish its own writing.
>
> (1987c: 36)

Barthes finds the process of asking directions so intensely pleasurable that he hopes to prolong the experience as much as possible. The inhabitants of the city make elaborate drawings when requested to give directions. They sketch in great detail the landmarks and routes of the city. This he describes as a form of 'delicate communication in which a life of a body, an art of the graphic gesture recurs . . .' (1987c: 34). Barthes elaborates his description of the sensuousness of the experience:

> it is always enjoyable to watch someone write, all the more so to watch someone draw: from each occasion when someone has given me an address in this way, I retain the gesture of my interlocutor reversing his pencil to rub out, with the eraser at its other end, the excessive curve of an avenue, the intersection of a viaduct; though the eraser is an object contrary to the graphic tradition of Japan, this gesture still produced something peaceful, something caressing and certain, as if, even in this trivial action, the body 'labored with more reserve than the mind' . . .
>
> (1987c: 34–5)

Here Barthes reminds us again of his earlier structuralist writings by employing the rural/urban distinction to contrast Tokyo, the exotic, Eastern city, with major cities of the West. He sees orientation in Tokyo as more akin to highly visual and sensuous way finding in the jungle or bush, whereas in the West it is a more abstract, cerebral experience sustained by the 'printed culture' rather than by 'gestural practice' (1987c: 36).

Just as the city of Tokyo has for Barthes no spiritual centre, neither have its neighbourhoods. The underground stations in which are found the subway, the urban trains as well as large department stores, are the centres of neighbourhoods. These he sees as spiritually empty. The station serves transition, commerce and departure, and thus, he argues, instability and emptiness. This station-centred underground Tokyo structures the urban morphology into neighbourhoods, each evoking, Barthes says, 'the idea of a village' (1987c: 42). Each village (neighbourhood), he claims in French structuralist (and orientalist) fashion, is 'furnished with a population as individual as that of a tribe, whose immense city would be the bush' (1987c: 42). Again we find the nature/culture distinction of which Barthes, the poststructuralist, has not rid himself.

'LA LUMIÈRE DU SUD-OUEST'

The final piece of Barthes's landscape writing that we will consider is a short article published in 1977 about Bayonne, the place of his childhood in the south-west of France. It is the last landscape essay that Barthes undertook. It represents the end of a literary journey that took him through the ideological monuments of the Spain of the *Blue Guide*, to the top of the Eiffel Tower, where he gazed at the structure that is Paris, to the streets where he lingered over the eroticism of walkers in that city, and finally to Japan, where he cruised the unmarked streets of Tokyo which swirl around that city's empty centre. Like *Empire of Signs*, 'La Lumière du sud-ouest' is about place, landscape, reading, pleasure and the body. And yet there are important differences between them, differences that stem from the fact that he is not only writing about different places but moving ever closer to his goal of reading with the body. *Empire of Signs* still contains traces of his earlier fascination with mythologies. These traces have been all but erased in 'La Lumière'. In this latter piece he claims that 'reading a landscape is firstly perceiving according to one's body and one's memory, according to the memory of the body' (1977a: 20). He continues:

> I have felt the south-west. I have already scoured the text which is generated by the light of the landscape, by the heaviness of a languid day under a wind from Spain, by a particular type of speech.
>
> (1977a: 20)

But what is this memory of the body? 'My body,' he replies:

> is my childhood, such as history made it. This history has given me a childhood which is provincial, southern and bourgeois. For me these three things are intertwined; the bourgeoisie is provincial, and the province is Bayonne, the countryside of my childhood.
>
> (1977a: 18)

The body, then, is something simultaneously located outside history (people's bodies have the same powers of sight and smell at different times in history) and within it; *my* body is as history made it (is located in a particular time, place and class: in Barthes's case, twentieth-century provincial and bourgeois). This dual nature of the body is exemplified, he argues, in the memories of childhood, which are the memories of the body, based on sight, sound and smell. 'These insignificant memories are like doors opening on to a vast region occupied by sociological and political knowledge' (1977a: 18). Having said this, he makes it clear that these are his own doors. 'I enter these regions of reality in my own way; with my body.' As an example of the memory of the body he describes his memory of the smells of the old quarter of Bayonne.

> All the objects of commerce mingle there to create an inimitable fragrance; the rope used by the old Basque sandal makers, chocolate, Spanish olive oil, old paper in the municipal library. All formed a kind of chemical formula of a type of commerce which has disappeared.
>
> (1977a: 19)

All that remains of this place is the memory of the body; the memory of childhood.

Let us now compare Barthes's experience of Japan with his experience of Bayonne. Japan is a place he encountered briefly in middle age, a place that was utterly foreign to him; a place that he first confronted as one of the foremost intellectuals of post-war France. Bayonne, on the other hand, was a place that he knew intimately, that he encountered as an unschooled child. The Bayonne that he wrote about was a place of memory for him, whereas Japan was devoid of memories, a place where memories were yet to be written in him. He attempted to experience Japan through the body by severing sociality, by severing speech. Let us re-examine a passage cited earlier on Tokyo:

> This city can be known only by an activity of an ethnographic kind: you must orient yourself in it not by book, by address, but by walking, by sight, by habit, by experience; here every discovery is intense and fragile, it can be repeated or recovered only by memory of the trace it has left in you: to visit a place for the first time is thereby to begin to write it . . .
>
> (1987c: 36)

This experience of place which he describes, through walking, sight, habit; this condition where 'every discovery is intense and fragile' is surely the condition of childhood. However, the wide-eyed, inquisitive European is not a child but an ethnographer. But the ethnographer does not experience through the body. Ethnography is perhaps the quintessential product of the Enlightenment. There are two intertwined ironies in Barthes's invocation of the ethnographer here. The first is that the ethnographer places primary importance upon linguistic competence, whereas Barthes celebrates his own ignorance of the language. The second and greater irony is that the ethnographer's job is to do what Barthes tries to avoid, that is, recuperate the experience of the Other by making it intelligible in our own terms.

Although he is an ethnographer uncorrupted by language, saved (or so he thinks) by its absence, he cannot escape the filter of the knowledge which his own history has put in place. He can return to the place of childhood only in memory; he can never go there for the first time again. He can never know another country with his body the way he knew Bayonne as a child. 'This is why childhood is the royal road by which we can best understand a country.' 'Essentially,' he concludes, 'there is no country but that of childhood' (1977a: 20).

CONCLUSION

Barthes's various critical approaches to the study of landscapes offer us some interesting insight into the way landscapes can be read as texts that are every bit as problematic as the literary text. Through his work we can see that the reading of landscapes is a political, albeit often passively political, act. Furthermore, unless one can penetrate well beyond the superficial obviousness into one or more of the secret histories of landscapes, it will be a naively conservative political act.

We might ask at this point what ideas about landscape interpretation can one take from Barthes's early work? His refusal to adopt a naive realist attitude toward landscapes and his insistent denaturalization of the landscape text and texts that represent landscapes provide us with some ideas which might be applied in many other contexts. For example, one might wish to look at tourist guides, travel advertising, Western novels, ethnographies and histories of the 'exotic' parts of the world as 'agents of blindness'. As Westerners we are burdened with layers of orientalist discourse which blind us to non-Western realities: tropes of time travel which invite us to 'step back in time' and tropes of emptiness which allow us to 'discover' new worlds as if they were either uninhabited or, if not, then at least 'uncontaminated' by Western culture. In other words, we seek out the traditional exotic culture and fail to see that which is modern or familiar. This literature effectively empties signifiers of all 'historical' (in Barthes's realist sense of true) meaning and refills them with Western constructions.

The idea of utter difference as Barthes presents it in *Empire of Signs* may be of use in undermining our own taken-for-granted cultural categories through which we view landscapes. What Barthes wishes to avoid above all is the recuperation of the categories of the utterly different Other (fictive or otherwise) into our own. By avoiding language he believes this can be achieved. While we may not necessarily advocate exactly this approach, we do believe that the concept of recuperation is a significant one for undermining taken-for-granted categories.

From his later work we might also take a sense of the crisis of representation as it has been felt by Barthes and his fellow poststructuralists. As social scientists we may wish to problematize our authorial role in representing Others and their landscapes. Barthes's search for an escape from the cultural filters imposed by language, however, leads us perhaps into a dark and nebulous realm where we may not wish to follow. In his search for unmediated landscape knowledge he has searched the memory of his own body for naive childhood perceptions, and we would argue that this is an unrealistic aim, at least for social scientists.

Finally, we wish to emphasize here that from the beginning to the end of his career Barthes's goal was radically to undermine power, whether it was found on the left or on the right. But Barthes felt that revolutionary acts cannot remain revolutionary for long and that his followers who attempt to create new methodologies based on his writings are in danger of 'recuperating' them. Thus in our writing/reading of the landscape text we may want to take our inspiration more from the spirit than from the letter of Barthes, to seek *jouissance* rather than to follow the more comfortable route of *plaisir*.

NOTES

1 As we will explain below, in his later work he comes to reject the implication of the terms 'decipher' and 'reveal' as he uses them here.

2 Eco (1976: 9–14), one of the foremost contemporary semioticians, further broadens semiotics to include such diverse subjects as zoology, proxemics, musical codes, visual communication, systems of objects, and cultural codes.

3 Culler (1982: 25–6) argues that there is no *sharp* break between Barthes's structural and poststructural periods. He quite correctly argues that elements of each intrude upon the other throughout his career. Having said this, we would nevertheless maintain that there was a shift of emphasis in his later work away from his 'scientific' structural work, a fact Culler (1983) himself notes in another work on Barthes.

4 There is a noteworthy similarity here between Barthes's notions of pleasure and bliss and Freud's notion of pleasure and unpleasure. For Freud, pleasure is a state of comfort, of relaxation, whereas unpleasure is excitation and tension. When confronted with the tension and disorientation of unpleasure, the mind seeks the safety of the known, the comfortable, of pleasure. For a discussion of the relation of Freud to Barthes and Lacan see Silverman (1983: 56–60).

ACKNOWLEDGEMENTS

We are grateful to Laurie Hovell and Maruja Jackman for their helpful comments on earlier drafts of this chapter.

3

THE IMPLICATIONS OF INDUSTRY

Turner and Leeds

Stephen Daniels

Turner's *Leeds* (Plate 3.1) is one of the few pictures of an industrial city in early nineteenth-century English water-colour. In this essay I want to show what a complex and knowledgeable picture it is: empirically precise, analytically intelligent and resonant with literary and political associations,.

The viewpoint is Beeston Hill, about a mile and a half south of the river Aire. Many of the buildings in view are located south of the river and the political boundary of Leeds in the out-township of Holbeck. Holbeck was functionally part of the city; in 1806 a local clothier described it as 'joining to Leeds like the suburb of a town' (Parliamentary Papers 1806: 75).

There are three preparatory sketches for the picture. They were probably done on the spot in September 1816 after Turner's sketching tour of the Dales, undertaken to collect material for illustrations to the *Richmondshire* volume of Thomas Dunham Whitaker's *History of Yorkshire* (Hill 1984: 104). Turner's *Leeds* was almost certainly intended to be engraved for Whitaker's volume on Leeds, entitled *Loidis and Elmete* (1816), and why it was not published there is an issue to which I will return.

The sketch which is the basis of the whole area depicted beyond the foreground is a sharply drawn, meticulously detailed and annotated panorama extending over three pages of sketchbook (Turner Bequest, CXXXIV, ff. 79v, 80r, 38r). The sketches of the foreground are much freer and made from a different viewpoint, farther down the road (Turner Bequest, CXXXIV, ff. 37, 38v, 45r). These two viewpoints are conflated in the finished picture.

Compared with the panoramic sketch the view in the water-colour is cropped and compressed; the road is severely foreshortened, the slope of the hill steepened, and they connect foreground and middle distance more directly and suddenly. The water-colour is more selective in its detail than the panoramic sketch. The effects of smoke and mist obscure many buildings and what emerge prominently are the large textile mills and churches, picked out by the colouring of their respective materials, red brick and pale stone. The emphasis on these features is reinforced by the realignment of the road to lead directly to the mill in

Plate 3.1 J. M. W. Turner, *Leeds*, 1816, water-colour heightened with a body colour over pencil, 292 × 425 mm (Yale Center for British Art, Paul Mellon Collection)

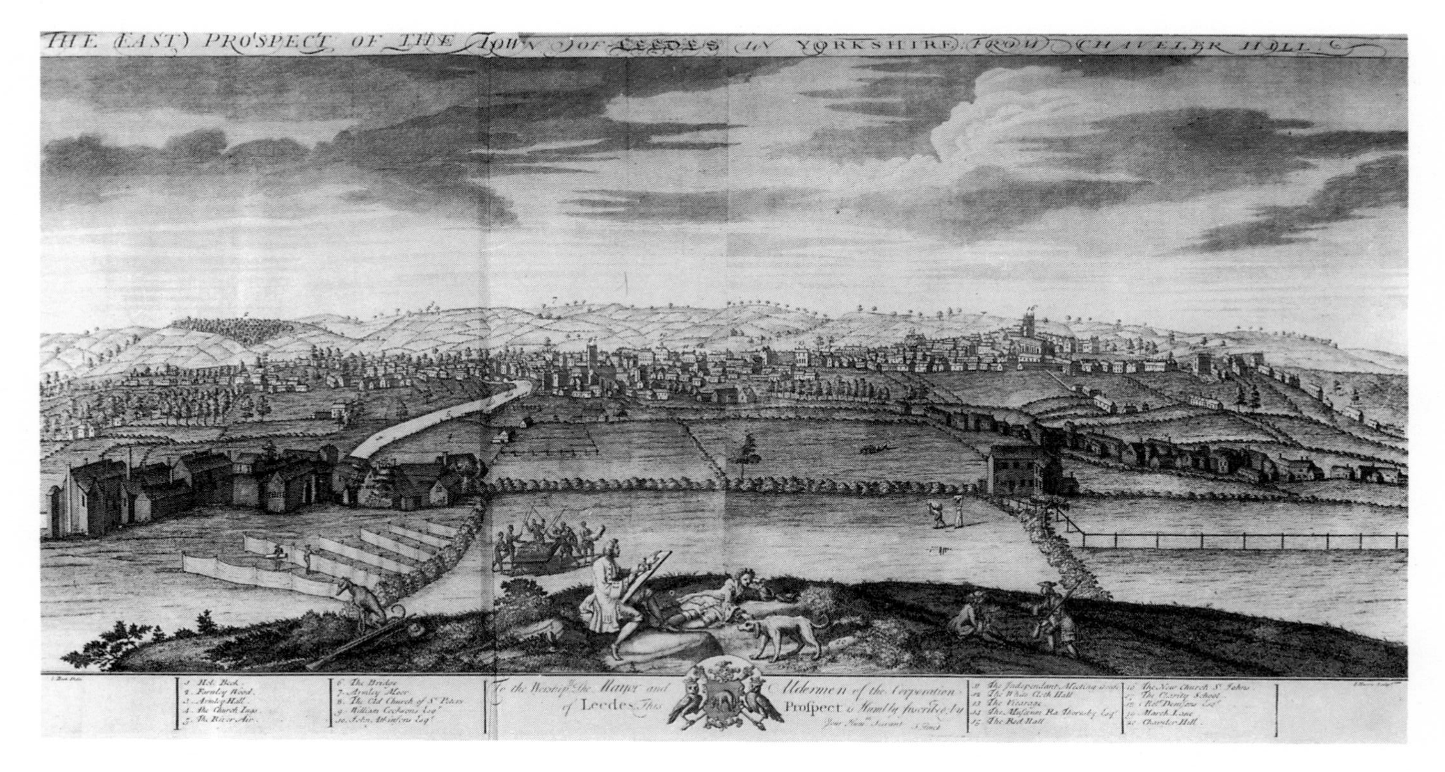

Plate 3.2 Samuel Buck, *The east Prospect of the Town of Leedes in Yorkshire*, 1720, line engraving by I. Harris, two sheets conjoined, 394 × 870 mm overall (Society of Antiquaries, London)

the centre and by the placement of two figures in the left foreground to frame the mill to the left.

In the foreground sketches there is just a shorthand indication of some of the human activity which appears in the finished picture. This does not mean that Turner was casual about the figures or that the figures are incidental to the meaning of the picture. On the contrary, he was extremely careful in the majority of his water-colours about details of the dress, disposition and activity of figures, their position in the landscape and role in its working. *Leeds* is no exception.

On the left of the picture tentermen are hanging newly woven and washed cloth to dry. It is the kind of morning – sunny and breezy – that local clothiers and mill owners welcomed. The large pile of cloth still to be hung confirms that the men are working at the end of a line of tenter frames crowning the crest of Beeston Hill. These may be indicated in one of the foreground sketches (Turner Bequest, CXXXIV, f. 45r) by a line of slanting strokes. A keen and knowledgable observer of industrial tasks, Turner portrayed industrial dyeing in two other pictures – the drying of cloth in his 1811 oil of Whalley Bridge and Abbey (Butlin and Joll 1984: No. 117) and the drying of paper in a water-colour of Egglestone Abbey and Mill, dating from 1816.

To the tentermen's right, and farther downhill, two figures are picking mushrooms. They occupy a similar position, and strike similar poses, to the couple picking mushrooms in Turner's view of Gledhow Hall. (This water-colour is exactly contemporary with *Leeds* and shows a landscape on the northern side of the city in a similar early morning atmosphere. The fact that it was published in *Loidis and Elmete* suggests that *Leeds* was intended to be too.)

At the focal point of the foreground are two masons who can be partially identified in one of the sketches (Turner Bequest, CXXXIV f. 37v). They stand at the apex of a wedge-shaped composition with its base at the bottom edge of the picture, one side described by the edge of the wall, the other by the edge of the shadow of the wall, a spade and a pile of mortar. Standing each side of the wall, the masons are about to lift a slab into place. The elbow of one and the turned head of the other direct attention to the figures coming uphill.

The leading figure is a clothworker shouldering a roll of cloth. Logically he has come from the row of workshops just down the hill (their weaving chambers visible on the upper storey) and he is perhaps about to turn into the field on the left to have the cloth tentered.

Turner would have been aware of the different stages of cloth manufacture and their organization in the landscape. He knew Leeds and its locality well. Since 1808 he had stayed there every year with his patron, Walter Fawkes, the local clothiers' radical political representative (Hill 1984: 18). In addition, informative descriptions and illustrations of each branch of cloth manufacture and other occupations associated with it were provided in *The Costume of Yorkshire*, by George Walker, published in Leeds and London in 1813 and 1814. Circumstantial evidence suggests that Turner knew Walker personally,[1] and it is likely that Turner based his depiction of four figures in Leeds directly upon illustrations in Walker's book. The milk carriers seem closely modelled on Walker's *Milk Boy*. The couple between the clothworker and the mason standing to the right of the wall resemble Walker's *Factory Children*. The milk carriers are returning from an early morning delivery to Leeds, a quickening activity at this period which Walker connects directly with urban industrial growth. If the two figures trudging uphill are factory hands they could be returning from a night shift in one of the large mills in the valley – the one in the centre was gaslit in 1815 (Connell 1975: 87).

Although the main concentration of industrial building is in the valley, there is nothing bucolic about the foreground. The stone pavement at the edge of the road and the rebuilding of the wall indicate that we look down an important economic artery. By the early nineteenth century fields in the southern out-townships of Leeds had been turned over to the grazing of milk cows, an intensification of land use that was, in its way, as 'urban' as the building of mills and houses. Indeed, so valuable was land for grazing milk cows or for tentering cloth that these activities inhibited residential development (Beresford 1980: 82). What Turner depicts is not a contrast between 'country' and 'city' but an integrated, wholly industrialized landscape. He illustrates various processes, some mechanized, some not, and their interconnection, from the spinning of yarn in the valley to the weaving and tentering of cloth on the slopes of the hill. In explicating the workings of the landscape Turner pays particular attention to the weather and the time of day. As much as the drift of dispersing mist and rising smoke,[2] the activities of tenterers, mushroom pickers, milk carriers and factory hands characterize the atmosphere of the early morning. It is a scene of concerted energy – meteorological, technological and human energy – harnessed to industrial expansion. The brightening sunlight, freshening breeze and rising smoke empower the scene and endow it with optimism.

Having specified the empirical and analytical aspects of Leeds, I now want to examine its political and literary associations. In doing so I will first return to the question of why it was not published in Whitaker's *Loidis and Elmete*. The picture was probably intended to be the frontispiece of the book and so complement Place's prospect of Leeds, the frontispiece to Ralph Thoresby's history of the city, *Ducatus Leodiensis* (1714), which Whitaker reissued to pair with his own. The contrast between Whitaker's appraisal of modern Leeds in his text and Turner's in his picture may have been sufficient for Whitaker to reject Turner's *Leeds* as an illustration. An emphatically conservative cleric, Whitaker deplored the effect of industrialization on the landscape of the Leeds locality and the morals of its inhabitants. Holbeck had been overrun in 'the general calamity' (p. 98). Recalling Holbeck as a village happily separated from the city by 'an interval of many pleasant fields planted about with tall poplars' (p. 98), Whitaker observed that 'verdure and vegetation are now fled'. He yearned for the more polite regime of the cloth industry a generation before, when it was controlled by the 'gentlemen merchants' of Leeds, a group who were withdrawing their capital from the industry. The only modern landscapes Whitaker admired were those of the villas that 'dread of smoke and desire of comfort' had prompted retired merchants to develop on the northern, more prestigious, side of Leeds (p. 87). This was precisely the kind of landscape that Gledhow exemplified, a landscaped estate with a fine house owned by the Dixons, formerly a leading merchant family in Leeds. Turner's view of Gledhow, which Whitaker did publish, shows to advantage the 'beautiful plantations' Whitaker singled out in his text. Quoting a phrase from Cowper's conservative poem *The Task*, Whitaker characterized grimy Leeds and its ring of parks as 'a swarth indian with his belt of beads' (p. 87).

The society which Whitaker admired, namely the Tory and Anglican oligarchy of gentlemen merchants, was losing ground at the time. When Turner depicted Leeds economic and political power in the city was shifting discernibly to Whigs and Dissenters (Wilson 1971; Frazer 1980). Their economic figurehead was John Marshall. It is Marshall's mill and that of his former partners, the Benyons, that are the dual focal points of Turner's view of the city. Leeds was the leading flax-spinning centre in England. When a Prussian factory inspector visited the city in 1814 he reported that the Continental blockade had stimulated the production of yarn and that six large new mills had recently opened. Many workers had been taken on, and employers and employees were prospering (May 1968). Marshall's mill had become a showpiece for progressive visitors to Leeds and the Benyons' mill was also renowned for its advanced and safety-conscious construction. The thread the flax mills produced was turned locally into rope and canvas, for which there was a quickening demand in the war years, notably by the navy (Rimmer 1960; Connell 1975). Marshall had weaving sheds attached to his mill but he also put yarn out to domestic weavers in Holbeck.

The prospect of Leeds prospering after five years of depression would have aroused Turner's patriotism – in August 1816 he was perhaps regaling the company at Fawkes's house with a song, 'Here's a health to honest John Bull', which he had copied down on the inside cover of his sketchbook (Turner Bequest, CXLVIII). The manufacturers of Leeds, notably John Marshall, had overcome the Continental blockade. At a time when British tourists were admiring scenes of Allied triumph over Napoleon abroad, Turner was admiring one at home.

Turner's wartime patriotism was directed as much to Britain's artistic achievement as to its industrial and scientific know-how, and in this respect also Leeds had proved itself worthy of Turner's enthusiasm (Lindsay 1966; 137). The formation in 1808 of the Northern Society for the Encouragement of the Fine Arts made Leeds the main provincial centre for the exhibition and marketing of modern painting. George Walker was instrumental in establishing the society, Fawkes assisted in running it, and it was probably Fawkes who encouraged Turner to exhibit at the society in 1811. Turner would surely have concurred with William Carey's praise of Leeds as the first city 'to form a Provincial society, upon the patriotic principle of patronizing the British School. The Artists of England owe much to Leeds, and Leeds to the Artists of England' (Fawcett 1974:

168). Exhibitions lapsed after 1811, apparently because of the adverse commercial climate, and so in 1816 Turner had even further reason to relish the prospect of Leeds prospering (Fawcett 1974: 89, 168–70). Turner would have seen a general cultural expansion where Whitaker saw only decline.

George Walker's *The Costume of Yorkshire* expressed a cultural outlook that accorded with Turner's own. Turner would have sympathized with Walker's introduction, which hoped that 'the British heart will be warmed by the reflection that most of the humble individuals here depicted ... contribute essentially by their honest labours to the glory and prosperity of the country' (Walker 1813 [1814]: introduction). He would also have been moved by Walker's affirmation that textile factories 'are essentially requisite for the widely extended commerce of Britain and furnish employment, food and raiment to thousands of poor industrious individuals' (Walker 1813 [1814]: 96). In Walker's illustrations and descriptions items of advanced technology (including a pioneering steam locomotive near Leeds) do not degrade workers, rather they enhance the dignity of labour in promoting industriousness throughout the county. His account of the woollen industry links skilled and unskilled, hand workers and machine operators, small masters and large employers, in concerted effort. There were tensions between various sectors of the industry which Walker acknowledges but his account overall is a harmonious one. It agreed with the consensus of testimony from the Leeds locality to the 1806 Commons Select Committee on Woollen Manufacture, which maintained that technology did not and would not disrupt the domestic system. By increasing the local supply of yarn mechanized mills dovetailed into the domestic system and arguably energized it (Daniels 1980: 55). It was still reasonable for political economists to argue that, rather than replacing skilled work, machinery and the division of labour would enhance it. The source of Britain's economic advance was still 'the wonderful skill and ingenuity of her artisans', working in concert with advanced technology (Berg 1983: 58). This, as opposed to the alarmist view of Whitaker, seems to me to be the political economy of Turner's Leeds.

In addition to the *Costume of Yorkshire* it seems the political economic imagery of *Leeds* is specifically influenced by an earlier picture and a poem: Samuel Buck's prospect of Leeds, published in 1722 and 1745 (plate 3.2) and John Dyer's poem *The Fleece*, first published in 1757.

Buck depicts Leeds from Chevalier [Cavalier] Hill to the east of the city, where cloth-making was in his time developing. The clockwise circulation of cloth in Turner's *Leeds* replicates the directional movement in Buck's prospect of the city. Like Buck, Turner positions weaving shops to the right and a figure shouldering a roll of cloth moving towards the two figures unrolling cloth on the tenter frames on the left. The direction and elevation of Buck's viewpoint enable him to complete his narrative of the cloth industry by highlighting in the city the cloth hall and merchants' mansions and the main export artery of the river Aire with its sailing barges. A second edition of the prospect, published in 1745, is glossed by a written account of the cloth trade of Leeds.

Buck's prospect belongs to an idiom of eighteenth-century topography which highlights features signifying the commercial progress of a locality and by extension the benign implications for the nation as a whole. The most famous example in prose is Daniel Defoe's description of the Yorkshire woollen industry. Completing his survey of the stages of cloth-making in Calderdale, Defoe observes from a height:

> almost on every tenter a piece of cloth, or kersie or shalloon, for those are the three articles of that country's labour, from which the sun glancing, and, as I may say, shining (the white reflecting its rays) to us, I thought it was the most agreeable sight I ever saw.
>
> (Defoe 1971: 492)

As Britain's most valuable industry, and one which linked agriculture, manufacture and trade, the woollen industry provided a paradigm for the nation's well-being. This appraisal of the woollen industry is expressed most forcibly as a political economic prospect in *The Fleece*.

In the tradition of Virgil's *Georgics*, *The Fleece* is an instructive poem.[3] It prescribes the most advanced procedures for the production and distribution of cloth. It does so as it traces the progressive transformation of 'Brittania's fleece' into cloth 'uphung/On rugged tenters, to the fervid sun ... it expands/Still bright'ning in each rigid discipline,/And gathered worth' (Dyer 1761: 131). It is a process that takes the poem from the hills of Herefordshire to the Yorkshire Dales, to the port of London, connecting all social ranks and a variety of specialized workers from artisans to factory hands. Having pictured Britain as a nation united through industry, Dyer goes on to chart its influence overseas. Packed into sailing barges 'with white sails glist'ning', then taken by 'tall fleets into the wid'ning main ... every sail unfurl'd', the cloth carries the benevolence of 'Britain's happy trade, now spreading wide/Wide as th' Atlantic or Pacific seas,/Or as air's vital fluid o'er the globe' (Dyer 1761: 188). Two things threaten Britain's moral geography: the rivalry of France and, more seriously, the allure of luxury, which undermined the empires of antiquity.

A zealous reader of English poetry, Turner possessed a copy of *The Fleece* in Volume 9 of his *Complete Edition of the Poets of Great Britain* (1794), edited by Robert Anderson (Falk 1938: 255). As it contained also the works of James Thomson and Mark Akenside (whose formative influence on Turner is well documented), this volume of Turner's must have been well read. *The Fleece* is glossed by Anderson's emphasis on the patriotism of the poem and his enlisting of Akenside's admiration of Dyer. There are echoes of *The Fleece* in Turner's own verse of about 1811–12. In Book II of *The Fleece* Dyer interpolates the legend of Jason and the Argonauts, noting the fashioning of the hero's wondrous ship, 'in th' extended keel a lofty mast/Up-rais'd and sails full swelling' (Dyer 1761: 100). Turner too applies the Jason legend to Britain, transferring to the construction of ships the narrative scheme that Dyer employs for the making of cloth, following the course of British oak from plantations to the building and launching of warships to engage the French fleet, 'ensigns broad displayed/Brittania's glory waved arrayed' (Turner Bequest, CVIII, f. 23v). In Book III of *The Fleece* Dyer describes the processing of flax and hemp in a mill in Calderdale by both children and adults, 'all employ'd/All blithe' (Dyer 1761: 135). In verses on Bridport in Dorset, a centre of the manufacture of naval sailcloth and rope, Turner explicates the flax industry again within the idiom of *The Fleece*, beginning with the 'breaking' (crushing) of flax in a scutch mill:[4]

Here roars the busy mill called breaks
Through various processes o'ertakes
The flax in dressing, each with one accord
Draw out the thread, and meed the just reward.
Its population great, and all employed,
And children even draw the twisting cord
Behold from small beginnings, like the stream,
That from the high raised downs to marshes breem [?]

First feeds the meadows where grows the line,
That drives the mill that all its power define,
Pressing . . .
On the peopled town who all combine
To throw the many strands in lengthened twine;
Then onward to the sea its freight it pours,
And by its prowess holds to distant shores,
The straining vessel to its cordage yields:
And Britain floats the produce of her fields.
Why should the Volga or the Rubicon
Be covetted for hemp? why such supply'd
The sinew of our strength and naval pride . . .

Plant but [?] the ground with seed instead of gold
Urge all our barren tracts with agricultural skill,
And Britain, Britain, British canvas fill;
Alone and unsupported prove her strength
By her own means to meet the direful length
Of Continental hatred called blockade

(Turner Bequest, CXXXIII, ff. 102v, 105v, 110v)

If Turner's appraisal of Bridport echoes *The Fleece*, Dyer's poem resonates more powerfully in his appraisal of Leeds. Turner had already transcribed the flax industry into the idiom of *The Fleece* before he depicted the city that was in 1816 the industrial centre for flax as well as wool. Leeds occupies a key position in the argument of *The Fleece*. Dyer builds up an image of the city that raises its status beyond any other in Britain, to the point where Leeds stands as the exemplar of British expansion and influence. Some of the features in Turner's view of Leeds confirm Dyer's imagery of the city from a similar, perhaps the same, vantage point, coming over the brow of the hills from the south:

> And ruddy roofs, and chimney-tops appear,
> Of busy Leeds, up-wafting to the clouds
> The incense of thanksgiving: all is joy;
> And trade and business guide the living scene,

The inhabitants of Leeds come out to work 'As when a sunny day invites abroad. The sedulous ants . . . O'er high, o'er low, they lift, they draw, they haste'.

> . . . th' echoing hills repeat
> The stroke of axe and hammer; scaffolds rise,
> And growing edifices; heaps of stone
> Beneath the chisel, beauteous shapes assume
> Of frize and column. Some with even line,
> New streets are marking in the neighb'ring fields,
> And sacred domes of worship. Industry,
> Which dignifies the artist, lifts the swain.

The imagery of building – 'beauteous shapes assume of frize and column' – is already classicized before Dyer connects the development of Leeds first with a city in antiquity, then with that of other British industrial cities:

> . . . Such was the scene
> Of hurrying Carthage, when the Trojan chief
> First view'd her growing turrets. So appear
> Th' increasing walls of busy Manchester,
> Sheffield, and Birmingham, whose redd'ning fields
> Rise and enlarge their suburbs.

Dyer's extended prospect not only glosses the smoking chimneys in Turner's *Leeds* as emblems of urban aspiration – a trope Turner had already employed in verse on 'the extended town' (Turner Bequest, CII, f. 4v) – but also endows the masons with a symbolic importance commensurate with their pictorial prominence. In magnifying the expansionism of Leeds by analogy with Carthage Dyer discloses a prospect of decline as well as prosperity. In Book II of *The Fleece* he had observed 'the dust of Carthage' in a survey of ancient cities ruined by luxury, monuments 'to those/Who toil and wealth exchange for sloth and pride (Dyer 1761: 118–19). This would have confirmed (perhaps it was the origin of) the cautionary analogy Turner himself drew between the fortunes of ancient Carthage and modern Britain (Lindsay 1966: 117), and it infuses his view of Leeds with the gravity of the scenes he painted a year before and a year after the building and the ruin of Carthage.

Masons are also prominent in a later water-colour of the Leeds periphery, *Kirkstall Lock on the River Aire* (Turner Bequest, CCVIII–L), which can be seen to complement Turner's prospect of the city. As in the view from Beeston Hill the masons flank a main artery into Leeds, the city being named on the barge. Sailing barges on the river Aire were an essential ingredient of the iconography of Leeds in topographical depictions of the city and its surroundings. Buck shows then, and Dyer exhorts, 'Roll the full cars down the winding Aire/Load the slow sailing barges' (1761: 137). In *Airedale*, a poem contemporary with *Kirkstall Lock*, John Nicholson celebrates 'the populous and flourishing town of LEEDS' (eclipsing Kirkstall Abbey, which once overshadowed it) in terms of the barges on the river Aire 'sails unfurling in commercial pride (Nicholson 1825: vi). Contrary to the title of Turner's picture, he has not actually depicted the river Aire (which flows in the background by Kirkstall Abbey) but the Leeds–Liverpool Canal (Shanes 1981: 32). He knew as much, because he had written the word 'canal' on the sketch (Turner Bequest, CCX, f. 61). The displacement of the name makes sense if we recognize that it was the river Aire that had the topographical associations, not the Leeds–Liverpool Canal.

Unlike Turner, practitioners of the picturesque could scarcely cope with industrialization, notably those who fashioned a conservative politics with the picturesque. For Humphry Repton a Uvedale Price textile mills were disturbing features in the landscape, storehouses of volatile social and economic energies which threatened landed interests and tastes (Daniels 1981). For the itinerant Tory John Byng textile mills were redeemed in wartime by their patriotic associations. In a striking analogy he likened them on those established emblems of British power, warships. The rows of windows on a mill reminded him of the ranks of gun ports on a first-rate man-of-war (Bryn Andrews 1935–6, III: 81–2; II: 194–5). Materially the analogy is not so fanciful. Naval dockyards were the largest industrial sites in the country and large warships harnessed as much human and natural energy as many a textile mill. The analogy may have been in Turner's mind too, not to redeem an industrial landscape but to enhance it. Warships were a favourite subject of Turner's, and in his verse, sketches and paintings he expressed their power and patriotic iconography. (*'Victory' returning from Trafalgar* (Butlin and Joll 1984: No. 59) he sold to Fawkes in 1806.) Considering also Turner's view that industry was ultimately realized in seaborne trade and that ships were the main vehicle of his views on industrialism, it is plausible that Turner's view of the industrial landscape of Leeds was infused with naval associations. The flax mills in his view of Leeds had overcome the continental blockade. What the tentermen are raising to billow in the wind might be sailcloth made from mill-spun yarn, and even if not in fact Turner would have been aware of Dyer's use of tentered cloth and spreading sails as dual ensigns of British dominion. To recall Turner's couplet:

> Urge all our barren tracts with agricultural skill,
> And Britain, Britain, British canvas fill;

In terms of Turner's water-colour *oeuvre*, *Leeds* can be seen to have as much connection with coastal scenes like Saltash and Devonport as urban scenes like Dudley and Coventry (see Shanes 1979: Nos. 1, 27, 66 and 58 respectively). Or, to enlarge the point, Leeds can be seen to occupy a central place in a series of over

350 water-colours that Turner produced between 1810 and 1827 which Turner himself saw as a coherent network of landscapes (Shanes 1981: 15).

If Leeds is an eccentric picture in terms of early nineteenth-century water-colour as a whole, it is not in terms of Turner's *oeuvre*. In many, if not most, of his topographical works Turner seems to have grasped the workings of a place and its implications in a style which interpenetrates general themes and particular observations (Shanes 1981: 13–16; 1979: 5–9). In his view of Leeds Turner concentrates the complex implications of local industry – economic, locational, social, moral and political implications – in a naturalistic impression of a late autumn morning. The range of research required to recover these implications testifies to the scope of Turner's topographical style.

NOTES

1 Walker knew Fawkes well. A neighbouring landowner, he shared Fawkes's love of hunting on the moors and painted panels in Fawkes's Farnley Hall with some hundred sporting scenes. Turner would surely have met him there (Nares 1954). I owe this reference to David Hill.
2 The naturalistic atmosphere of *Leeds* resembles another panorama which may have influenced it, Girtin's *Eidometropolis*, described by the *Monthly Magazine* (XIV: 254–5): 'he [Girtin] has generally paid particular attention to representing the objects of the hue with which they appear in nature, and by that means, greatly heightened by illusion. For example, the view towards the East appears through a sort of misty medium, arising from the fires of the forges, manufacturers, &c. which gradually lessens as we survey the western extremity.' I owe this reference to Scott Wilcox.
3 My reading of *The Fleece* owes much to John Barrell's (1983).
4 This process, along with other details of flax-growing and processing in the Bridport area and the importance of the industry are described in Stevenson (1812: 294–303, 446–7). Stevenson collected his material in 1811 and so his appraisal is exactly contemporary with Turner's as well as agreeing with it in most particulars. Turner may not have known of the sequence of twelve illustrations by William Hincks of the Irish linen industry, published in London in 1782 and 1791, which follows the progress of flax processing from the sowing of seed to the baling of linen for export and includes a perspective view of a scutch mill showing the machinery for breaking the flax.

ACKNOWLEDGEMENTS

This essay was first published, with fuller illustration, in *Turner Studies* 6, 1 (1986): 10–17.

4

READING THE TEXTS OF NIAGARA FALLS

The metaphor of death

Patrick McGreevy

Many visitors have looked on the landscape of Niagara Falls as a kind of text to be interpreted. For example, in 1843 the American writer Margaret Fuller came to Niagara 'to woo the mighty meaning from the scene' (Fuller 1856: 21). Particularly in the nineteenth century, many believed that God – the author of this text – had written a special message in Niagara's landscape, and since then visitors have read a bewildering array of meanings in it (Adamson 1985; McGreevy 1985, 1987; McKinsey 1985). This essay will examine a sub-set of these readings of Niagara Falls – those that find in Niagara's landscape a metaphor of death. The evidence for this investigation is drawn primarily from poems, novels, travel journals and other texts Niagara's visitors have written. However, the essay goes beyond merely examining these texts, for the metaphor of death is also manifest in the landscape that humans have created around the waterfall. At Niagara Falls there are tangible institutions and behavioural traditions related directly and indirectly to the symbolism of death. These features of the cultural landscape are themselves a kind of text that can be understood in conjunction with the written texts. There is a complex intertextuality here. On the one hand, because people understand Niagara partly through written texts, their understanding and actions towards it are based to some extent upon the symbolism of death elaborated in those texts. On the other hand, because people now view and act towards Niagara on the basis of the death metaphor, it is further impetus for others to write about it in those terms.

The essay is divided into three sections. The first traces the manner in which death came to be symbolically associated with Niagara Falls through the presence of actual danger and death. Archival evidence and visitors' written accounts, demonstrate that many treated Niagara not only as a symbol of death but also as a stage on which to act out individual responses to death. But these responses can be understood only within a broader historical and cultural matrix. The second section explores the place of death in this broader matrix. A fervent public discourse on the meaning of death raged throughout the Victorian period. Its terms of reference were certainly traditional, but, as the evidence at Niagara shows, this discourse allowed a remarkable degree of latitude for imaginative speculation. Nevertheless, both the visions of death that visitors saw at Niagara and the way people shaped Niagara's landscape in accordance with those visions must be interpreted in terms of prevailing cultural beliefs about death. The third section examines the death metaphor in detail. Certain physical features of Niagara Falls – the brink, the plunge, the abyss, the rising mist and rainbow – have been consistently understood in terms of prevailing notions of the after-life. Understanding the individual elements of the metaphor makes it possible to interpret certain alterations of the human landscape at Niagara Falls.

DEATH AT NIAGARA FALLS

Niagara Falls is literally a place of death. There is a family of daredevils at Niagara Falls who claim that they have pulled more than 500 corpses from the lower river since the turn of the century (Williams 1976). This 'River of Death' (Seibel 1967: 395), as some have called the Niagara, began claiming victims at least as long ago as the eighteenth century, when – we learn from the diary of Mrs Simcoe, wife of Upper Canada's first lieutenant-governor – eight British soldiers were accidentally swept to their death (Marden 1932: 113). As many as fifty-three victims in a single year have been 'rolled and hurled', as William Dean Howells (1888) put it, 'headlong on to the cataract's brink, and out of the world'. Here was a place that made people aware of their mortality. It was also a place that aroused their curiosity about the after-life.

As early as the 1830s death had become part of the lore of Niagara. Guidebooks repeated the gruesome details of accidents, suicides, murders and narrow escapes (Holley 1883; DeVeaux 1839). A favourite tale was that of a young man who, while playing with a child near the brink of the falls, pretended to throw her in. She accidentally slipped through his arms, and when he tried to save her, both were 'hurried into eternity', as one visitor described it (Bird 1856: 245). In each of the two earliest novels which feature Niagara the hero dies at the falls. Chateubriand's *Atala* (1801) reaches its climax when the young Indian heroine takes poison and dies in her lover's arms. Similarly, in James Fenimore Cooper's *The Spy* (1821) the patriotic hero dies in battle at Niagara Falls and is buried there. The theme of death even entered the popular fiction of Niagara Falls. A series of five-cent detective mysteries, published near the turn of the century, used the falls as a backdrop for murder (Deputy 1888; Francis 1895) – a device repeated more recently in Henry Hathaway's film, *Niagara* (1953), which starred Marilyn Monroe. Part of Niagara's attraction, it seems, is based on a fascination with death itself.

Partly because of the stories of death that made their way into popular lore, many people have found it difficult, while standing at the brink of Niagara Falls, not to be aware of death. One nineteenth-century writer described the brink this way:

> And there every islet, every rock, every point has its legend of terror; here a boat lodged with a man in it, and after a day and a night of vain attempts to rescue him, thousands of people saw him take the frightful leap, throwing up his hands as he went over; here a young woman slipped, and was instantly whirled away out of life; and from that point more than one dazed or frantic visitor had taken the suicidal leap. Death was so near here and so easy.
>
> (Warner 1897)

People felt the nearness of death not only because of the stories but also, as Mrs Jameson (1838) commented, because of 'the immediate danger, the consciousness that anything caught within its verge is inevitably hurried to a swift destination, swallowed up, annihilated'.

Given the presence of real danger and the accumulation of stories and legends of death, it is not surprising that the falls should eventually be treated as a metaphor for death itself. James K. Liston's *Niagara Falls: a Poem in Three Cantos* (1843) presents the basis of the metaphor:

> The slope of life, where none can retrograde
> A single step, but must each moment live;
> Live but to move right onward to the brink
> Of Death's huge precipice invisible,
> Unfenced, uncharted in the maps of life
> Because of its most strange ubiquity.

The metaphor appears not only in poetry but also in prose, non-fiction and even an 1888 insurance advertisement:

> The water does not flow over 'Niagara Falls' more steadily nor more resistlessly than the stream of living humanity passes over from this world to the abode of the silent majority; therefore be worldly wise and provide a future support for those who depend on your now strong arm. Insure you life in Massachusetts Mutual Life Insurance Co.
>
> (*How to see Niagara*, 1888)

Niagara Falls became a grim reminder of the ultimate limitedness of the human condition. 'Here is the spot of all others upon the broad earth,' wrote Thomas Rolph (1836), 'where the nothingness of human pride comes home upon the heart; where ... its hard wrestlings with the doom to which it is fated, sink into their native insignificance.'

But the human heart does not always respond the same way to 'the doom to which it is fated'. Niagara is a stage where many responses to death have been played out. There were some who fought against death at Niagara. Using all their strength and ingenuity, they pulled themselves or others from the brink of death. One of these was Edward Di Ruscio, a twenty-year-old Chicagoan who, in 1951, swam fifty feet into the river to rescue a woman just a hundred feet above the American falls. Literally hundreds of people, most of them attempting suicide, have been rescued at the brink (Continelli 1988).

There were others who defied death at Niagara. They walked or rode bicycles on tightropes stretched over the river; they trod its waters in barrels and giant rubber balls. Philippe Petit, the briefly famous French aerialist, planned a trip on a wire over the falls in 1974. When asked about the dangerous stunt he replied, 'In the world of the possible, it is just before the impossible' (Allen 1974: 12).

Others attempted feats close enough to the impossible to lose their taste for defiance. Annie Edson Taylor, the first person to survive a trip over the falls in a barrel, warned after her 1901 ordeal:

> If it was with my dying breath I would caution anyone against attempting the feat. I will never go over the Falls again. I would sooner walk up to the mouth of a cannon, knowing it was going to blow me to pieces, than make another trip over the Falls.
>
> (Baker 1975: 12)

In the mid-nineteenth century Joel Robinson was known as the most intrepid and resourceful of Niagara rivermen. He had rescued so many from the vicinity of the falls that people called him the 'Navigator of the Rapids'. When presented with the opportunity to pilot the then unused steamer *Maid of the Mist* through the lower rapids in 1867, Robinson felt an 'irrepressible desire' to take up the challenge. When the ship entered the turbulent water it was battered and slashed with such fury that Robinson feared it would immediately disintegrate. The ship survived but Joel Robinson, according to his wife, 'was twenty years older when he came home that day than when he went out' (Holley 1883: 94–5). A contemporary local historian further described this transformation:

> He decided to abandon the water and he advised his sons to venture no more about the rapids. Both his manner and his appearance were changed. Calm and deliberate before, he became thoughtful and serious afterward. He had been borne, as it were, in the arms of a power so mighty that its impress was stamped on his features and on his mind. Through a slightly opened door he had seen a vision which awed and subdued him.
>
> (Holley 1883: 95–6)

There were many others who, in their defiance of death, knocked too loudly on the door that Robinson had peeped through. One of them was an English barber named George Henry Stephens. In 1919 his oak barrel 'smashed like an egg' beneath the Horseshoe Falls; only his tattooed arm was recovered. Another was the Greek chef George Stathakis: his wood and steel barrel became his coffin after a twenty-two-hour pummelling beneath the falls in 1930. Finally, in 1957, a man claiming to be God arrived at Niagara Falls; as several of his followers looked on, he attempted to 'walk on the water' above the American falls. Needless to say, he found the footing hazardous and was swept over the brink in seconds (Seibel 1967: 395).

Many have come to Niagara not to defy death but to seek it out. For those who have felt beaten down by this world, Niagara has provided a rapid and certain exit. A recent study concludes that Niagara Falls has seen far more suicides than any other place in North America, including the Golden Gate Bridge; 'year after year, (at least) one person a month uses the falls to kill himself' (Continelli 1988: 8).

But many who ended their lives at Niagara had not arrived with that intention. Seized by a sudden impulse at the brink, they left this world without apparent revulsion or remorse. In a recent documentary film, a woman who works at a concession stand near the brink of the falls spoke of the numerous suicides she had witnessed: 'The strangest thing about people who jump over the falls is that they don't even think. Like I said, the one woman was waving at everybody as she floated on her back over the falls' (Garey and Hott 1985). A Toronto woman recently pulled from the river at the brink of the falls explained that she had been captivated by the 'overpowering roar that became louder and louder in my ears' (Continelli 1988: 8). The chief of Canada's Niagara parks Police says that many who are rescued 'tell us that they had no intention of committing suicide, but they still had this feeling of the attraction of the water' (Continelli 1988: 8). Until recently local authorities have hesitated to acknowledge the suicide problem, fearing that publicity would affect the tourism industry. Now the special blue phones have been placed at the most popular suicide sites along the American side of the river (plate 4.1). These phones offer a

Plate 4.1 Suicide prevention measures at Prospect Point, Niagara Falls, New York (photo by the author)

direct connection to a suicide prevention counselling centre. In addition, an eight-foot-tall suicide fence stands at the observation tower at Prospect Point.

These suicides were not the only ones who felt 'the attraction of the water'. Many travellers have described an attraction, even an urge to jump in (Chapman 1875: 208; Barham 1845; Buckingham 1845: 44). T. R. Preston, a Torontonian who visited the falls in 1842, nearly succumbed to 'an attracting influence' which he compared to 'the fascination exercised by the loadstone or the eye of a rattle-snake' (Preston 1845: 84). George William Curtis tried to analyse this hypnotic phenomenon: 'It flows so tranquilly, is so unimpatient of the mighty plunge, that it woos and woos you to lay down your head upon its breast and slide into dreamless sleep' (Curtis 1852: 91). In 1834 Harriet Beecher Stowe visited Niagara Falls and recorded this often expressed sentiment:

> I felt as if I could have *gone over* with the waters; it would be so beautiful a death; there could be no fear in it. I felt the rock tremble under me with a sort of joy. I was so maddened that I could have gone too, if it had gone.
> (Fields 1897: 89–90)

Stowe's fascination with death would have been unthinkable in an earlier age. Niagara was, in many ways, a lightning conductor for nineteenth-century enthusiasms, and death became one of that century's most spirited obsessions. The next section examines the development and characteristics of this obsession. To understand the Niagara texts that deal with death, it is necessary to place them within the nineteenth-century context in which they were written.

THE CHANGING FACE OF DEATH

Death is something that people of all eras have in common. As Arthur Koestler has put it, we are all 'faced with the intractable paradox of consciousness emerging from nothingness and returning to nothingness' (Koestler 1976: 238). But the way people in the Western world have viewed death has changed dramatically through time. In the twentieth century there is a horror associated with death and consequently a desire to separate the dead from the living. Cemeteries, for example, are clearly demarcated from the sphere of normal life. Philippe Aries has found that the medieval cemetery, in contrast, was a distinctly public place – a site of dancing, gambling, juggling, sport and business. Death itself in the Middle Ages was a simple public ceremony. In Aries's words, 'death was both familiar and near, evoking no great fear or awe' (1974: 13).

Beginning in the twelfth and thirteenth centuries, a new emphasis on individuality altered both the perception and the practice of death. This change began with the idea of a personal judgement at the moment of death. By the fifteenth century the personal judgement had become a trial, a final temptation. In the last moment, it was believed, one's life flashed before one's eyes. An individual's 'attitude at that moment would give his biography its final meaning, its conclusion' (Aries 1974: 38). Several other developments in the Western way of dying parallel this emphasis on the individual's last moment.

Burial customs gradually became more individualized. From the fifth century until the thirteenth, burial had been anonymous. Inscriptions reappeared first on the tombs of saints and royalty. By the eighteenth century they had spread to the middle class. In the art and literature of the fourteenth to the sixteenth centuries there appeared the figure of the partly decomposed cadaver. Appreciation of the individual life, in Aries's view, had transformed death from something natural to something strikingly horrible (Aries 1974: 39–50).

This idea that death was not a natural part of life – that it represented a rupture, a break from the familiar world – helps to explain a further development: the association of death and eroticism, which emerged in the art and literature of the sixteenth century. Aries writes:

> Like the sexual act, death was henceforth increasingly thought of as a transgression which tears man from his daily life, from rational society, from his monotonous work, in order to make him undergo a paroxism, plunging him into an irrational, violent, and beautiful world.
>
> (Aries 1974: 57)

The erotico-macabre themes in this art and literature prepared the way for what Aries calls the 'romantic death' to be found, for instance, in the writings of Twain and the Brontes (Aries 1974: 58). 'The very idea of death moved them,' but their 'morbid fascination', Aries suggests, 'may merely be a sublimation ... of the erotico-macabre phantasms of the preceding period' (1974: 61).

The growing awareness of, and fascination with, individual death culminated in the nineteenth century. This was exhibited in a number of ways. First, there was the rise of spiritualism in the 1840s, beginning in the United States but soon spreading to Europe (Garrison 1978: 11; Rowell 1974: 11). Second, there was the fervent public debate, which raged between 1830 and 1880, concerning the meaning and the existence of hell (Rowell 1974: 17). Third, there were numerous books on the subject of death which enjoyed great popularity; they bore titles such as *The Grave*, *Meditations against the Tombs*, *Death-bed Scenes* and *The Future State* (Rowell 1974: 6–7). This obsession even affected Britain's royal family; Prince Albert and the Queen read together a volume called *Heaven our Home* which described heaven as an 'etherialised, luminous, material habitation' (Rowell 1974: 120). Fourth, there were the immensely popular 'Judgement Pictures' of John Martin. The first two canvases depicted scenes of apocalyptic terror and the third portrayed a 'celestial landscape in iridescent blues and golds peopled with white-robed figures' (Rowell 1974: 10). These paintings were first exhibited in London in 1855 as 'the most sublime and extraordinary pictures in the world'. Later they drew large crowds throughout Britain and North America.

A good deal of nineteenth-century literature seems to be informed by this fascination with death and what lies beyond death. This is particularly true of the Gothic tradition and what has been labelled 'Dark Romanticism' or 'Negative Romanticism' (Thompson 1974). The romantics made no attempt to hide their fascination with death, nor even their fascination with hell itself. They read Milton's and especially Dante's depiction of the next world from a new perspective. Clearly Dante, Milton and their respective readers had been fascinated with the diabolical as well as the angelic, but they had always evaluated the former in the traditional negative terms. The Romantics had no such traditional obligations; Lucifer's defiance, moreover, appealed to their Promethean urges. 'The Dark Romantic hero,' writes G. R. Thompson (1974: 6), 'by working in and through evil and darkness, by withholding final investment of belief in either good or evil, ... perhaps attains some Sisyphean or Promethean semblance of victory.'

Edgar Allan Poe was perhaps more obsessed with death than any other nineteenth-century writer. Harry Levin describes this preoccupation as 'his morbid curiosity, his restless desire to penetrate the ultimate secret' (1970: 101). What Poe really desired has been suggestively characterized by George Poulet as a state of 'posthumous consciousness' (Levin 1970: 130). Poe's immense international popularity suggests that, to many people, this desire did not seem alien. Freud once commented, 'Whenever we make the attempt to imagine our own death ... we really survive as spectators' (Bardis 1981: 72). Similarly, Rowell notes that, among those most vehemently committed to the existence of hellfire, not one 'seriously considered it, in all its horror, as a possibility for himself' (1974: 30). Death can be fascinating if one is still here looking over at it. A certain separation is necessary; it must be remote. Many of those who travelled to Niagara, particularly in the nineteenth century, seemed to share Poe's fascination with death.

Geoffrey Rowell claims that the growing stature of science caused many to abandon belief in a sequel to death. With characteristic turn-of-the-century optimism, William Gladstone commented that the doctrine of hell had been 'relegated ... to the far-off corners of the Christian mind ... there to sleep in deep shadow as a thing needless in our enlightened and progressive age (Rowell 1974: 212).

The problem, according to some twentieth-century interpreters, is that not only has hell been relegated to some far-off corner, but so also has death itself. The intense nineteenth-century awareness of death has not been overcome but repressed (Becker 1973). Toynbee argues that the loss of traditional beliefs has made it more difficult to accept 'the hard fact of death's inevitability' (Toynbee 1976: 17). The British anthropologist Geoffrey Gorer theorizes that the final collapse of belief in an after-life has made 'natural death and physical decomposition ... too horrible to contemplate or to discuss (1970: 79). The result is that in the twentieth century we have a 'pornography of death' similar to the Victorian pornography of sex. This new pornography, in Gorer's view, is expressed in a taste for violent death in fiction, drama and film. Death from natural causes titillates our curiosity as little as sex between married people titillated the Victorians. One might object, however, that the people of the twentieth century have no monopoly on this fascination with violent death: it is difficult to forget the success of the Roman gladiatorial shows, or the popularity of public executions in nearly all periods.

Given the breadth and intensity of this discourse on death in the nineteenth century, it is not surprising that the issue should appear at the site of another nineteenth-century obsession – Niagara Falls. And, indeed, at Niagara nearly the full range of the discourse finds expression.

THE METAPHOR OF DEATH

The death metaphor appears at Niagara Falls partly because of the presence of actual danger and the record of suicides and accidental deaths. But the elaboration of the metaphor derives its impetus from the vigorous discourse on death bubbling in Niagara's wider cultural context. This section explores the elements of the death metaphor in detail.

Those who have read a metaphor of death in Niagara's natural landscape have used their experience at the falls to explore the nature of death. Gazing from the brink of Niagara Falls has moved them to imagine what the experience of dying would be like. A good part of the descriptive literature they have produced could be characterized as imaginative speculation on the afterworld. Although these visitors have read the landscape like a text with a definite meaning, the metaphor has allowed this text to serve as a focus for imagination, a catalyst for exploring the nature of death. What these visitors have brought to the landscape, rather than something inherent in it, is obviously key to their readings. They have brought, first of all, the knowledge of stories and legends of death and perhaps some familiarity with the writings of others who have seen the metaphor of death in Niagara's landscape. Their readings, therefore, are complexly intertextual. And many of their speculations on the after-life take their structure from the prevalent literary and theological sources of Victorian Christian culture, but often, as we shall see, they imaginatively develop that structure in highly unorthodox ways. Like most texts, Niagara's landscape text cannot be isolated from its readers, and its readers cannot be isolated from their world.

This exploration of the metaphor of death will proceed sequentially through the elements of the river landscape. There is a fair amount of agreement among a whole host of writers on the symbolic meaning that should be attributed to the brink, the plunge, the abyss, the rising mist and the rainbow – all in relation to the theme of death. The remainder of the essay, therefore, presents a sort of geography of Niagara's metaphor of death.

The brink

If life is a river that will ultimately plunge over a precipice, the brink of that precipice is the boundary line between life and death, a place where one could perhaps look beyond if only in imagination. 'There in the solitude and on the brink / Of thy abyss,' wrote William Chambers Wilbor (1907), 'I seem to stand, upon the / Verge of time, alone with thy creator.' James K. Liston's *Niagara Falls: a Poem in Three Cantos* (1843) is especially rich in these images of Niagara's brink which he variously names 'Death's huge precipice', 'Death's dark door-posts', the 'frontiers of Life's bounded territories' and 'that mysterious line/That separates eternity from time'. Liston argues that people ought to be more curious about:

> . . . what is going on
> On further side of that most narrow line
> Between the visible and the unseen worlds!

For Liston, apparently, a trip to Niagara would necessarily inspire this curiosity. It would set one wondering whether to expect 'the palm of victory, or the blasts of flame' (Liston 1843: 38). Many writers, especially those with a religious purpose in mind, have emphasized the soul-searching which the brink inspires. Archbishop Lynch of Toronto, for instance, wrote in 1876 that Niagara's brink should recall to the individual 'his own last leap' when 'the soul shall tremble on the precipice of eternity', and he reminded his reader that 'death holds many a deep secret of a good or ill-spent life'. The symbolism of the brink as the final moment of life connects the metaphor with the numerous examples of real deaths at Niagara. The eight-foot fence on the observation tower and the emergency phones are further manifestations of this connection.

The plunge

If the brink of Niagara is the final moment of life, then the plunge is a fall from life, a fall into death. Some visitors have called it an endless fall, a fall into eternity:

> I saw, I heard. The liquid thunder
> Went pouring to its foaming hell
> And I fell,
> Ever, ever fell
> Into the invisible abyss that opened under
>
> (Anonymous 1879)

Death is a moment of transformation. The plunging waters make an appropriate metaphor of transformation as, for example, in this description by Henry James:

> Even the roll of the white batteries at the base seems fixed and poised and ordered, and in the vague middle zone of difference between the flood as it falls and the mist as it rises you imagine a mystical meaning – the passage of body to soul, of matter to spirit, of human to divine.
>
> (James 1884)

There is an unusual story which should be mentioned here because, in terms of the dominant symbology of death at Niagara, it freezes the plunge just above the abyss. Linda Fulton's *Nadia: the Maid of the Mist* (1901) is a tale of death, but there is little horror to it. At the moment of death, victims are snatched away to a special Niagara limbo. Fulton's main character is a young man who comes to

Niagara thinking of suicide. While there, he has a dream that a beautiful maiden is calling for help in the river. When he attempts to save her she reveals that she has been acting, and leads him behind the falls to a spacious chamber ruled by Aeolus, god of winds. There he discovers Indians, voyageurs, soldiers, youths and maidens – 'the spirits of all who have gone over the great falls, or who have been drowned in the river or the rapids below, and thus become vassals of the water king until the Judgement Day' (Fulton 1901: 22–3). The young man questions several Niagara victims about the circumstances of their death and their state of satisfaction under the water king. One of the voyageurs responds that he 'cannot complain', and adds, 'This is much better than the purgatory I had looked forward to as my just desserts' (Fulton 1901: 24). The victims of the river have somehow managed to escape the abyss that was seen as inescapable by nearly all who commented on death at Niagara. Within the context of this literature, therefore, the story seems anti-climatic. The effect would have been the same if Dante had included such a limbo in his vision of the after-life just before the sign that warns: ABANDON HOPE, ALL YE WHO ENTER HERE! (Dante 1948: 5). In terms of the wider metaphor, it is significant that Fulton locates her limbo behind the falling water. The plunge is still a moment of transformation, a falling out of this world, but the fall and the transformation are both arrested short of entry into another world. This tale also illustrates that individuals could sometimes imaginatively transcend predominant images of the after-life and create unorthodox alternatives.

The abyss

Entering the abyss, we break into a new dimension:

> Looking over into the abyss, we behold nothing below, we hear only a slow, constant thunder; and, bewildered in the mist, dream that the Cataract has cloven the earth to its centre, and that, pouring its waters into the fervent inner heat, they hiss into spray, and overhang the fated Fall, the sweat of its agony.
>
> (Curtis 1852: 85)

Symbolically the abyss is out of this world and the limits of this world no longer apply. We can see this in two ways. First, the abyss is unmeasurable: 'fathomless as Hell' (Lord 1869: 19), a 'boundless, bottomless eternity' (Liston 1843: 36). Second, we find that things can change into their opposites. A 'hell of waters' (Woods 1861: 241) becomes a hell of fire. The earth's 'fervent inner heat' mysteriously appears beneath the water, creating 'an unearthly cauldron heated by a hidden volcano' (Tudor 1834: 239).

If, when we enter the abyss, we have left this world, the question remains: what kind of place have we entered? Clearly it is not a good place, even to those who assign it the ultimately beneficial role of purgation. People *experience* it as horrible: what they imagine may derive much of its content from religious and scientific ideas they have learned, but the horror they feel is real and personal. Although many visitors have remarked that Niagara's abyss is something only a Dante or a Milton could describe (Tudor 1834: 240), these visitors clearly could imagine that abyss themselves. Consider this experience, recorded in the journal of Henry Tudor, an English traveller who visited Niagara Falls in 1834:

> when I turned my eyes to the curling masses of ever-rising vapour issuing from the turbulently-boiling surface, that seemed pendent over a hidden volcano, that awful passage from Revelations was immediately and most forcibly brought to mind: 'And the smoke of their torment ascendeth up for ever and ever, and they have no rest day or night.'
>
> I was literally overwhelmed by the unequalled grandeur of this stupendous landscape, by the solemn and absorbing train of thoughts which it had called forth, and by the pitch of overexcitement to which my imagination was wrought; and I felt a chill of secret horror creep through my veins, and curdle, for the moment, my very heart's-blood.
>
> (Tudor 1834: 266)

Although Tudor borrows a sentence from Revelations to describe the abyss, he has clearly had an experience of feeling a 'secret horror' himself.

The abyss at Niagara has been described as 'that deep torture-dungeon' (Blake 1903), a 'valley of darkness' (Houghton 1882: 119), the 'hell of the lost' (Appleton 1872) and the 'bottomless perdition of Milton's fallen angels' (Tudor 1834: 261). The power to imagine terrors seems almost unlimited: the abyss, in Anthony Trollope's view, was 'so far down – far as your imagination can sink it' (1862). Even for those who saw this abyss as a purgatory which the soul would ultimately leave for a state of eternal bliss, the prospect of death usually remained terrifying. Death, it seems, inevitably meant a taste of hell. Even Jesus had his descent into hell (1 Peter 3:18–22).

But people who imagine death with bone-chilling horror may still be attracted to it. If it is someone else's death, it may arouse as much curiosity as fear. This fascination with death, even in its most gruesome aspects, is directly illustrated by the numerous museums of horror which have clustered at Niagara Falls, successfully playing on the terror which the falls, and in particular the abyss, inspire, for the abyss, understood metaphorically, is the locus of death's deepest horrors. A thorough tourist could begin with a visit to Castle Dracula (plate 4.2) to see 'the recreation of DANTE'S INFERNO, FRANKENSTEIN, a haunting EGYPTIAN MUMMY and many more exceptionally strange characters' (Castle Dracula, n.d.). At the Haunted House the visitor will find 'a Ghost in Every Corner, a Skeleton in Every Closet, Every Kind of Ghoul Imaginable' (Haunted House, n.d.). Next one might try the Boris Karloff Wax Museum, the House of Frankenstein (plate 4.3) or the Criminal's Hall of Fame. At Ripley's 'Believe it or not!' Museum the tourist may 'wonder at the fearsome ancestor skull, a prize exhibit from New Guinea' and observe the 'world's most unusual graveyard' (Ripley's, n.d.). Then one might 'look danger in the eye ... from a place of safety' at the Niagara

Plate 4.2 A landscape of death: Castle Dracula, Niagara Falls, Ontario (photo by the author)

Plate 4.3 The appeal of Niagara seems to be related in part to a fascination with death: the House of Frankenstein, Niagara Falls, Ontario (photo by the author)

Serpentarium or visit Louis Tussard's English Wax Museum, which features 'such vivid spectacles as the slow death of Nelson, whose chest actually heaves as he expires on deck' (*Time*, 18 June 1965).

These businesses are not a new phenomenon at Niagara. The Niagara Falls Museum, according to advertisements, is 'the oldest museum in North America' (Niagara Falls Canada, n.d.: 42). Founded in 1827, this establishment was worth a visit, in the opinion of one early traveller, if one wished 'to sup full of horrors' (Bonneycastle 1849). The museum's leading attraction since before the American Civil War has been the collection of Egyptian mummies. As late as 1946 the museum claimed it had the largest such collection in North America (Abbot 1946). More recently the operators have added several naturally preserved American Indian 'mummies' (*Niagara Falls Museum News*). To link the horror museums with the death metaphor is to connect visitors' readings of Niagara's natural landscape with the construction of Niagara's cultural landscape. For the purposes of this essay they form a single text.

One of the most terrifying depictions of the after-life comes from the journal of Fitz Hugh Ludlow, a nineteenth-century traveller who described himself as a 'hasheesh eater'. The falls and rapids deeply impressed Ludlow, but, for months after his visit, Niagara haunted his dreams with visions of a watery death. He found himself 'in every variety of posture, helpless, friendless, frequently deserted utterly of every human being, ... hung suspended over the bellowing chasm'. One 'most agonizing vision', he wrote, 'stamped itself upon my mind with a vividness lingering even while I was awake'. In this nightmare he sat at the edge of the cliff next to Niagara Falls, mesmerized by a 'singular fascination'. A woman sitting beside him repeatedly directed him to gaze into the scene below. Suddenly he felt the rocks slipping beneath him, nearly dragging him down. Realizing that the woman meant to do him harm, he sprang to his feet and grabbed her by the arm:

> I glared into her beautiful icy eyes; I cried out, 'Woman! accursed woman! is this your faith?' Now, casting off all disguise, she gave a hollow laugh, and spoke: 'Faith! do you look for faith in hell? I would have cast you to the fishes.' My eyes were opened. She said truly. We were indeed in hell, and I had not known it until now. Wearing the same features, with the demoniac instead of the human soul speaking through them – wandering about the same earth, yet aware of no presence but demons like ourselves – lit by the same sky, but hope spoke down from it no more.
>
> I left the she-fiend by the river bank, and met another as well known to me in the former life. Blandly she wound to my side as if she would entrap me, thinking that I was a newcomer into hell. Knowing her treachery, as if to embrace her I caught her in my arms, and, knitting them about her, strove to crush her out of being. With a look of awful malignity, she loosed one hand, and, tearing open her bosom, disclosed her heart, hissing hot, and pressed it upon my own. 'The seal of love I bear thee, my chosen

> fiend!' she cried. Beneath that flaming signet my heart caught fire; I dashed her away, and then, thank God, awoke.
>
> (Ludlow 1857: 268–9)

Ludlow's fear of falling into the abyss is shockingly realized when he discovers that Niagara's entire landscape has become 'the hell of the lost'. This is another example of how the dominant symbology of Niagara's death metaphor could be imaginatively elaborated in quite unorthodox ways.

Some visitors, in spite of their Christian beliefs, have expressed grave doubts about the after-life their religion promises. A poem written by Thomas Gold Appleton in 1842 provides an example. Appleton conveyed his doubts by structuring the poem into two contrasting stanzas. The first stanza gives a daylight view of the falls, full of beauty and 'majestic calmness', with 'a spirit of mist' returning 'to the Heaven it came from'. The second stanza begins with an abrupt difference:

> As deepens the night, all is changed,
> And the joy of my dream is extinguished;
> I hear but a measureless prayer,
> As of multitudes wailing in anguish;
> I see but one fluttering plunge,
> As if angels were falling from heaven.

Heaven is now uncapitalized. The stanza concludes:

> As deepens the night, a clear cry
> At times cleaves the boom of the waters;
> Comes with it a terrible sense
> Of suffering extreme and forever.
> The beautiful rainbow is dead,
> And gone are the birds that sang through it,
> The incense so mounting is now
> A stifling, sulphurous vapor,
> The abyss is the hell of the lost
> Hopeless falling to fires everlasting.
>
> (Appleton 1872: 29–30)

Appleton emphasized this dark picture of the abyss by placing it last. 'The beautiful rainbow is dead': can we really expect mercy from a God who, in Noah's time, found only eight worthy out of the whole human race? But even Appleton – again through the form of his poem – retained an element of ambivalence toward the after-life.

For many interpreters of Niagara's symbolism the ambivalence has been weighted towards an optimistic view of death. For two late nineteenth-century poets even the abyss itself held a positive meaning. Richard Lewis Johnson (1898):

Nor let thy eyelids ever close,
In Neptune's arms in sweet repose,
Till all the nations shall disclose,
Like thee, Niagara,

A charity as broad and deep
As is thine own encircling steep,
Or as thy vortex where we peep
Thro' azure mists to heaven.

and Frank B. Palmer (1901):

Behold, O man, nor shrink aghast in fear!
Survey the vortex boiling deep before thee!
The hand that ope'd the liquid gateway here
Hath set the beauteous bow of promise o'er thee!

For these two poets the abyss held no real terror: it was a vortex that became a tunnel, a passage leading to paradise.

The rising mist and rainbow

The mist that ascends before the falls has nearly always been interpreted as a sign of hope, an indication that there is more to the after-life than the abyss. These lines are typical:

Though 'Hope's bright star' is sometimes pale,
Let Hope, not fear in man prevail,
The misty ghost within the veil
Proves life's resurrection.

(Palmer 1901)

But this resurrection is not a complete one: only a purified soul survives. As James Warner Ward (1886) put it:

Frivolous things are cast aside disdainfully;
Nothing the brink can pass but heaven-lit purity.

And, similarly, Harriet Beecher Stowe wrote, 'that beautiful water rising like moonlight, falling as the soul sinks when it dies, to rise refined, spiritualized, and pure' (Fields 1897: 89).

Some have described the mist metaphorically as a 'column of ascending incense' (Grinfield 1845). These descriptions were often part of a larger metaphor in which the falls were an altar, a temple or a church. 'The cataract of Niagara has been called "Nature's high altar": the water, as it descends in white foam, the altar-cloth; the spray, the incense; the rainbow the lights of the altar' (Lynch 1876). The rising mist has sometimes inspired a different metaphor when, instead of an altar, the falls are compared to 'earth's grandest cathedral', as in these lines by George Houghton (1882):

> Tall above tower and tree looms thy steeple
> builded of sunshine,
> Mystical steeple, white like a cloud,
> upyearning toward heaven.

The steeple and the incense are both images of yearning for connection with another world. Another symbol of connection – explicitly between heaven and earth – is the rainbow. That there should be, after the abyss, a rainbow – a bridge to heaven – is of special significance to those purified souls whose 'writhing ghosts' are 'out from the vortex flung', as Henry T. Blake (1903) put it:

> Swift through the glittering archway overspread
> Like radiant portal of immortal hope
> Upsoaring, vanish at the gate of Heaven.

The image of the rainbow as a bridge or portal appears often in the Niagara literature (Wellsteed 1849), although the rainbow is not always explicitly a portal to heaven. Henry Austin, for instance, used the rainbow as a less literal portal in his poem 'Niagara':

> . . . eternal rainbows crown the rocks,
> Halos of Hope, charmed circles of high Faith,
> Commanding entrance through the chasms of Doubt
> To deeps of nobler knowledge and soul-strength.
>
> (Austin, 1900: 2827)

The rainbow is a sign of hope and of connection between earth and heaven largely because it is also a sign of covenant. In the Genesis account of the flood, God promises, after the waters have subsided, that 'there shall be no flood to destroy the earth again'. And then, in a passage repeatedly alluded to by Niagara visitors, God gives Noah a sign of this promise:

> Here is the sign of the Covenant I make between myself and you and every living creature with you for all generations: I set my bow in the clouds and it shall be a sign of the Covenant between me and the earth. When I shall gather the clouds over the earth and the bow appears in the clouds, I will recall the Covenant between myself and you and every living creature of every kind. And so the waters shall never again become a flood to destroy all things of flesh. When the bow is in the clouds I shall see it and call to mind the lasting Covenant between God and every living creature of every kind that is found on the earth.
>
> (Genesis 9: 1–16)

Many writers have interpreted the rainbow at Niagara with explicit reference to this promise. After the deluge to the abyss, the rainbow signifies the re-establishment of order, of peace between heaven and earth.

For some visitors, peace has been more than symbolic at Niagara: it has been an experience. Just as the experience of terror that many have reported at Niagara has its symbolic counterpart in the images of hell and the abyss, so the rainbow as a symbol of peace has its experiential counterpart in the feelings of peace and repose which many visitors have also reported. These feelings of repose have not always been consciously connected with death, redemption and the symbolism of the rainbow. An English visitor, Greville J. Chester, travelling shortly after the Civil War, found peace in the horseshoe itself:

> Standing on the brink of the Canada Fall gazing into the center of the great 'horseshoe', where monotony and continuity seem to strive with ever-varying progress, the mind is affected with the deepest sense of peace and repose, and seems to catch the reflected image of Eternity itself. Deep, too, and deeply impressive as are the voices of these many waters, painful and oppressive they nowhere are; and these, too, speak peace to the soul.
>
> (Chester 1869)

One might postulate that even the Horseshoe Falls is connected to the comforting rainbow: it is, after all, a rainbow-shaped cataract.

The experience of peace has led some visitors directly to thoughts of death and eternal rest. Charles Dickens was one of these. In his *American Notes for General Circulation* he wrote:

> It was not until I came on Table Rock, and looked – Great Heaven, on what a fall of bright-green water!– that it came upon me in its full might and majesty.
>
> Then, when I felt how near to my Creator I was standing, the first effect, and the enduring one – instant and lasting – of the tremendous spectacle was Peace. Peace of mind: tranquillity: calm recollections of the Dead: great thoughts of Eternal Rest and Happiness: nothing of Gloom and Terror.
>
> (Dickens 1843)

The peace that Dickens and other writers saw in Niagara Falls and expressed in words also found expression in various concrete proposals for the Niagara landscape.

Archbishop Lynch of Toronto, in his pastoral letter issued at the opening of a monastery at Niagara Falls in 1875, described the feelings of peace and hope that he believed the Christian pilgrim would discover at Niagara:

> He looks upon that broad, deep and turbulent volume of water, dashing over a precipice ... and thinks of the awful power of Him who speaks in 'the voice of many waters', and of his own last leap into eternity. In hope he raises his eyes and sees quietly ascending clouds formed from the spray, bridged in the center by the beautiful rainbow. Again he cries out: 'let my

> prayer ascend as incense in Thy sight. Let my last sigh be one of love, after making my peace with God and the world.'
>
> (Lynch 1876: 104)

Several years earlier, Lynch had proposed to Pope Pius IX that a pilgrim shrine be established at Niagara Falls and dedicated to 'Our Lady of Peace'. The archbishop made the proposal at the beginning of the American Civil War, when he felt:

> moved with sorrow at the loss of many lives and the prospect of so many souls going before God in judgement, some it is to be feared, but ill prepared, and at the sight of the beautiful rainbow that spanned the cataract, the sign of peace between God and the sinner.
>
> (Lynch 1876: 99)

Pius IX officially dedicated the shrine in 1861 'to avail for all future time' (Lynch 1876: 100).

There have been several other proposals to establish institutions dedicated to peace at Niagara Falls, but none of them has succeeded. For example, Mr Leonard Henkle, a successful inventor from Rochester, proposed the construction in 1896 of an enormous meeting hall – averaging forty-six storeys in height – to be built out into the river just above the falls. Henkle's description of the hall's purpose is reminiscent of the United Nations charter, but it is hardly so secular, despite its claims:

> This hall shall be forever free for nonsectarian international religious and social and festive purposes. And in it shall be taught the paternity of God and the brotherhood of man; that there is only one true God, one humanity, one country, one religion, and one common ultimate destiny for man.
>
> The nations of the world will be welcomed to assemble in this hall, to be taught to cease the conflicts of war and to love one another. The social distinctions between poverty and wealth shall be therein destroyed.
>
> (*New York World*, 9 February 1896)

Henkle's idealistic scheme was not the last to go unrealized at Niagara (Gillette 1894). In 1945 an international committee proposed to site the new United Nations headquarters on Navy Island, a Canadian possession about three miles up river from the falls. The following year, when it became clear that only an American site would be seriously considered, a similar proposal was made for an adjacent island in American territory. The promoters called Navy Island 'the ideal site for the world peace capital' because, as they put it, 'here there is peace'. Peace was a 'spiritual attribute' of the site. But it was more than that, for 'here too', they pointed out, was 'international peace, long established, enduring'. This, they felt, made the Niagara region particularly appropriate:

> Surely, to those who will implement the purpose of the United Nations, it

> will be inspiring to execute their high duties in a locality steeped in traditions of peace and good-neighborliness, among peoples of various ancestry who have forged indissoluble bonds of international good will and co-operation, and who have made peace work.
>
> (Executive Committee to Promote Navy Island 1945)

The international peace along the Niagara has moved many to express sentiments like these. Sir Angus Fletcher, for instance, who served as the British Consul at Buffalo during World War II, once remarked, 'We have won, as it were, a local victory over war' (Graham 1949: 6). These sentiments have helped to inspire at least one successful international venture: the construction and dedication of the Peace Bridge in 1927 (Graham 1949: 293).

There have also been a number of peace conferences at Niagara Falls. During the American Civil War several prominent Confederate statesmen sailed around the Union navy and up the St Lawrence to Niagara. Horace Greeley, convinced that these southerners wanted to sue for peace, became their self-appointed intercessor with President Lincoln. After considerable exchange of messages, Greeley persuaded Lincoln in 1864 to send his personal secretary, Major John Hay, to Niagara. It seemed for a time that the two parties were close to an agreement to end the war. Greeley believed that the Confederacy was willing to agree to the abolition of slavery in exchange for some $400 million in compensation. No agreement could be reached, however, and Major Hay left Niagara quickly (Severance 1914: 79–96).

Canada was the host of a second peace conference at Niagara Falls in 1914, when an incident involving the landing of American marines at Vera Cruz created international tension. Mexico was already in the midst of revolution, and it now appeared that an American invasion was imminent. The governments of Argentina, Brazil and Chile offered to mediate in the dispute. Montreal and Havana were suggested as neutral sites, but the parties eventually agreed upon Niagara Falls. After two months of negotiation the tension between Mexico and the United States was successfully defused, although Mexico's internal strife continued (Severance 1914: 3–78).

It is difficult to find explicit connections between these peace conferences and the view of death as 'Eternal Rest and Happiness'. Nevertheless, it seems safe to posit at least an indirect link. The first official recognition of Niagara Falls as a place of peace – the shrine to Our Lady of Peace – was indeed inspired by such a vision of death. Archbishop Lynch tied his proposal to a symbolic interpretation of individual experience at the falls. The success of this proposal, along with the dissemination of travellers' accounts which described the falls as a peaceful spectacle, helped to create a tradition of associating Niagara with peace, which in turn made it seem an appropriate site for a peace conference.

Peace at Niagara has nearly always been connected with prior violence. This peace is not the absence of wrath but its aftermath. The rainbow arrives after the fury of the flood has spent itself. The bliss of heaven follows the purifying hell of the abyss. The Christian can conceive of the after-life as peaceful only because Christ:

> Endures the Father's curse and braves the storm
> Which threaten'd on our guilty heads to burst,
> And leave us deluged everlastingly
> Below the surges of a sea of wrath.
>
> (Liston 1843: 89)

And the Christian can interpret the rainbow as a sign of peace only for that same reason:

> That overwhelming cataract of wrath,
> Which on my Savior pour'd to rescue me
> Thus may I gaze upon the bow of mercy
>
> (Dowling 1848: 31–2)

Even the peace that Archbishop Lynch had in mind when he proposed the peace shrine had this connection with divine wrath. Lynch felt that Niagara Falls itself would inspire people to seek an alternative to this wrath: 'In this holy retreat of Niagara Falls many will find the road to heaven, and the true pleasure of serving God, and the real joy of having escaped the terrors of the world to come' (Severance 1914: 106).

Comparing Niagara Falls to an altar, even one with a foamy white altar cloth, has violent implications too. An altar is a place of sacrifice; the altar that Noah built after the flood was a place of blood sacrifice.

Lynch proposed the peace shrine at Niagara after blood had been spilled in the Civil War. Similarly, the peace conferences in 1864 and 1914 were prompted by prior bloodshed. The United Nations proposals came in response to the massive destruction of the Second World War. The first sentence of the United Nations charter recognizes this: 'WE THE PEOPLES OF THE UNITED NATIONS determined to save succeeding generations from the scourge of war, which twice in our lifetime has brought untold sorrow to mankind. . . .' (United Nations 1945).

The often-noted peace along the Niagara frontier that helped to inspire the United Nations proposals and the dedication of the Peace Bridge is actually a recent development. Like the rainbow after the deluge, it is a peace that follows a great deal of bloodshed. Conflict on the Niagara may be due in part to its geographical situation as a point of narrow separation along the natural barrier of the Great Lakes. Even before Europeans arrived, this was a place of conflict between the Iroquois who lived south of the lakes and the Hurons who lived to the north (Mau 1944: 8). Later, Niagara became the scene of a series of bloody battles between the British and the French, culminating in a British victory in 1759 (Severance 1917, 2: 303–28). The worst violence along this river came in the conflict of 1812–14. Indeed, Niagara was the only continuous theatre of battle during this period. Every major settlement on both sides of the river was destroyed. In the battle of Lundy's Lane alone – fought within earshot of the falls – nearly 2,000 perished (Howard 1968: 190). Although this was the last declared war on the Niagara frontier, open conflict was to flare up twice more. During the Canadian rebellion of 1837 the Upper Canadian rebels – abetted by American sympathizers – set up camp on Navy Island and prepared to invade Canada. International tension reached a peak in 1838 when a Canadian force crossed to the American shore and burned a steamship which had been ferrying supplies to the rebels and sent it drifting towards the falls. Support for the rebel cause gradually diminished, in part because the United States government repeatedly stated its neutrality (Guillet 1938). Finally, in 1866, 1,500 members of the Fenian Brotherhood crossed the Niagara river, intending, through the conquest of Canada, to force Britain to free Ireland. In a disorganized series of bloody clashes the invaders were eventually repulsed. Only after this point can we say that real peace came to the Niagara frontier (Howard 1968: 206).

It was this history of violence Lord Morpeth had in mind when he composed these lines in 1841:

> Oh! may the waves which madden in thy deep
> *There* spend their rage nor climb the encircling steep;
> And till the conflict of thy surges cease
> The nations on thy banks repose in peace
>
> (Holley 1883: 162)

Nearly all the images of peace at Niagara are, like this one, images of spent fury. Paradoxically, then, Niagara may have become a symbol of peace precisely because it had been the scene of so much violence, human and natural.

CONCLUSION

The nature of the possible worlds beyond death's door, so elaborately described at Niagara Falls, may seem simply irrelevant today. In the nineteenth century death was obviously more on people's minds. There was an increasing emphasis on the individual as a subject with unique experiences and as a competitive economic actor, particularly among the growing ranks of the middle class. This may have increased interest in death, the ultimate individual experience. The meaning of death was being negotiated, partly in response to the impact of Darwinism and other new ideas on religious beliefs. Some of Niagara's visitors questioned the traditional Christian view of death. But the idea that there might be no sequel to death – be it heaven, limbo, purgatory or hell – was still unthinkable. Those who wrote about death at Niagara assumed an after-life, and the possibilities of its nature fascinated them, not just because they each believed they would one day experience it but also because exercising their imagination in this way was exhilarating. The opaque face of death, and indeed the landscape of Niagara Falls itself, was a screen on which they could project their deepest fears and their highest aspirations. But Niagara's landscape was not completely opaque: it had a certain structure which could be interpreted in a way that matched the prevailing Christian understanding of death. Those who articulated the metaphorical meaning of Niagara's landscape produced written texts. In addition, certain behaviour patterns, traditions and elements of the human landscape at Niagara also form a series of texts that further articulate both the metaphorical meaning of the landscape and the meaning of death itself. This essay has presented an interpretation of all these texts in terms of the metaphor of death.

5

THE SLIGHTLY DIFFERENT THING THAT IS SAID

Writing the aesthetic experience

Jonathan Smith

A notable geographer and expositor of landscape visited Syracuse University, where I was a graduate student, and asked to see the local landscape. As his scholarship betrayed no particular interest in North American cities or the history of New York State, we sought to identify the source of his curiosity. He replied, frankly, that he loved landscape: without extensive knowledge of past events or present circumstances, he was pleased to gaze upon its shapes and surfaces. Furthermore, he expressed surprise that a geographer should be asked to articulate a motive that was so obvious or to justify a desire that was so natural. He simply took pleasure in sightseeing, as others take pleasure in symphonies, paintings or professional wrestling tournaments. Ignoring the technical problem of the 'virtual artist', which theoretically precludes landscape from consideration as art, most readers will recognize this pleasure (Dipert 1986; Zumbach 1984).[1] Certainly this is largely a matter of the sensuous enjoyment of light and shade, colours and textures, movements, sounds and smells; but it also follows a narrative pattern of confinement, exposure, foreshadowing, expectation, surprise and conclusion (arrival). In short it has to do with the fact that landscapes are stimulating, and even though we are well aware of the role of imagination in our perceptions it remains true that castles on hilltops are more stimulating than castles in the air.

Now this pleasure can be naive or sophisticated, but, because among professional geographers it is a rather guilty pleasure that seems slightly frivolous, we have tremendous difficulty conveying whatever subtleties of sophistication we may have discovered to one another, to our students and to the general public. When we ask, like Major Younghusband seventy years ago, for 'liberty to extend our knowing up to the extreme limit where it merges with feeling' we encounter an obstinate block (Younghusband 1920). The origin of this block is not, strictly speaking, institutional. Some of the field's most prominent individuals have endorsed an enlargement of our sensibility in a series of programmatic statements. The origin is neither philosophical, theological nor psychological; neither is the block symptomatic of a popular malaise. The root of the problem is what could be called, in the broad sense of the word, a linguistic difficulty that has arisen with changes in the way we read the landscape and in the way we write about it.

Let's begin with a characteristic analysis of our reputed insensitivity to the beauties of the natural world. Peter Fuller (1988) has described this as the result of a sturdy wedge called Modernism which has split the study of nature from the pursuit of art. The split in which this wedge lodges was left, Fuller argues, by the erosion of natural theology, or the conviction 'that nature, rightly seen, revealed moral and religious truths'. The subversion of this conviction was accomplished by the constituent belief that these truths were accessible through 'a relentless empirical naturalism'. Paradoxically, far from providing access, this very scrutiny forever closed the door on moral and religious truth and reduced nature to a 'grey lifeless monotone', which signifies nothing but moral relativism. The theologians of nature were not unaware of this danger. In poetics it is called the pathetic fallacy, and its dispiriting presence was linked to empiricism by William Wordsworth when he rued:

> That we should pore, and dwindle as we pore,
> Viewing all objects unremittingly
> In disconnection dead and spiritless.

This disconnection is of course the infamous world of broken images, where 'the image of nature as a harmonious garden ... fails to command the finest imaginative minds'.

In addition to empiricism, Fuller argues, this structure of feeling is the consequence of 'the continued retraction of religious belief, rise of uncreative factory work, and estrangement from nature'. Of these causes the second lay beyond the experience of 'the finest imaginative minds', while the second and third exerted their influence primarily within this self-appointed elect. So what, we may very well ask, is the state of mind of their quotidian counterparts? This we can only infer from behavioural evidence, but if facts as mundane as National Park attendance or National Geographic subscriptions are admitted, we must conclude that the masses do not find landscape beauty problematic.

Atheism, Fuller's personal conviction, has hardly carried the day. Indeed, if we include vernacular varieties of religious belief, there is considerable evidence to suggest that religious sensibility is resurgent. A great deal is said about the rise of uncreative factory work, and a century ago the observation may have had some validity; but very little is said about the decline of uncreative factory work, an observation that seems today more pertinent. And there has been, I believe, an overall gain, it being highly doubtful whether the neolithic farmer was more creative than the humanistic scholar, or even the hamburger vendor. It was Hawthorne's discovery at Brook Farm that, far from the clods being etherealized by the thoughts of their transcendentalist tillers, 'our thoughts, on the contrary, were fast becoming cloddish'. Finally, I find it difficult to impute the mass of people with a sensed estrangement from nature. Again, if we look to the vernacular interpretation of Darwin and Freud we find a doctrine of animalism. Indeed, Reinhold Niebuhr characterized the modern attitude as one of benign naturalism.

Fuller errs, I believe, when he assumes that the majority of people live in the modern world that has been defined by the grim and anaemic sensibilities of the historical avant-guard, and that those who do not are the possessors of 'a sentimental residue'. If they did, the avant-guard would not deserve its title; if they are, the attitude of the majority is made strangely negligible. This attitude, predictably modified by mass appropriation, is the attitude of the nineteenth-century avant-guard. It is, as it was, sentimental. But it is not a residue. It is a new way for the majority to regard their world.

Of course scenic values have never before existed outside that group which Fuller credits with the 'finest imaginative minds'. In the United States of the early nineteenth century the mode of regard with which the landscape was viewed was instrumental rather than sentimental. In the newly occupied Genesee Country, Isaac Weld was shocked to learn that 'the Americans . . . seem totally dead to the beauties of nature, and only to admire a spot of ground as it appears to be more or less calculated to enrich the occupier by its produce' (Handy and McKelvey 1940: 7). Eight years later, in 1804, Timothy Dwight (1969: 20) reiterated the point when he wrote that 'the phrase beautiful country, as used here, means appropriately and almost only lands suited to the purposes of husbandry, and has scarcely a remote reference to beauty of landscape'. Even the domicile, which we continue to sentimentalize by referring to the object as a home, was designed as an instrument of comfort. In 1862 Edward Dicy (1971: 9–11) marvelled at 'how little artistic taste there must be amongst a people who, with such incomes, are content to live in dwellings of such external unattractiveness'. Nowhere had he seen 'show so much sacrificed to solid comfort . . . substantial comfort, without external decoration'. Indeed, whether the object was personal enrichment, cultivation or solid comfort, the popular national aesthetic was founded on what Neil (1975) has called instrumental functionalism (Kouwenhoven 1982). From the beginning the popular preference was for buildings that were, as Louis Sullivan would say in 1892, 'well formed and comely in the nude' (Pevsner 1975: 28). Imaginative minds 'discovered' this, as Columbus 'discovered' America, rather after the fact, and they called the doctrine modernism. This was in spite of the fact that they ridiculed the American preoccupation with instrumental values, particularly as they were codified by Benjamin Franklin; but their criticism of the American's pragmatism ignored its fundamental relativism and multivocality, two more ideas that the elect have recently 'discovered'.

A second error is to assume that religious (pseudo-religious) or aesthetic sensibilities will necessarily recommend environmental prudence. In John 1 it is plainly stated, 'Love not the world, neither the things that are in the world.' As documented by Ann Bermingham (1986), for the aesthete 'the countryside exists less in its own right than as a fantasy solution to dilemmas formed elsewhere'. Indeed, those who seek beautiful landscapes are often, as Zaring (1977) puts it, 'studiously blind' to environmental destruction. Aesthetics has nothing to do with ethics. Hitler was an artist, and he would not have been a better artist if he was a better man.

If the deficiency exists, the fault lies elsewhere. We have little difficulty seeing beauty in the landscape, but, when we have, we find it awkward, or even embarrassing, to say so. The sensibility has not become 'sentimental residue', but the expressions, as Wolfe (1982: 67) puts it, fail 'to communicate meaning other than the imagery of their own collapse'. This situation is alarming, because the way of being in the world to which these expressions refer is made manifest and realized only in these expressions (Taylor 1980). If the expressive tradition peters out, the values and sentiments it presents will be lost. They are not like animals that roam the forest with or without a name.

PROBLEMS OF READING

When I say reading, I mean reading the landscape. This is said within the trope of landscape as a text. The trope is not in a formal sense different from other tropes common to geography. These include the earth as a spaceship, or as a mother; the city as a machine or as an organism; culture as a drama or as a growth; the economy as an engine or an athlete; houses as spoors; roads as a circulatory system, and so forth. When we look at the social history of reading the landscape we see that taste has varied, much as taste in literature varies. Also, we see that much of the modern landscape, like much of modern literature, is not to the general reader's taste, although this reader manages, as Tom Wolfe (1981) puts it, 'to take it like a man'. What we are witnessing is a divergence of taste and art, with taste remaining comfortably settled in the old and enduring pastoralism, while art, including the art of urban design, has been compelled by social and impelled by logical necessities to move outside, and come into competition with, this image of Arcadia.

Of course, when we say 'taste' we are talking about matters of consumption, in this particular case the consumption of landscape. By switching our mode of expression to the textual metaphor we have begun to talk about something we will call reading. The switch of terminology is useful at this point, since the metaphor directs us to a small but important literature on the social history of reading. George Steiner (1978) has based a series of provocative arguments on a tentative outline of such a social history.[2] He observes that private, silent reading was not the norm until the sixteenth century, a century which, incidentally, saw a revolution in the design of private gardens. The designs of Ligorio and Vignola, such as the Villa d'Este at Tivoli and the Villa Lante at Bagnaia, brought landscape into a new conjunction with the individual personality of its owner. The private library, the culturally sanctioned demands that one could make on others for silence and solitude, the luxuries of leisure and education: each of these was fundamental to the classic experience of reading. That these four necessities were attainable attests to the fact that within privileged groups reading was regarded as a worthwhile activity – indeed, that it was regarded as supremely worth while.

It was given its own space and time, and it was prepared for with a daily conservation of strength and a lifelong process of education. Individuals began to regard the landscape with an attitude that was similarly studious, and similarly serious. To gauge the change that has occurred, we need only ask when and where the bulk of reading, both popular and studious, is done today. We need only consider the degree to which the book that is open before us serves as a sign to others that they are not to disturb us for any reason more trivial than, say, a fire alarm. We have only to reflect, in other words, on the priority of reading to discover that it has become an activity of waiting places, like airports and toilet stalls, and of marginal scraps of time – over coffee and before bed, and that, far from serving as a 'do not disturb' sign, the book in our hands has become a homing beacon to the garrulous. In short: we read, or we are presumed to read, when we have nothing better to do, the list of these better things being extensive.

Outwardly, this shift in the priority of reading in the literal sense is paralleled by a shift in the metaphorical reading of landscape. Gardens, which in this metaphor are the counterpart of the library, have become a functional ensemble of recreational spaces and vegetable patches yielding satisfactions equivalent to those found on a shelf of self-help and how-to books. When this is not the case they are called (in the United States) yards, and they are nothing but an economical reference to a dwindling pastoral tradition, which are preserved and left unread for the same obscure reasons that the Harvard Five Foot Shelf of books is preserved but left unread. At the same time there are shifts in the time we devote to reading the landscape. The composition of a photograph is not necessarily easy, but it is always quicker than sketching or painting, activities that Ruskin (and more recently Peirce Lewis) deemed essential to proper seeing (Ball 1971: Lewis 1986). Hawthorne came to Niagara Falls 'haunted with visions of foam and fury, and dizzy cliffs, and an ocean tumbling down out of the sky'. It took him several days to rid his mind of this literary hyperbole, and realize that it was a scene that 'Nature had too much good taste and simplicity to realize'. It was only after looking for a long time that he could write, 'I came to know, by my own feelings, that Niagara is indeed a wonder of the world, and not the less wonderful because time and thought must be employed in comprehending it.' Didn't Carl Sauer admonish geographers to 'sit on vantage points and stop at question marks'? A fact about the English Romantic poets that draws our attention is the length and frequency of their rambles over the countryside, a ten-mile walk to fetch the mail not being uncommon. In addition to the sheer amount of time that was once spent looking at the landscape, which when compared with our own abbreviated and desultory attention seems to parallel the difference between reading Sir Walter Scott and Donald Barthelme, we find this time, like that devoted to reading, marginalized to those times of the day or week or year when we are exhausted. In a final parallel, we find that looking at landscape is an activity which our culture does not protect from invasion by that ubiquitous menace, the bored and garrulous person.

With this sketch of the external parallels between what is literally reading and the metaphorical reading of the landscape behind us, we can turn briefly to an internal parallel. This is, obviously, the issue of privacy. The question is, how much space and time are we allowed to hold apart from the various forms of collective social and economic activity? The answer is clear. No one today can claim the private time or space which the small group who turned landscape into art claimed and received from their companions.

I will draw one last parallel with reading, this time as described by C.S. Lewis (1988). In an attempt to define the 'good' book, Lewis said that good books were read by 'good' readers. Answering the question that this begs, he said that good readers were those readers who, first of all, give reading priority in their lives, and who, second, read books more than once. Reading a book for the first time, Lewis wrote, the reader exhausts his 'lust' for the plot. Reading a book for the second time, this novelty is removed and, remaining in the trope of concupiscence, the qualities of the themes of the text are revealed. Under this stricture most of us must confess to being poor readers. We are all guilty of saying, 'I read that,' and meaning that I read it and got the point and consider the matter finished. The parallel with reading the landscape is, once again, striking. In travelling we are satisfying a 'lust' for new places. Our pleasure, being a pleasure of surprise, is very much like the pleasure we derive from the plot of a story. Our boasts are 'I've been there, I've seen that,' as if we, once again, considered the matter finished.

This external similarity is rooted in an internal similarity. Available books by their sheer numbers strongly discourage rereading, unless we have failed to get or have forgotten the point. Likewise, cheap and swift transport has drastically enlarged the number of places and landscapes that are available to us, and lessened the likelihood that we will return, unless we have failed to get or have forgotten the photographs. This can only leave us, like Van Wyck Brooks's Oliver Allston, 'baffled by the multiplicity of all those doors past which one can see only through the key hole' (1941: 100). Coming to admit the futility of this sort of travel, Allston concedes:

> For travel, I wish only to see the countries that I have seen before, into which I have a degree of initiation. The rest, much as one may know about them in the casual way of modern men, can never be more than moving pictures, mere shells and outsides, like so many animated postcards.

Let's consider this image of the postcard, scrawled, short and speaking of haste. This we should place beside the index card, bearing at its head a title and in its body a summary, but serving like the postcard as a sort of visa stamp testifying to the owner's having been there. Both these texts are paraphrases or glosses of a larger text, a text of landscape and a text of literature. They are messages for readers who have not been there or for readers who do not expect to return. Going back to Lewis's trope of concupiscence, I would add that decks of index cards and boxes of postcards share with collected photographs of old lovers the drab melancholy that follows all lusts. The paraphrases of the index cards, the postcards, the photographs, seem from any distance futile; and they invariably prompt the questions 'Why did I read this?' 'Why did I go there?' 'Why did I desire her?'

In the work cited above, George Steiner defined a 'text' as something which was read with the intention of submitting a response. The book is a text when the reader is, at least potentially, a writer. The same definition is helpful when we begin to puzzle over the metaphor 'the text of landscape'. Landscape becomes a text when the reader intends to respond (whether as a civil engineer or as a writer). In the writer's case, as we have seen, the social conditions of reading have changed. The reading, and the text that is created in this reading, is, owing to the social devaluation of reading, subject to frequent irrepressible interruptions. At the same time the new reading is a series of glosses or paraphrases that is not unlike a notice board covered with postcards. It is driven by the 'lust' for new places that modern technology allows us to indulge, and it yellows as quickly as newsprint.

Of course, none of this should apply, at least in theory, to scholars, who are given the space and the time to read (or do fieldwork), and who are guided by the restrictions of their discipline to something like a comprehensive understanding of a certain limited canon of books (or places). Scholarly reading represents an exception, an anomaly in the setting I have just described. But this reading, and the text it creates, is plagued by its own problems, and these are the problems to which we will now turn.

PROBLEMS OF WRITING

Scientific and historical knowledge has forced on us a new and troublesome awareness of the symbolic. This knowledge is, by its own claim, literal. The claim can be denied, but the effect of the claim on other forms of descriptions cannot. It not only denigrates certain descriptions as poetic, figurative or subjective, but it also selects certain poetic, figurative or subjective descriptions for which it can offer a literal alternative while it leaves untouched in a second category descriptions of subjects which it does not know. The selected descriptions are explicitly debunked. To the extent that this debunking discredits the language of figurative descriptions generally it carries an implicit condemnation of all figurative descriptions but, except for the most doleful positivists, it does not demolish figurative descriptions altogether. The descriptions that are not selected are set apart in a separate category. The distinction appears to stem from ontological subtlety when it is in fact a consequence of linguistic subterfuge or what Andrew Clark might have called one of our 'esoteric taxonomic myths'. What it does is create a sense of two worlds, one the domain of literal description, the other accessible only to imagination and figurative language. This segregation of rhetorical styles, with figurative language herded into the overcrowded homelands of adolescent longing and middle-aged introspection, precludes perception of what Marcus and Fischer (1986) call 'the ironic conditions of knowledge'. This segregation into two separate texts transforms difference into distance, and the valuable irony of the 'slightly different thing that is said' is, quite regrettably, defaulted.

If we look at the discourse on scenic values that became popular in the nineteenth century, we find a series of terms which are to our minds figurative. However, there is every reason to believe that those who expressed themselves in these terms found it necessary to believe that their meaning was literal (Bevan 1938: 254–74). For example, when they said that a chasm was awful or that a cataract was terrible they believed that the adjective was properly attributed to the chasm or the cataract. We on the other hand believe that they were interpolating a reference to the psychological state that these features produced in their own minds. What their language exhibits is, in George Santayana's words, 'the survival of a tendency originally universal to make every effect of a thing upon us a constituent of its conceived nature' (1955: 31). The same tendency is evident in respect of beauty, which Santayana defined as 'pleasure regarded as a quality of a thing'.

To give some substance to this idea of the qualitative attributes of objects it may be helpful to retell a story about the romantic poet Samuel Taylor Coleridge. During a walking tour of Scotland, Coleridge stopped to admire the falls of Clyde. Taking a seat, he began to puzzle whether the sight before him was grand, majestic or sublime. Just as he settled on an ascription of majesty, the poet was joined by another spectator. Without hesitation the newcomer blurted out that the falls were surely majestic. Led by this exclamation to suspect the stranger gifted with rare perception, Coleridge addressed the man, who promptly affirmed his initial opinion, and then added, to the poet's dismay, that they were sublime and beautiful too (Watson 1970: 111). He could as well have said, 'However it strikes you, mate, it's all the same,' for he was certainly a relativist. Persuaded that such terms were at best matters of opinion, he exhibits an attitude very much like our own. He rejects the proposition, characteristic of traditional thought, that emotions and valuations of objects, in this case objects in the landscape, can be correct, proper or ordinate (Lewis 1947).[3]

To read any landscape is to read a text of metaphors, or more precisely a text of synecdoches (Duncan 1990). The falls in the story I have just related are a synecdoche for the idea of majesty. In them majesty is manifested and realized just as the falls of Reichenbach are a synecdoche which manifests the idea of dread surrounding Sherlock Holmes's last battle with Professor Moriarty. To understand the difference between figurative and literal language, we must ask the following question. Are such metaphors considered proper, or are they considered common? If the metaphor could be, at least in theory, proper, or expressive of a property of the object, then majesty and dread become a part of literal description. The falls are no less so than they are wet. On the other hand, if majesty is common, and as objectively correct when attached to an anthill, the phrase is merely psychological interpolation. It is unverifiable and therefore inadmissible in scientific discourse, which considers such interpolations as corruptions.

On a map a dot is a conventional sign that can stand for anything that is not a line or an area. Its meaning is stabilized by the key. The silhouette of a pig is an iconic sign that can stand for a handful of things. Its meaning is held in place by the general shape of pigs. It can only represent places where pigs, pork or perhaps piggish habits are found. Both signs are representations that make some part of reality manifest on the map, and both signs are meaningful; their difference lies in the key on which they are based. The key to the conventional sign is on the map. If it is mutilated the map is illegible. The key to the iconic sign is a certain experience of the world in which pigs are not wholly alien. In this case the sign is its own key, and if you have the sign you have the meaning. If you pick up a shred of a map that depicts the emblem of a pig beside a thin blue line, you will be able to construct some meaning. If you pick up a shred of a map that depicts two dots, you will be unable to construct any. Now are the metaphors of the landscape conventional and common like a dot, or are they iconic and proper like the emblem of a pig? If they are iconic and proper their expression is not an interpolation and a corruption but a verity and an enrichment.

To answer this question we must locate the key by which the affective cues of the landscape are made legible. There is, obviously, a high degree of textualization, in the form of guidebooks, and reification, in the form of parks and scenic overlooks. But these do not fully explain the pleasure that we take, as Emerson put it, on our 'necessary journey'. On the contrary, it is in these unguided pleasures that the institutions originate. Just as the manuscript represents the idea and the bound volume the manuscript, the view represents the falls, and the guard rails and road signs represent the view. In either case the object is to some extent constituted by the image of its ultimate form, be it the published work or the developed site, but no amount of later packaging will compensate for the absence of the original seed. There are books without ideas and sites without scenery, but even slick commodification can not constitute from nothing what they inherently lack.

Some time ago Merleau Ponty (1964: 2–21) observed that human actions 'reveal a sort of prospective activity in the organism', an 'orientation towards the meaning of certain elementary situations' that suggests 'familiar relations ... which predispose the organism to certain relations with its milieu'. Arguing against the 'pure exteriority' of behaviourism and the 'pure interiority' of idealism, he stated that the body is 'the place where the spirit takes on a certain physical and historical situation', and in so doing 'it applies itself to space like a hand to an instrument'. Perception is the result of motor, affective and sensory functions. The first two are learned, and thus 'the structure of the perceived world is buried under sedimentations of later knowledge'. Martin Heidegger (1962: 190) described this 'prospective activity' as interpretation:

> In interpretation, we do not, so to speak, throw a 'signification' over some naked thing which is present-at-hand, we do not *stick* a value on it, but when something within-the-world is encountered *as such*, the thing already

> has an involvement which is disclosed in our understanding of the world, and this involvement is one which gets laid out [made manifest] by the interpretation.

Thus an individual's perception of water depends on the ability to swim, and on cultural attitudes towards swimming, as much as it depends on the sound of splashing, the feeling of wetness, the sight of liquidity or the smell of seaweed. Given these 'sedimentations of later knowledge', we have a key, and with this key a pond is objectively a swimming place and on a hot summer's day it is literally beautiful.

Of course, our encounters with these 'elemental situations' may be rather limited, and few things are encountered 'as such'. Everywhere we look we encounter a pre-interpreted landscape, or a landscape made legible. Michel de Certeau (1984: 97) describes built form as a metaphor that makes legible those actions that it pretends to facilitate, but in his eyes the metaphor is deceptive. Drawing the ratio that walking is to the city as speech acts are to language, he notes that 'the trace left behind is substituted for the practice'. Furthermore, with a debt perhaps to Bergson, who described the activity of consciousness as the 'interval between representation and action', he finds the representation false and deceitful. Being visible, he writes the landscape 'has the effect of making invisible the operations that made it possible', and this obstruction 'exhibits the (voracious) property that the geographical system has of being able to transform action into legibility, but in doing so it causes a way of being in the world to be forgotten'. This argument is, of course, dialectical. The most legible representation becomes the most illegible representation. A similar argument is encountered below in the notion of hyperreality, a term, apparently coined by Umberto Eco (1983) and employed by Jean Baudrillard, which refers to the utterly unrealistic nature of the most realistic representations.

Jean Baudrillard (1988) did what he claims ideas can never do. He crossed the Atlantic. He made this journey:

> in search of astral America, not social and cultural America, but the America of the empty, absolute freedom of the freeways, not the deep America of mores and mentalities, but the America of desert speed, of motels and mineral surfaces.

He found what he was looking for:

> Where the others spend their time in libraries, I spend mine in the deserts, and on the roads, and I get to know more about the concrete social life of America from the desert than I ever would from official or intellectual gatherings.

What Baudrillard gets to know while he is truant from academic meetings is, in a sense, what Robert Frost learned when he informed the star that 'to be wholly taciturn is not allowed'. He learns, like Edward Abby, that the desert means nothing; its significance is silent. You will recall the star's response to Frost: it said only that 'I burn'. Likewise, in the desert, things stand for nothing but themselves. They refer to nothing human, but in fact mutely critique the whole pretension of humanity, significance and culture. Reviving the theme of Paul Horgan's novel *The Return of the Weed* (1980), Baudrillard writes that the desert signifies 'the mental frontier where the projects of civilization run into the ground'; while at the same time it 'designates human institutions as a metaphor of that emptiness'.

What Baudrillard learns about the concrete social life of America is contained in what he calls this 'irony of geology', but it is amplified and completed by 'the profusion of sense' that seems to flourish undesiccated by this desert. The extravagant gesture of desert culture splashed on the pathos of the stones is a metaphor for the American individual and the American collective splashed on the pathos of history. This 'electric juxtaposition' is the pivot of the extended metaphor, modernity is the American landscape, which structures most of Baudrillard's book. The scheme of his narrative is to use manifest refutations of environmental determinism, or a discontinuous geography, to illustrate the denial of historical determinism, or the historical discontinuity, that lies, he argues, at the heart of American modernism.

America broke with Europe, and with history, when it undertook to make utopia real, a pragmatic design one would expect of Americans, since, as Baudrillard writes, 'it is not conceptualizing reality, but realizing concepts and materializing ideas, that interests them'. Realization and materialization yield simulation; conceptualization and ideation yield ideology. As America eschews the latter it embraces the former. In Baudrillard's words, here:

> Everything is destined to reappear as simulation. Landscapes as photography, women as the sexual scenario, thoughts as writing, terrorism as fashion and the media, events as television. Things seem only to exist by virtue of this strange destiny.

This 'strange destiny' of simulation is therefore the aesthetic form of 'our ulterior, asocial, superficial world', and the pleasure that seems to be a quality of that form is 'the charm of disappearance'. This pleasure is every day realized, he claims, in the experience of driving over 'excessive, pitiless distance', and in this act he discovers a metaphor of endless futurity and an obliterated past which serves to illustrate all that is modern, this contestable term being, in Baudrillard's account, 'an original break with a certain history'. It is leaving the past behind, becoming anonymous, becoming amnesic. Simply driving, with your eyes on the road and your mirrors askew, 'you learn more about this society than all academia could ever tell you'. The question he poses is 'how far can we go in the extermination of meaning . . . in the non-referential desert form without cracking up; and, of course, still keep up the esoteric charm of disappearance?'

Our fugitive journey may very well end in our 'cracking up' if our only navigational aid is a map with a mutilated key, a map from which the meaning has disappeared. This conclusion is certain if we grow sufficiently enamoured of the map, and find its more legible simulation of the landscape the only image we dare trust. On the other hand there is an undeniable charm in letting the slightly different thing that the map says stand in for the real thing. But the hyperreality of the map, which exterminates the meaning of waterfalls and swimming holes, is charming only so long as we maintain our human bond with these metaphors and can thereby enjoy the slightly different thing that is said when the metaphor is referred to one or the other key.

A PROPOSITION

On the surface it appears that we have on the one hand what Mircea Eliade (1959) called 'a shattered universe', and on the other what Cecil Day Lewis (1946), following T.S. Eliot, called 'a heap of broken images'. These are, of course, metaphors for Nietzsche's proposition 'God is dead,' a proposition which I have argued is not, at least in pragmatic terms, valid (James 1902). What Day Lewis and countless other artists have suggested is that 'the soul of things' is lost, but that some intimation of the soul of things may be recovered in poetry. In his image, the poet 'includes both the object and the sensation connecting him to that object', and thus, in accord with Santayana's definition, he compensates for the holy with the beautiful. The same point was made by D. N. Jeans (1979) when he called for us to describe a place with 'an attempt to show forth its nature as deriving from its absorption into mind'.

In 1928 Owen Barfield produced a slender book entitled *Poetic Diction: a Study in Meaning* in which he presented a theory of language which made metaphor the source of all meaning, and the root of all being.[4] It is the source of meaning because, when they are newly invented, metaphors register the role of human intention and imagination in perception. Like his ever present mentor, Coleridge, who called Primary Imagination 'a repetition in the finite mind of the eternal act of creation in the infinite I Am', Barfield concluded that 'the mind can never even perceive an object, till the mind has been at work combining the *disjecta membra* of unrelated percepts into that experienced unity which the word "object" denotes'. It is the root of all being because, in those ancient metaphors that make up the tissue of our language, ancient similarities long since divided by logical thought remained linked. Barfield's favourite example was the Greek word *pnuma*, a word standing for a fundamental entity that we have divided into the concepts of wind, breath and spirit (Adey 1978). In many respects the book is a product of its time, and I am not about to take up its anthroposophical agenda, but it does make a point highly relevant to this discussion. Barfield observes that the central role of metaphor is strangeness, or what we now call defamiliarization. Strangeness is 'the pleasure of realizing the slightly different thing that is said' when an idea is expressed in a foreign language, an unfamiliar word, a poetic or an archaic phrase. It is the word 'slightly' rather than the word 'different' that should be stressed. When we write, as we do not at present, of our aesthetic enjoyment of the landscape, the validity of our interpolation rests on the claim that the landscape is our subject. It is as if we were sitting on a hilltop with a map, alternately looking at the presentation and the representation, and coming to some understanding with the pleasurable exercise of crossing and recrossing that difference.

NOTES

1 Dipert notes that although the landscape may contain 'sensuous beauty' it is difficult to regard as intended other than by those with a religious impulse who regard God as a 'virtual artist' who 'intends that certain intentions be attributed to him'. Clearly he is referring to the natural landscape. Clark Zumbach on the other hand has argued that art by ascription succeeds by converting imagined design functions into interpreter's use functions. It is the charge of interpretation to show how the parts add up to the suggested whole.

2 This begins to follow a suggestion made by the historian Brian Stock at the conference entitled New Directions in Cultural Geography, which was held at Syracuse University in March 1989.

3 I draw the idea of 'objective values' from Lewis, but must admit that his use of this same story to illustrate his point appears on my reading to be incorrect.

4 Barfield's ideas are, broadly speaking, similar to those presented far more comprehensively by Ernst Cassirer.

6

LINES OF POWER

Gunnar Olsson

For Jacques Lacan the unconscious was structured like a language. For me power is structured like a knowledge.

As I now move from this double beginning, I quickly find myself in the company of René Girard and his theory of mimetic desire. I am driven there by my conviction that knowledge is by definition an exercise in translation. But just as desire desires not to be satisfied, so translation translates into distortion. And, in this light, communication shows itself to be what it really is: a form of collective violence designed to neutralize the deviant by sacrificing it on the altar of social cohesion. The unknown must submit to the known just as the world is divided into reasonable and unreasonable. Subjects are produced through subjection, objects through objection.

Boiled down to its essentials, telling the truth is to claim that something is something else and be believed when you do it. Others trust me when I say that this is thus or that $a = b$. To succeed in this tricky business is not so easy, however, for after Nietzsche everybody knows that this is this, not thus, that *a* equals *a*, not *b*. The similar and the same are similar, not the same; there is never presence, only proximity. And with this reminder I have already demonstrated how perfect knowledge is as impossible as perfect translation, how truth has more to do with trust and conviction than with what here and now happens to be the case. In addition, I have suggested that, even though *a* and *b* are striving to become doubles of one another, this desire can never be fulfilled. Both knowledge and language require a defect or a fault, an *alter ego* who keeps his distance. The dynamics of power are rooted in difference.

What remains is a play of make-believe. To this end, I shall attempt a deconstruction of the word IS – the epistemological marker *par excellence.* Nobody can do without this sign of mimetic desire in concentration, for without it there would be neither understanding nor communication. Without IS nothing would be, not even Nothing.

At the same time, IS would not be IS if it did not have many meanings. It is indeed this ambiguity that turns the IS into a key concept in the vocabulary of power. Its nature is to be supplementary and constantly shifting; if captured in one context, it promptly escapes to another. As it performs its juggling tricks at Polity Fair power stays in power because of its evasive anonymity; by showing itself it conceals itself; in hiding it shows. The word is the manifest non-existence of what it designates.

To keep track of myself, I shall concretize some various meanings of the IS by constructing a set of straight lines. It is in the turning and twisting of these objective correlates that I shall then catch a glimpse of power with its pants down. A blue eye meets a brown, analysing and experiencing in the same glance; the law itself as an instance of Epimenides' paradox of the right to say I: all Cretans are liars, for power is in that miniscule yet insurmountable space between speech and silence.

My first drawing of the IS consisted of two parallel lines, an equals sign of the type:

=

During the first half of the 1970s I spent some of my best years in the company of this symbol. Every evening we went to bed, every morning we had breakfast, every day we conversed. Bewitched, bothered and bewildered. And yet. The emptiness between the two lines eventually melted away, the two became one, turned on their side, transcended themselves and became the slash of:

/

In the mid-1980s the potent angle changed once more as the line found a resting place in the Saussurean bar of:

—

While '=' demarcates what is identical to what and '/' stands for the penumbra of a mutual relation, the '—' is the rendezvous of signifier and signified. While '=' has its roots in logics and '/' comes from dialectics, '—' is central to semiotics. Regardless of these differences, however, the three signs are all possessed by the same kind of mimetic desire; what stands on one side of the line wants to merge with what stands on the other. But, here as elsewhere, the desire is defined by its impossibility.

The story of the straight line contains everything I know and everything I have not yet understood. This includes fragments of power. To me it is in the richness of these symbols – signs for the simultaneous splitting apart and joining together – that power reveals itself in its most clear, most elementary, most beautiful form. In comparison, the king's sceptre, the general's baton or the bishop's crozier appear as vulgar as the Danish emperor without clothes. It was Georges Bataille (*Visions of Excess*, 1985: 5) who noted that 'the *copula* of terms is no less irritating than the *copulation* of bodies'.

If the equation is the wheel of positive science, then the equals sign is the nave of that wheel. The statements on each side of the sign should represent the same quantity, express the same amount or say the same thing in different words. Clear enough. Yet there is a serious problem: since every reformulation is of necessity supplementary and therefore strictly speaking a distortion, a given supplement can sometimes be accepted as a truth, sometimes rejected as a lie. The issue is further complicated by the requirement that the reformulation should be not only true but also informative; whereas a tautology is by definition true but not informative, a metaphor is often informative but not true. Once again, it is not sufficient to note that something is the case. I must also be believed when I point it out: truth is not a private conviction but a social convention. Truth is not truth unless shared. Hence the statement $a = b$ becomes credible when it refers to phenomena that are open to common inspection, a condition which legitimizes the power of logical empiricism and grounds the metaphysics of presence. It is nevertheless an argument of postmodern deconstruction that such ability to share the world must not be taken for granted. The pronouns 'I' and 'he' are radically different. It is not surprising that truth is often invoked as an excuse: 'Objectively, comrade!' 'Unconsciously, sister!'

Indeed, it now seems clear that every truth contains an element of distortion, of lying, of them overruling me. Every truth is inevitably formulated around a nucleus of difference, for knowing what something *is* involves knowing what it is *not*. In the public congregation of categorization new truths are being blessed and old prejudices sacrificed. Ecstacy is that brief moment of conversion in which one set of taken-for-granted is replaced by another. But what does it mean to change one's taken-for-granted?

To change one's taken-for-granted is to change one's central beliefs. It is in the middle of this leap that epistemology turns into ontology, problems of existence turn into issues of being, reasoning turns into power. And thus it is that both knowledge and power, resistance and legitimation, rest on a foundation of convention and agreement. The frightening alternative is madness, for as Ludwig Wittgenstein remarked in his *Zettel* (1967: 393):

> If I were sometime to see quite new surroundings from my window instead of the long familiar ones, if things, humans and animals were to behave as they never did before, then I would say something like 'I have gone mad'; but that would merely be an expression of giving up the attempt to know my way about.

Does this mean that to be mad is to approach the limits of meaning? Perhaps! For in such situations I am left completely alone, for then I am not like anybody else; if a denotes myself, then there is no b to go with it. Not knowing my way about is another way of saying that I am completely lost, with no fixes to keep me steady, with no contexts to share. It is exactly at this moment of horror that the equals sign shows itself in its imperial nakedness: a blessing in disguise, an instrument of socialization, a standardized tool for making you and me normal, predictable and interchangeable; logic is not merely a matter of form, even though the copula is the real subject of speculative thought.

The basic assumption of this reasoning mode is in the Leibnizian principle of *salva veritate*; truth is preserved when two propositions are interchangeable in every context. Put differently, what I say about an object is assumed to be true regardless of what the object is called. In reality, however, two contexts are never exactly the same, for then there would not be two contexts but one. Likewise, two synonyms are never totally overlapping. My credibility is therefore a function of how I distinguish one context from another and of how I name the resulting categories. As expected, this is a privilege held by God himself. Let there be! And there was! Light and darkness, air and water, fish and fowl, Adam and Eve. In pairs of two all flowed out of his mouth. The Creator did not build the world, he uttered it. Context creates order, power creates context.

Even the naming is nameable. And this possibility poses again the Epimenidean question of whether a particular Order of Context is to be trusted. Is God – the key figure of power – logical? Is God himself true to the principles of *salva veritate* and to the laws of consistency that are symbolized by the equals sign? Of course not! Especially in the Old Testament, God instead exercises his power by creating paradoxes, predicaments and double binds; first he demands obedience and then he changes his commands. Abraham provides the paradigmatic case, and that is why he was so crucial both to Søren Kierkegaard and to Franz Kafka. But the man in question was already ninety-nine years old when the Lord made the covenant with him that he was to be the father of a multitude of nations. This altered context was then codified in a decree which proclaimed that his name should be changed from Abram to Abraham. New worlds require new labels, for what is true about an object is not independent of what it is called.

But the story continues. *Genesis* 22. The Almighty designs a test of Abraham's obedience. As a sign, he must sacrifice Isaac, his only son, whom he loves. No regrets, everything ready, the son tied to the altar, the knife on his throat. Then, suddenly, from the sky, the voice of the Lord's angel: 'Stop! Stop! For Heaven's sake. Don't kill. Your master knows that you fear him. He has changed his mind.' The murder is no longer necessary, for in everything Abraham has already demonstrated his submission. Yet it is at this very moment of apparent relief that Abraham experiences limitless terror. God – the incarnation of power – is not to be trusted, for if he has changed his mind once he can do so again. Only logic is predictable, and God is not logical. Indeed, 'God' is nothing but a proper name for everything we sense is too important to ignore and too evasive to specify. 'God' is a pseudonym of power.

To logic paradox is an enemy, to power predicament is an ally. Later dictators have become experts in the same form of institutionalized double bind and uncertainty. The torturer who by mistake kills his victim has eliminated this sense of uncertainty and thereby the whole point of his business; the torture is for the audience, not for the victim. The only defence may indeed be in the deployment of irony, satire, humour and poetry; even though revolt is often necessary but ineffectual, refusal is always possible and irritating. As a consequence, dictators fear laughter more than tears, lonely poets more than organized protesters, happiness more than sorrow. But whereas humour and poetry engage in a play of logical types, deduction is characterized by consistency and a form of reasoning which moves from an axiomatic beginning to an inevitable end. Under the pressure of *salva veritate* even the equals sign transforms itself into an arrow, the key symbol of mimetic desire.

The best example is of course in the causal model. Here the equals sign serves as a selective filter which lets through more influences in one direction than in the other. At the same time, causal relations are often dressed up as logical relations, even though the two types of implication are drastically different; whereas cause-and-effect involves time, power and responsibility, logic is without time, without freedom and without guilt. In both cases, however, the argument strives for acceptance, for like all philosophy also logic is inherently rhetorical; the trick is to be believed, to reason in a manner that increases one's credibility. Belief is the instrument of power *par excellence.*

In this context the most interesting rhetorical trope is the metonymy. Its strategy is to create trust by making the reader or listener recognize himself; its tactic is to let the concrete rule over the abstract, the specific over the general. This practice is itself part of the metaphysics of presence which permeates all empiricism. There is a high correlation between credibility and concretization, power and thingification, order and communication; the anatomy of power says that the way to the mind goes via the body, that discipline is anchored in details. Privilege and exploitive power have always masked themselves as duty and responsibility.

Perhaps there is nothing more powerful than the power of the example; speech is not only indicative but also imperative. The current challenge is to subvert that power. Indeed, I have come to believe that our very survival depends upon improved ability to be abstract enough; the most radical point is the point of insolvability. And with this call to abstractness I have already begun to move from the ordering parallels of the equals sign to the devious slant of the slash.

As the equals sign belongs to logic, so the slash stems from dialectics. While the former dominates positive science, the latter permeates deconstruction. It is in the former to symbolize knowledge as the restatement of identities, in the latter to denote the inseparability of identity and difference. The former searches for the certain in the ambiguous, the latter for the ambiguous in the certain.

Even though the two tropes are thoroughly intertwined, the equals sign gets its convincing power mainly from metonymy, the slash mainly from metaphor. It follows that the equals sign points the way to standardization and thingification, whereas the slash eludes all attempts to catch it. In both cases, however, the structures of substitution reflect the forms of power and the forms of power the structures of substitution. The ensuing problem has its roots in Plato: what makes us see must itself be invisible, what makes us understand must itself be non-understandable; the mirror is a mirror not for the mirrored but for the mirroring, not for the object in front of the glass but for the tain behind it. It is with the writings of Jacques Derrida that the discussion is now returning to a level of abstractness worthy of Hegel, Kant and Descartes: what is reflection and what makes reflection possible? What can I know about what, except that epistemology never left the mirror stage and that thirsting for knowledge is an instance of unquenchable narcissistic desire.

It is this desire for the invisible that itself becomes visible in the slash, for to me the slanting line serves as a concrete symbol of an abstract relation; with T. S. Eliot, I conceive of symbols as objective correlatives of human feelings. This may sound as if I try to re-enlist the forces of metonymy, but then I seek consolation in Jacques Lacan's observation that the symbol manifests itself as the killing of the thing. Lacan's remark is important, because it is the very essence of a relation to be extremely abstract, invisible and untouchable; like silence, every attempt to capture it fails, for every attempt destroys it. It follows that relations cannot be defined, only experienced. Relations should not be confused with what is related, just as the desire for love should not be confused with the loved one.

Through the slash everything hangs together with everything else in a maze of internal and self-referential relations. Even algebra may serve as an example, for three times three would not be nine unless three times seven were twenty-one and seven times seven forty-nine. And so it is easier to understand both why dialectics has been called the algebra of revolution and why all revolutions fail. The concept of implication is a subjunctive which in empirical operationalization is perverted into an indicative. In the need to speak there is nevertheless a desire for silence, an idea of infinity, a sense of neither identity nor difference. The future is future because it is ungraspable, not because it is manipulable.

It follows that dialectics is not a predictive mode of reasoning or a metonymic system of causal models. It is rather a form of epistemology specialized in the discovery of the hidden in the apparent. As a consequence it can never look into the future, only into the past, an insight which was born with Hegel, lived with Kierkegaard and died with Stalin: dialectics is not a language of commands but a language of understanding, especially of the relations between repression and submission, master and slave.

The dialectic nature of the slash becomes especially evident in confrontations between the individual self on the one hand and the collective taken-for-granted on the other. Thus there are many who are shocked when they experience how society issues orders which to them seem unfair or how it makes claims which to them are false. In these situations the analyst usually asks whether he should put more trust in the individual or in the collective. The dialectician instead answers that neither is to be trusted, for the two terms are defined in terms of one another and are hence in constant and unresolvable conflict.

It is nevertheless the void between individual and society that constitutes the realm of political power. The deep insights which fill this social space of silence must not be revealed, just as God the Father could not permit his children to eat of the tree of knowledge or to mention his name. It is indeed through the automatic recital of commandments that society protects itself from its members, and individuals guard against the collective. It is this that the slash symbolizes, that which the angels fear: relations beyond the related.

But how can I know that I as a subject conceive of the world in the same fashion as you do as an object? Assuming that this question lacks an answer (for there is no Other of the Other), by which right do you then engage in political action, an activity which in practice always means that the will of the Other is subordinated to your own? Is sociality to desire your neighbour? If so, how do you handle the problem that the collective 'we' can never be anything but a majority; 'social democracy' is an oxymoron. If the personal is political, then perhaps the political is pathological. For how small must a minority be before it is forbidden? And how large must a majority be before it becomes such an integral part of the taken-for-granted that it turns silent? For the state to whither away, it must be everywhere. Fascism is a tendency to homogeneity. After Gulag and Auschwitz critical thought can be nothing but plural.

Political power is structured as a slash: internal, self-referential, insatiable as the silence of desire itself. Of all political concepts, intentionality is the most central and the least understood. Thingification is the price we pay for not accepting that there is a beyond beyond the beyond of expression, an absence inscribed without a trace in every discourse. The slash tries to be a symbol of this excluded third which is neither either–or nor both–and but something entirely outside the realm of naming; the slash is not what it first might have seemed – a bridge between opposites – but the void of categorial limits itself. And through the silence breaks the voice of Samuel Beckett's Unnamable (*Beckett Trilogy*, 1979: 352):

> perhaps that's what I feel, an outside and an inside and me in the middle, perhaps that's what I am, the thing that divides the world in two, on the one side the outside, on the other the inside, that can be as thin as foil, I'm neither one side nor the other, I'm in the middle, I'm the partition, I've two surfaces and no thickness, perhaps that's what I feel, myself vibrating, I'm the tympanum, on the one hand the mind, on the other the world, I don't belong to either. . . .

And so it is that the slash perhaps can serve as the signifier of that constellation in which nothing takes place except the place. Maybe it is even at this Mallarméan place of silence that truth and power hold their secret meetings. To stave off all trespassing into this sanctuary, God himself spoke these words to Moses, for him to bring down to the priests and the people (*Exodus* 20:4–6):

> You shall not make for yourself a graven image, or any likeness of anything that is in heaven or above, or that is in the earth beneath, or that is in the water under the earth; for I the Lord your God am a jealous God, visiting the iniquity of the fathers upon the children to the third and the fourth generation of those who hate me, but showing steadfast love to thousands of those who love me and keep my commandments.

And so it is that issues of representation and the sublime may have more to do with ethics than with logics and aesthetics. Power is structured as a knowledge:

'And the Lord said to Moses, "Go down and warn the people, lest they break through to the Lord to gaze and many of them perish"' (*Exodus* 19:21). And in doing as he was told Moses became the first *politruk*. The eye shows itself to be more powerful than the voice, the gaze more violent than the word.

Inherent in the issues of representation is a heightened awareness of the fundamental difference between word and object. This awareness is itself a part of that crisis of the sign which became acute in the second half of the nineteenth century, especially with writers like Baudelaire, Rimbaud and Mallarmé. Once again these artists were driven by a desire for the words they did not possess. In realizing that every utterance of necessity is indirect, they experienced how conventional language did not furnish the means to express what they most urgently wanted to express. Even when they said 'stone' and meant 'stone' they were not stone. Even when they said 'you' and meant 'you' they were not you. One can share what one has, not what one is. As a speaking subject I have no choice but to live in a language which is common and social. The limits of my language mean the limits of my world.

It is this unfathomable problem of representation which to me finds an objective correlate in the Saussurean bar, i.e. in that horizontal line which simultaneously splits and unites the two fractions of the sign of:

$$\frac{S}{s}$$

where *S* stands for 'signifier' and *s* for 'signified'.

It is of course tempting to tie the capital *S* to the touchable physicality of the sign and the small *s* to its non-touchable meaning. Such an interpretation would be a serious oversimplification, however, for signs are always threaded together into braids of desire and justification. In a rythmic dance of creativity, nominator and denominator constantly change positions.

Illustrations of such turn-abouts are already in the *Odyssey*, especially in Odysseus's tale of how he was caught in the cyclopes's cave (ninth song). When Polyphemos demanded the name of the intruder, he got the answer Ovtis, *Nohbdy*. The beast in turn replied that 'Nohbdy the last one will be that I eat,' upon which he fell asleep, full of red wine and human meat. It was at this crucial moment that the trickster saw his opportunity, rammed the red-hot pike of olive into the drunkard's eye, leaned on it and turned it as a shipwright turns a drill. And the pierced eyeball hissed broiling and the roots popped. The cyclope howled in pain and when his like gathered outside the cave to learn what was wrong Polyphemos roared in reply, 'Nohbdy, Nohbdy's tricked me, Nohbdy's ruined me!' To this his friends responded, 'Ah well, if nobody has played you foul there in your lonely bed, we are no use in pain given by great Zeus.' And Odysseus was filled with laughter to see how like a charm the name deceived them.

Thus sounds the original tale of the crisis of the sign, of power structured as a knowledge, of identity statements as lies, of thingification and the sense of meaning. But this *ur*-tale has interesting connnections also with the Greek concept of mimesis and thereby both with postmodernism and with René Girard. Inherent in this notion is the idea that art is an imitation – a re-pre-sentation – of reality. But neither the true artist nor the hegemonic ruler is satisfied with the mere copying of the outer appearance of a phenomenon. They also wish to touch its inner soul. Only with the sublime do they become fully fledged forgers. The bar eliminated, word and object united, mimetic desire completed.

Constitutive of the concept of mimesis is the assumption of a privileged original, a Holy Script, a Book of Nature. With this text as a starting point, the truth-teller's task is to provide a perfect translation, a mirror image, a narcissistic reflection. At the same time, I have already noted how perfect translation is impossible. The supplementary nature of language in fact implies that any reference to an original is highly misleading; the copy is not a copy of an original but a simulacrum, a copy which lacks an original. The Book of Nature is itself a simulacrum. As such it is untranslatable.

And thus it must be repeated: meaning does not emerge from the identity of signifier and signified, but from the difference between them. Indeed, it is difference alone that allows a signifier to signify. Without signs no thoughts, without splits no signs. Referring to a once-and-for-all beginning is therefore not to refer at all, for without tenses and cases we would be literally lost. It follows that the ideal of total and perfect representation does not guarantee the stability of truth. Instead it vouches for its indeterminacy.

Filling the ontological gap between signifier and signified is the Saussurean Bar of Power. But whereas power is power because it transforms categories, analysis is analysis because it keeps them apart. As a consequence the '–' can function not only as the wand of the ruler but also as the crowbar of the critic. This explains why also Marxian thought may be considered an early response to the crisis of the sign; the use value of the commodity corresponds to the signifier, the exchange value to the signified. To fetishize is in fact to see only the physical and be blind to its meaning; in reality, though, a commodity is a very queer thing, abounding in metaphysical subtleties and theological niceties. No society can do without its religion of ontological transformations.

Similar attitudes permeate the theories of psychoanalysis and the practices of psychotherapy. Since the unconscious after Lacan is structured as a language, human crisis can be interpreted as a crisis of the sign, more specifically a crisis locatable to the bar. The therapeutic strategy is first to discover a repressed signified and then to kill it with an explicit signifier. The patient speaks herself well, for, in the power-filled act of naming, anguish becomes graspable. The horrible and non-communicable looses its frightening grip once it is caught in shared categories and domesticated in common expressions. Desire is tamed when subverted into desired; the phallus becomes a symbol of lack, desire thingified. Penis envy speaks clearly, the castration complex as well. But just as Saussure's bar blocks the road to the fulfilment of desire, so the way to self-understanding always goes via the Other. Fort-da. Fort-da. Fort-dada. But what does a woman want? What does it mean to be a woman in an age of iconography, at once a reflexive consciousness and a social product?

Here, as before, the processes of thingification and alienation seem to be governed by a mimetic desire. This drift to imitation expresses itself less in metaphorical condensations than in metonymic displacements. It lies within René Girard's theory that we are striving to make signifier and signified into doubles of one another. But when this desire approaches satisfaction the search for identity turns back on itself and changes into hostility. The desire can be kept alive only through the sacrifice of one of the antagonists. A scapegoat is selected, burdened with the collective guilt and driven out of the community.

To me, the scapegoat carries many of the same traits as the rejected hypothesis, the power of the example, the prohibition against graven images. The desire is a desire for perfect communication, a desire impossible to satisfy. Put differently: power is a desire, not a need. It is this desire that I have symbolized as lines of power. It is the truth of this desire that no ruler can bear to hear.

And now, towards the end of the beginning, it should be clear what fascinates me. It is that strange transition point at which the light or sound waves of my textual performance hit your eyes or ears, move into the grey substance inside your head, stir around and become meaningful words, powerful enough to change both our understanding and our action. This abstract penumbra is to me in the intersection of a set of lines, all of which serve as condensations of knowledge, hence of communication. It is not sufficient to say how something is; I must also be believed when I say it. This is not to argue, however, that anything goes. On the contrary, for to be believed is to tread a dance with the taken-for-granted. To be believed is to have power. Power is almighty, the aparatjik its prophet. Not to philosophize is still to *philosophize*. Power is a desire to control meaning. The prime symbol of meaning is the copula IS, a verb designating an event.

Finally: is this geography? Of course it is! For what is geography, if it is not the drawing and interpretation of lines? The only quality that makes my geography unusual is that it does not limit itself to the study of visible things. Instead it tries to foreshadow a cartography of thought. To practise this art, however, is incredibly difficult, for any attempt must face the challenge of being abstract enough.

In this context, it is with some dismay that I note how my thoughts of power have been unthinkable without the correlates of the straight lines. Even though this technique of fixing ideas is a minimalist approach, it is an approach nevertheless. But how do I write a theory *à la* Brancusi, a theory in which form and content are indistinguishable? And how do I practise a writing *à la* Beckett, where I do not write *about* something, but where the writing *is* that something itself? Mallarmé's example was to paint not the thing but the effect it produces.

Perhaps I enter this social space of silence by living in the world as I found it. A world where the unconscious is structured as a language, a world where power is structured as a knowledge, a world where lines are taken to their limits.

Mondrian.

> all these words, all these strangers, this dust of words, with no grounds for their settling, no sky for their dispersing, coming together to say, fleeing one another to say, that I am they, all of them, those that merge, those that part, those that never meet, and nothing else, yes, something else, that I'm something quite different, a quite different thing, a wordless thing in an empty place, a hard shut dry cold black place, where nothing stirs, nothing speaks, and that I listen, and that I seek, like a caged beast born of caged beasts born of caged beasts born of caged beasts born in a cage and dead in a cage, born and then dead, born in a cage and dead in a cage, in a word like a beast, in one of their words, like such a beast . . .
>
> (*The Unnamable*, 1952:356)

7

THE ARCHITECTONIC IMPULSE AND THE RECONCEPTUALIZATION OF THE CONCRETE IN CONTEMPORARY GEOGRAPHY

Michael R. Curry

INTRODUCTION

The past few years have seen the rise and fall of a wide range of approaches to the practice of geography. If in the 1960s we saw the quantitative revolution and a series of humanistic backlashes, the 1980s have seen the renaissance of quantitative work, in the form of geographical information systems, and a new set of alternatives, knit loosely around the rubric of postmodernism. Superficially, of course, the parallels between the present situation and the one twenty years ago seem striking, even if in the current situation the competing groups appear the more likely merely to dismiss one another's work. But there is a deeper sense in which these movements and counter-movements reflect a fundamental tendency within geography, and elsewhere. I term it the 'architectonic impulse'. This impulse, this desire to create an ordered, hierarchical system, is by no means new; like much else, it can certainly be traced at least to the Presocratics. Indeed, the *Oxford English Dictionary* cites examples from as far back as 1801 of the term 'architectonic' being applied to this desire. And in 1877, in his *Critical Account of the Philosophy of Kant* (from which I take the term), Edward Caird described 'the architectonic impulse of reason, which seeks to refer all sciences to one principle'. Himself a Kantian, Caird can hardly be described as critical of the process which he saw as arising from the action of reason, and neither ought he to be faulted for failing to be critical. Yet, for those of us who engage in social scientific research, the architectonic urge can lead in quite the wrong direction. Perhaps more important, the impulse is often masked, and we see it operating in areas in which, explicitly, at least, it is at the same time being roundly criticized.

Today an understanding of this architectonic impulse is of especial importance. Advocates of some positions – postmodernism, for example – have criticized 'meta-' or 'master-narratives', large-scale accounts of the way in which the world operates. They thereby appear to be fundamentally criticizing architectonic thinking. To many it now appears that the alternatives must be universalism, essentialism and architectonic thinking or relativism and anti-essentialism; that one must be a system builder or critical of all systems. But an understanding of the development of the architectonic urge, and of its critics, can suggest a way of cutting through those dualisms, and of attempting to develop a more ethnographically honest approach to philosophical problems and, by implication, to the study of human life.

Although a comprehensive history of the manifestations of this impulse is well beyond the scope of this chapter, it is possible to begin to see the lineaments of it through a limited surrogate, the concept of 'forms' or 'ways of life'.[1] Since the turn of the century a variety of authors, working in geography, sociology and philosophy, have developed works based on these conceptions. Although developed in widely different fields, in each case the concept has moved along a similar trajectory, beginning as a descriptive term, then gradually being systematized, and ultimately being offered as the basis of a systematic ontology.[2] That ontology, in turn, characterizes the nature of the 'concrete'; it determines what is to be counted as a concrete existent, and what is to be seen, rather, as an abstraction or a construction. This chapter is about the relationship between these two features of modern thought, the architectonic impulse and the nature of the concrete.

It proceeds by examining the development of the concept of a way (or form) of life in three domains. In each of the first two, traditional regional geography and urban sociology, the concept of a way of life has been central to a body of work. In the late nineteenth and early twentieth centuries the concept of *genre de vie*, or way of life, was central to the regional geographies of Paul Vidal de la Blache. Over the course of the first half of the twentieth century Vidal's students and followers adapted the term, until, in the end, it had developed a meaning very different from the original. A second domain, and one of more immediate interest to many contemporary geographers, begins with Wirth's classic article 'Urbanism as a way of life' (1938). Wirth's work is still a subject of debate in sociology, but it has also been central to the development of contemporary urban geography. And, as in regional geography, in urban sociology the term has undergone a substantial reinterpretation over the last fifty years.

The third domain, in philosophy, is rather more complicated. There, in the 1930s, Wittgenstein began to develop a critique of certain traditional philosophical notions, a critique which relied in part on a conception of forms of life. Rather than offering a full-scale analysis of his work, or a historical study of accounts of his work, I shall limit my remarks to two related areas, both within geography. First, Peter Gould and others have recently attempted to develop a new and better approach to geography, centered on R. H. Atkin's use of Q analysis. Gould explicitly invokes Wittgenstein as a source of authority for what turns out to be an explicitly and fundamentally architectonic project. And, second, there has more recently been, also within geography, the development of a postmodern movement which both appeals explicitly (often via Lyotard) to Wittgenstein's conception of forms of life and appears critical of the kind of explicitly architectonic project on which Gould has embarked. In fact it will become clear that the postmodern project, too, is at least implicitly architectonic in intent, and the two superficially different uses of Wittgenstein's work share a great deal, among which is an attitude towards the nature of the concrete.

As with so much else, it would be quite easy to begin an enquiry into forms or ways of life by looking at the Greeks; one might, for example, see as its intellectual forebear the term *ethos*, and search for changes in the meaning of that term through the decline of Athenian civilization. In a different way the term actually *has* been discussed by geographers, who have seen elements of a consideration of the concept in writers like Hippocrates and Strabo (Lukermann 1961). Certainly the search for its intellectual roots could, with less stretching, extend back as far as more recent authors like Vico and Herder and Dilthey (Berlin 1980). One of the first serious attempts to apply such a concept was in the latter part of the nineteenth century, in geography, in the work of Vidal de la Blache (1911, 1928). And Vidal is a good place to start, because he not only attempted to work out a geographical system based on the concept, but also became the father of a school of geography.

THE VIDALIANS

Before Vidal there was little serious academic geography in France; on his death he left a thriving discipline in which virtually every Chair in every department in France was occupied by one of his students. Students of these students are alive today and have continued implicitly and explicitly to redefine the term 'ways of life'. So simply to look at the development of the French school of geography is to see the nature of changes in beliefs about ways of life.

Unfortunately the enquiry into Vidal's work is rather perplexing. Some historians of geography have attempted to understand the French school by simply situating it temporally, seeing it as following on the heels of geographers of earlier in the nineteenth century, like Friedrich Ratzel and Carl Ritter and Alexander von Humboldt (Holt-Jensen 1982). But in reality Vidal's work defies such an easy apprehension. Indeed, on the evidence it seems fair to say that his work is much more closely related to the work of philosophers like Boutroux and Poincaré and Ravaisson (Lukermann 1965; Berdoulay 1976) and hence ultimately to that of Kant[3] than to that of earlier geographers. And, while this has implications for the understanding of his conception of the concrete, it at the same time suggests one way in which his successors have diverged from his position as they saw it as part of another intellectual tradition.

Vidal's discussions of ways of life refer fundamentally to means of livelihood, and especially to hunting and fishing, pastoralism and various forms of agriculture. In his article 'Les genres de vie dans la géographie humaine', for example, ways of life are divided by climatic region: there are ways of life in dry lands, ways of life in the tropics, ways of life in the European forests, and so on. There Vidal argues that ways of life 'arise from diverse local circumstances, which had developed their own natural patterns before man arrived' (1911: 212). Notwithstanding this, ways of life are created through the marshalling of those pre-existing forces and the use of one against the another. Indeed, ways of life are often characterized in terms of these materials: one speaks of a way of life based on the use of adobe, or of forest materials, or perhaps, with Lewis Mumford, of iron and steel.[4]

But if this is how we typify a way of life, of what is that way of life itself constituted? For Vidal – and here he revealed his interest in the natural sciences – ways of life are groups of human habits, habits that in important respects are related to tools and machines. At the same time – and here Vidal really parted ways with Durkheim and the social morphologists (who were interested in developing sets of atemporal laws by which to account for human behaviour) – he argued for the importance of time in the relationship between these ways of life, these historical accumulations of habit, and the places in which they develop (Berdoulay 1978).

Vidal began his *Tableau de la Géoraphie de la France*, itself the introductory volume of Lavisse's *Histoire de la France*, by arguing, after Michelet, that 'France is a person'[5] (Vidal de la Blache 1928: 13) But – and this later had important implications – he continued:

> A geographical individuality does not result simply from geological and climatic conditions. It is not something delivered complete from the hand of Nature. . . . [A] country is a storehouse of dormant energies, laid up in germ by Nature but depending for employment upon man. It is man who reveals a country's individuality by moulding it to his own use. He establishes a connection between unrelated features, substituting for the random effects of local circumstances a systematic co-operation of forces. Only then does a country acquire a specific character differentiating it from others, till at length it becomes, as it were, a medal struck in the likeness of a people.
>
> (Vidal de la Blache 1928: 14)

So for Vidal ways of life which have developed in particular places may, over the course of time, become so ingrained that the place and the people become almost one; the place acquires a personality. Some of his students, it should be added, argued that in France Brittany cannot be seen as having had time to develop such a 'personality', since it underwent structural changes in the fifth century, and that for a great many, Vidal included, Germany was totally without personality.[6] At the same time, though, it is not this personality of place, but rather the development of ways of life, which is absolutely essential for human survival; for Vidal the idea of a human group that lacked a way of life made no sense.

By contrast, though, it was only a few years later that one of Vidal's students, Max Sorre, adopted a very different point of view towards ways of life. Sorre argued that:

> with the progress of industrialization (and also urbanization), modes of livelihood based solely upon manufacturing cut their ties with agriculture and were located according to their own needs.... [T]he perfection of the steam engine by James Watt inaugurated a new era in the history of *genres de vie*.... [I]n the older *genres de vie* the activity of the group satisfied all of its needs, including food supply, tools, shelter, and clothing.... A period began when functional specialization caused the differentiation – one might say dismemberment – of *genres de vie*, and reduced the activities of the group as such.
>
> (Sorre 1962: 405)

So today, in contrast with the situation in Vidal's time:

> [T]he concept has been transformed while being enlarged.... In its general sense, the expression is applied both to individual behavior, determined by personality, social status, and professional mores, and the habits of groups.
>
> (Sorre 1962: 411)

Hence changes in society, and especially in the nature of transport and communication systems, have meant that society itself has changed. In a way reminiscent of the work of such recent agrarians as Wendell Berry (1970), Sorre has (albeit perhaps not clearly) seen in the twentieth century a new world, one radically different from the traditional world of even the nineteenth century. Now some of us have ways of life, and some have not. In a way this change of heart was quite striking; it was as though a cultural anthropologist had suddenly asserted that some of us live in cultures and some of us don't.[7]

If Sorre saw ways of life as still related to sets of concrete activities in particular places, twenty years later Anne Buttimer took a different view:

> [A]s situational influences predominate over localized group influences, *genres de vie* tend to adopt a spatial distribution (morphology) which is more closely associated with economic forces.
>
> (1971: 188)

Here ways of life are no longer strictly associated with places; one begins to see a society in which they are, rather, associated with places only through the medium of 'economic forces'.

For both Buttimer and Sorre, the twentieth century has ushered in a new world, in which tradition has lost importance: the group has been replaced by the isolated individual, and custom by deliberation and rationality. From there it was only a short step to Paul Villeneuve's 1984 'Pour une géographie des genres de vie urbains', where the geographical study of those ways of life involves all manner of techniques, from participant observation to the gathering of census and other quantitative data. Whatever the technique, for Villeneuve the concept is a means of ordering or typologizing human action. It has become a sort of catch-all phrase, one that refers less to a concrete reality, or even to an element in a theoretical schema, than to a belief that the study of urban life ought to be comprehensive, and ought to allow for all possibilities. Hence, for Villeneuve, ways of life are not themselves concrete existents; rather, they are conceptual aggregations of concrete elements themselves not determined by perceptibility, for example, but rather defined in terms of a specific research agenda.

URBANISM AS A WAY OF LIFE

To many geographers modern communciations and transport have brought about the dissolution of ways of life, and hence there can by definition be no urban way of life, no way of life strictly associated with and definable in terms of urban places; there can be only a variety, ultimately, of different ways of life located indifferently with regard to cities. For some sociologists, ironically enough, this has not always been thought the case. Nowhere is it more obviously the case than in the work of Louis Wirth.

As Herbert Gans (1968) has put it, 'The contemporary sociological conception of cities and of urban life is based largely on the work of the Chicago School, and its summary statement in Wirth's essay "Urbanism as a way of life"' (95; see also Wirth 1938). The lineaments of Wirth's argument are familiar; he argued that there are three aspects of the city that affect the urban way of life: the size of the resident population, which is much larger than that in towns and villages; the density of population, which is also much greater; and the heterogeneity of people there, the fact that so many people act and think in different ways. For Wirth these facts suggest that the city, whatever its positive features, will be the seat of a variety of problems. In the urban way of life people replace primary group contacts with more superficial secondary ones; their relationships with others are characteristically brief and impersonal, and life in the city is, ultimately, anonymous. At the same time, people live more segmented lives, defined more in terms of roles; actions once guided by tradition now come to be inspired by the spirit of rationality.

Many of the footnotes to Wirth's work have consisted of criticisms of his imputed location of the urban way of life in cities and the rural way of life in the countryside. Gans, for example, has argued that such a way of life is characteristic only of a small portion of the inner city; at the same time, he has attempted to characterize a third geographically based way of life, that of the automobile-oriented suburbs, while attempting to show that there are a variety of non-geographical ways of life – those of cosmopolites, the unmarried, ethnic villagers, the deprived, and the downwardly mobile – which have a great deal more to do with the stage of a person in the life cycle than with geographical differences (Gans 1968: 97).

But it ought to be noted that, if for Wirth ways of life were not abstract, technical concepts in the current sense, neither did he see them as actual, concrete entities. Rather, for him the urban (and rural) way of life lay somewhere in between; it was an ideal type, not necessarily associated with any real place.[8]

For example, he argued in a later work that:

> City ways of life have in some respects taken on a rural cast, particularly in the suburbs. On the other hand, industry, which hitherto was characteristic of cities, has gone into the countryside.
>
> (Wirth 1969: 165)

For Wirth what was at issue as an ideal type of human behaviour that was associated with certain physical aspects of industrialized society and the city. None the less, his critics have failed to exploit this difference between his work and both Vidal's and their own, and have moved away from Wirth's original formulation.

While Gans was not averse to speaking about such ideal types, for him ways of life were more accurately, and usefully, seen as explanatory concepts in the sense in which those concepts are used in the physical sciences; one reduces urban and rural ways of life to more fundamental, underlying factors, to what he called 'characteristics':

> The way of life of the deprived and trapped can be explained by low socio-economic level and related handicaps. The quasi-primary way of life is associated with the family stage of the life-cycle, and the norms of child-rearing and parental role found in the upper working class, the lower middle class, and the non-cosmopolite portions of the upper-middle and upper classes.
>
> (Gans 1968: 111)

Gans was not alone in arguing that there are a variety of other ways of life. Raymond Pahl, for example, went so far as to argue that for a great many people choice is a way of life (1968: 272). But with the exception (perhaps) of Pahl, by the mid-1960s sociologists, like geographers, had come to see ways of life not as real, concrete elements of reality, nor even as ideal types, but rather as scientific, theoretical constructs. As Fischer (1972) put it, what Wirth had really created (had he only been self-conscious enough to understand it) was a:

> middle-level – a dynamic model ... [that] assumes that structure operates on behavior through the mediation of cognition and is itself the aggregation of individual behavior ... [and that can be] presented in the form of a block-recursive model and ideally should be subjectable to path analysis.
>
> (Fischer 1972: 188–9)

Just as in geography, real concrete ways of life have long since disappeared from this view, and so even have those ideal types that were at once abstract and particularistic; what remain are explanatory concepts, there to be used or not on the basis of their explanatory power. What remains is the 'aggregation of individual behaviour', and the power of the model rests on its ability to account for the possible variations in that behaviour.[9]

Both the Vidalians and the followers of Wirth have explicitly used the term 'way of life', and we have seen that within both groups there has been, over the course of the last fifty to seventy-five years, a movement from seeing forms or ways of life as concretely existing entities to seeing them as abstractions within systems which define the concrete in other terms. Among contemporary geographers we find much the same process in Peter Gould's work (1981); it is relevant here because in calling for geographers to let the data speak for themselves Gould refers to the work of Wittgenstein, and suggests that one ought to appeal to some commonly available set of phenomena; to, for example, human ways of life.

Portions of Gould's earlier work, on mental maps, rested on an image or metaphor, wherein it was imagined that people have, perhaps quite literally, mental 'maps' of the world, images of the world in terms of which they act (Gould and White, 1974). In the hands of geographers schooled in behaviourist psychology, that initial insight, first expressed in prints of, for example, 'The New Yorker's view of America', was quickly recast in terms of a more abstract methodology, where mental maps became merely statistical summaries.

Although turning away from the issue of mental maps, Gould's later work mirrors this process, as he has developed a view derived from the Q analysis propounded by the mathematician R. H. Atkin (1974; 1981). Gould, of course, is not alone in his admiration of Q analysis; in geography it has been applied in different ways by Couclelis (1983), Johnson (1981) and Gatrell (1983) and has received favourable attention by Pickles (1985).

According to Atkin the basis of Q analysis is the notion of hierarchy; in our (or any) language there is a hierarchy of terms, where the ones at the top of the hierarchy cover, or include, those below. As Atkin put the matter:

> [T]his word *cover* really contains the whole essence of what we mean by moving up or down a hierarchy of sets of ideas, and all our intuitive experience as expressed in our idiomatic language seems to be an elegant illustration of this very idea.
>
> (1981: 23)

When we speak of hierarchies, Atkin continues, we must not be misled (as Christopher Alexander was in *Notes on the Synthesis of Form*, 1964) into believing that hierarchies are necessarily tree-like. Rather, we must see that there can be an overlap of the sets that constitute their elements, that the language we use may be inherently ambiguous.

On the basis of these notions of hierarchy and of sets Atkin attempts to establish the groundwork of a new and better social science, one that will be more rigorous in its methods and more fruitful in its results. Basic to that new social science will be the notion that sets and hierarchies (or, for example, people and behaviour) can be seen in terms of n-dimensional spaces. Those spaces define the nature of the events that can occur or the actions that can be undertaken, and can best be seen as the backcloth of those events and actions, while the events and actions themselves constitute a form of 'traffic' on that backcloth.

The backcloth itself, then, defines the set of existing possibilities for human action; changes in the backcloth may well result (in fact will by definition result) in a change in those possibilities. Previously those changes were too complicated to be clearly understood, but with Q analysis, claim Atkin *et al.*, it will finally be possible to keep track of the nature of those possibilities, and the potential effects of their being changed. For example:

> Not long ago, the Mianus Bridge collapsed on the main New York to Boston Thruway. Here was a dramatic example of structural change in the backcloth changing patterns of traffic and altering transmissions. But a description of the system in entropy-maximizing terms would simply recalibrate all the λs, and tell you Portchester was now less accessible – something the inhabitants, police, and truck drivers had already figured out for themselves. A Q-analytic description would tell you that you had to start redirecting the trucks fifty miles away to alleviate the 'shock waves' working out from the break.
>
> (Gould 1983: 387)

According to Gould, the advantages of this sort of conceptualization of empirical research are manifest. The statements of the physical sciences are all functions, in the form of $Y = f(X)$; they all are either deterministic or probabilistic. By contrast, the backcloth described in Q analysis is 'an allowing and forbidding, *but not requiring*, geometry. This appears much more suitable – that is, capable of capturing more truthful characteristics – for the human world' (1983: 383).

Both Gould and Couclelis have subjected Q analysis to at least some criticism. Gould (1983: 385–6) has noted that the backcloth–traffic analogy lends itself to mechanistic interpretation, and further that the emphasis on the dimensionality of connectivity (as in Atkin) may hide more important features of the situation in question. Couclelis (1983) has noted perhaps more fundamental problems; she has suggested that the initial premise of Q analysis, that there are well defined sets, may be incorrect. Further, she has noted that the traditional view of a set as a collection of items may be inadequate; accurate analysis may require the establishment of an order within sets themselves. As she put the matter:

> By viewing higher-order 'things' as collections or aggregates of lower-order 'things', Q analysis misses the fact that, in most nontrivial cases, higher-order objects correspond to more or less complex *structures* defined on, rather than mere concatenations or disjunctions of, their lower-order elements.
>
> (1983: 431–2)

Her solution, though, is not to reject Q analysis, but rather to attempt to define that order through the use of Boolean operatives.

Here Gould and Couclelis are undertaking an architectonic project, and one of a particular type. It can perhaps best be characterized as concerned with the nature of possible worlds. It thereby shares a good deal with the earlier work of Wittgenstein, particularly his *Tractatus Logico-philosophicus*. In both cases the world is imagined to consist of a set of objects and attributes that *do* combine in a variety of ways, that *might* combine in others, and that for various reasons *cannot* combine in still others. In the philosophical literature the notion of 'possible worlds' arises directly from the work of Leibniz, in the following way. Traditionally philosophers have asserted that there are some statements that are necessarily true and some that are not. It is necessarily true that red is a colour, or that $2 + 2 = 4$, or that if $a > b$ and $b > c$, then $a > c$. At the same time it is only contingently true that today is Saturday or that I have brown hair or that Washington, D.C., is the capital of the United States. According to the most common interpretation, Leibniz defined the terms 'necessary' and 'contingent' in the following way: if a statement is necessary, then it must be true in all possible worlds, while if it is contingent, then there is at least one possible world in which it is not true. Similarly, in Q analysis there are some combinations that are just not possible, while there are some that are necessarily so.

In both cases it is assumed that the language that describes the world is composed of a specific set of elements – subjects, predicates and logical operatives. Within the confines of traditional logic the subjects may be any object (concrete or otherwise); the predicates are qualities or actions (to name two); and the operatives are typically 'and', 'or' and 'not'. In Q analysis the vocabulary is somewhat more complicated, but ultimately it rests on notions like 'set ' and 'inclusion' and 'exclusion'. In either case, though, the world fundamentally consists of objects in relationships, and, whether one takes the objects themselves or the objects-in-relationships[10] as basic, the presumption is that the world consists of a catalogue of simple elements, and that our knowledge of the world derives from our ability to combine and concatenate those elements in appropriate ways. Hence this implicitly architectonic project contains explicit definitions of the concrete; that which is concrete is seen as merely the most basic of elements, and abstractions – or at least meaningful abstractions – are built from those basic elements.

POSTMODERNISM AND THE REJECTION OF TOTALISM

It seems reasonable to believe that a calculus that allows any possibility definable in terms of 'and', 'or' and 'not' will be one which can define the very limits of possibility, and that therefore an approach that begins with such a notion of possible worlds is operating at the greatest possible level of possibility, that at that level nothing is excluded. In fact this ability of possible-worlds theories to operate at the greatest degree of generality appears to be presumed by those recent postmodern geographers who criticize it for that very reason, as they reject the possibility of a 'totalizing discourse'.

Certainly their work appears in important respects like quite the opposite of that which Gould and Couclelis have been propounding. It will turn out, though, that postmodern geography shares a great deal with the architectonic project of Gould and Couclelis, and that that is true despite the fact that it appeals for authority to the later work of Wittgenstein, work which on the face of it is strongly critical of the possible-worlds theory of Gould, Couclelis or, indeed, the earlier Wittgenstein.

According to Michael Dear, 'The postmodern challenge is to face up to the fact of relativism in human knowledge, and to proceed from this position to a better understanding' (1988: 271). The primary aim of this relativism is to work 'against the potentially repressive power of theoretical metalanguages which can act to marginalize a nonconforming discourse' (Dear 1988: 265). This, on the face of it, suggests that postmodernism attempts to move away from the sort of architectonic project that we have seen so far. Indeed, with regard to the issue of whether one such structure or another can claim to be the 'right' one, 'The essence of the postmodern answer is that all such claims are ultimately undecidable' (Dear 1988: 265–6). This, in turn, is related to Dear's criticism of realism, where 'our conceptual orderings ... do not exist in the nature of things, but instead reflect our philosophical systems' (Dear 1988: 266).

Language is central here, for two reasons. First, there appears today, according to postmodern geographers, to be a crisis of the sign, occasioned by the realization that language does not derive its meaning by 'pointing at' reality. And, second, language now appears to be a fundamentally social phenomenon; Lyotard compares the using of language to the playing of a game of chess (1984: 10). Because language is now seen as centred in individual practices, it appears that our knowledge is fundamentally 'local'. 'There are many different language games – a heterogeneity of elements. They only give rise to institutions in patches – local determinism' (1984: xxiv). And that, again, implies that grand schemes no longer have the power to capture reality.

The postmodern use of the term 'language games' comes, of course, from Wittgenstein. And postmodern geographers have at the same time appealed to his notion of forms of life, as where Pred asserts that:

> To know a form of life,
>   a geographically and historically specific form of life,
>   a *genre de vie*,
> is to know language.
> To actively engage and understand speech acts,
> to become embedded in discourse....
>
> (1988: 138)

As we see reference in Dear to an attack on foundationalist epistemologies, an incredulity towards metanarratives, and a rejection of the search for ultimate foundations, it may appear that in this movement we have found a group of social scientists who have been able to stave off the tendency for terms like 'language game' and 'form of life' to be transformed into abstract and technical terms.

However, there are aspects of this work that suggest that it has not succeeded at this task. Consider the very use of the term 'language game'. Both Lyotard and Dear use the term as though it refers to a discernible set of linguistic practices. For Lyotard a language game is very much like a game of chess, and in using this analogy he refers directly back to Wittgenstein's original conceptualization. In language, as in chess, he says, 'each of the various categories of utterance can be defined in terms of rules specifying their properties and the uses to which they can be put' (1984: 10). And this appears to lead directly to a linguistic or conceptual relativism, where 'science plays its own game; it is incapable of legitimating the other language games' (1984: 40).

We find much the same in Dear's 'The postmodern challenge is to face up to the fact of relativism' (1988: 271). This relativism appears to follow directly from a view in which language games are separable, isolable units, ones which we can take seriously as having a real, factual existence, and which can be characterized in coherent factual and conceptual discourse. But to adopt this relativism is immediately to be led away from Wittgenstein's understanding of the concepts that he was using.

If in Vidal the concept of forms of life was of questionable parentage, that is also the case with Wittgenstein. Janik and Toulmin have presented evidence to the effect that, in the latter part of the nineteenth century in Vienna, talk of forms of life was all the rage. 'One of the most successful works of popular neo-Kantian literature, published shortly after the First World War, was a contribution to characterology [entitled *Forms of Life*] written by Eduard Spranger' (Janik and Toulmin 1973: 230). They immediately add, though, that Wittgenstein's use of the term was radically different from that of Spranger or his teacher, Dilthey. At the same time, Haller (1988) argues that Janik and Toulmin are clearly in error, and that for the source of the concept we need to look to Austria, to W. Fred, Otto Stoessel and Paul Ernst.

There has been substantially more interest in understanding Wittgenstein's use of the term than in understanding his sources for it; indeed, the issue is so complex that for a time a virtual cottage industry was involved in interpreting it.[11] How did Wittgenstein use the term? The answer is not as clear as one might wish. Of course some have criticized even the very use of the concept. Hilary Putnam, for example, argued that for Wittgensteinians the 'fondness for the expression "form of life" appears to be directly proportional to its degree of preposterousness in a given context' (1977: 114). And others, especially those interested in religious questions, have rejected the use of the concept on the grounds that it is less meaningless than dangerous, because it implies a fideist interpretation of religious discourse in which that discourse is purely conventional, and without any ultimate groundwork.

This last view appears to be supported by one of the few statements that Wittgenstein made on the term: 'What has to be accepted, the given, is – so one could say – *forms of life*'[12] (1968, II; s. 226). One might well take this to mean that there are certain conventional aspects of human life, and that those lack any further support or justification. Indeed, one might take this to be evidence either that Wittgenstein was a thoroughgoing relativist (Munz 1987) or that he was a political conservative (Nyiri 1982).

Further, he asserts that 'to imagine a language means to imagine a form of life' (1968, I: s. 19). This appears to suggest the sort of conceptual or linguistic approach to culture that we find in Sapir, where linguistic categories are seen to be the organizing elements of human life; there too we appear to find support for a radically relativist attitude.

Later he asserts that 'Here the term "language-*game*" is meant to bring into prominence the fact that the *speaking* of language is part of an activity, or of a form of life' (1968, I: s. 23). And 'It is what human beings *say* that is true and false; and they agree in the *language* they use. That is not agreement in opinions but in form of life' (1968, I: s. 241). Here we might begin to see language as an activity, and forms of life, too, as sets of activities. Finally, though, he asserts that 'Can only those hope who can talk? Only those who have mastered the use of a language. That is to say, the phenomena of hope are modes of this complicated form of life' (1968, II: s. 174). And this, somewhat perplexingly, has suggested to many that for humans there are not many forms of life, but only a single form of life, the all-encompassing human form of life.

Wittgenstein's spare use of the term – and his aphoristic writing style – have left his work open to an extraordinary range of interpretations; if Munz has been able to argue that he is a radical relativist, Haller (1988) has argued that he is a 'praxeological foundationalist'. It might seem from Wittgenstein's assertions that he believes forms of life in some sense to be the 'ultimate constituents' of the human world. But in fact, as Mason (1978: 335) has suggested, in respect of language games, Wittgenstein 'uses the notion of a language game primarily to call attention to the circumstances and peculiarities of our use of philosophically problematic expressions', as a 'kind of pointing device'.

Hence it would be a mistake to conclude that there must be some set of criteria by which to identify forms of life. Indeed, Wittgenstein's position is a decidedly anti-essentialist one – that the criteria that are used to determine that a certain sort of activity is an element of a form of life vary from time to time and from place to place, that they are, indeed, different with each form of life.

If we now turn back to the earlier work on possible worlds (including his own) which Wittgenstein rejected, his later position is merely this: the notion of possible worlds that he was adopting failed because it was not general enough.

And it was not general because it assumed a realist position toward the elements of logic, one in which the operatives of logic (or of mathematics) really exist, and further, are unambiguously given, once and for all. Yet in his discussions of the following of rules Wittgenstein produced compelling arguments to the effect that we must see notions as seemingly unambiguous as 'addition' and 'subtraction' as socially defined. Indeed, 'To obey a rule, to make a report, to give an order, to play a game of chess, are *customs* (uses, institutions).... To understand a language means to be a master of a technique' (1968, I: s. 199).

In understanding this we begin to understand the reasons why we need to construe Gould's and Couclelis's work as involving moves away from a notion of 'forms of life' to one of 'possible worlds'. In one sense this occurs because they appear unaware of the existence of the third central element in the study of language, the pragmatic. If semantics deals with the relationship between words and their meanings or referents, and syntactics with the relationships among words, pragmatics deals with the relationship between words and their users. The important thing to note here is that to fail to recognize the existence of the pragmatic dimension is *ipso facto* to fail to be able to distinguish among verbal signals ('Ouch!'), reflective symbols (the intentional use of scientific models, or of forms of argumentation) and typological metasigns (dictionaries, or the theoretical laws of a branch of science). It is to fail, that is, to recognize differences among the ways in which language is used. Moreover, it is to fail to see the close connection between verbal signs ('Ouch!') and non-verbal ones (the reaction of a football crowd to the referee's throwing both arms above his head). And finally, and perhaps most important, it is to fail to understand the relationship between the use of language and the community within which it is used. And it is this last feature of language that becomes most central in Wittgenstein's work, and in his concept of 'forms of life'.

Yet it is important here to understand that to conclude therefrom that one can somehow develop a better, more accurate or more rigorous social science by resting that social science on a conception of forms of life would be a mistake. Indeed, it would be quite counter to the way in which Wittgenstein conceived of the term and to the use which it served in his work. For far from offering it as the basis of a better architectonic project, he offered it as a means of criticizing all such projects. It was not a means of systematically characterizing what exists, but rather of pointing to the inevitable existence of a stopping point to ruminations about language and social life. And at the same time it points to the fundamental failings of attempts to develop positions based on some radical form of relativism, for it shows that those projects ultimately rest within some architectonic project. Rather, in a world where there are only various levels of locality, the question of relativism can arise only within those levels, and not across all localities.

If Wittgenstein's view of language games – and by implication forms of life – was not that they were technical terms requiring well defined sets of criteria, what can we say about his position on relativism? Although some have argued that his work must be seen as supporting a thoroughgoing relativism, this appears not to be the case, and why it is not the case should be clear both from his explicit statements and from the nature of his project. The first is nowhere more clear than when one raises issues about human nature. Wittgenstein's later works more and more demonstrate his attention to the importance of the need to see at least something as deeply and fundamentally characteristic of human life. These works explicitly emphasize the importance of the non-conventional, the natural, the limits of human nature. He came finally to attend to the ways in which, as we look at other human groups (and, indeed, at groups of animals as well), we see many features similar to those of our own. We see certain similarities between the experiences of others and those of ourselves. None of these singly provides us with an adequate criterion for establishing the 'humanness' of an individual or a group; indeed, there is no such criterion, and the border between the human and the non-human cannot be drawn with any certainty. At the same time, though, we find as we look at human activities that our understanding of the most remote of them relies on very general notions of the common behaviour of humankind. He put the matter in the following way, in discussing Frazer's interpretation in *The Golden Bough* of the Beltane festival. Frazer had described a ritual in which children put bits of cake into a hat, and the one who draws the black piece is considered the 'devoted' one. Wittgenstein asked:

> What makes human sacrifice something deep and sinister anyway? Is it only the suffering of the victim that impresses us in this way? All manner of diseases bring just as much suffering and do *not* make this impression. No, this deep and sinister aspect is not obvious just from learning the history of the external action, but *we* impute it from an experience in ourselves....
>
> [T]hat which I see in those stories is something they acquire, after all, from the evidence, including such evidence as does not seem directly connected with them – from the thought of man and his past, from the strangeness of what I see in myself and in others, what I have seen and have heard.
>
> (1971: 39–41)

At the same time, his position on relativism is clear if we reflect on the nature of his project. For Wittgenstein one of the central reasons for philosophical problems arising is that language comes to be used in ways that are not natural to it. Hence the ontological argument and the confusions that surround it arise because the concept 'to exist', which has multiple meanings, is thought by some to refer in the last instance to a single attribute. Similarly, the argument that the nature and meaning of all human actions must be relative to some culture or community or place or other arises when the commonplace notion that we need to understand the meanings of terms by looking at the contexts in which they are used, or the notion that moral judgments ought to be considered in terms of the beliefs and commitments of the holders of those beliefs, is inappropriately generalized to the extent that it appears self-referential and thereby paradoxical. For

Wittgenstein the arguments that lead to radical relativism are merely the result of a misunderstanding and misuse of language.

THE CURRENT SCENE

At the outset it was noted that in a variety of disciplines there has been both an expected and an unexpected turn in the use of the concept of a way of life. In the nineteenth century the concept typically functioned to refer to what those who used it took to be 'real' ways of living. For sociologists it referred to social groups, for cultural anthropologists it referred to cultural groups, and, as one might expect, for geographers it referred to patterns of human activity fundamentally tied to particular places.

For geographers, as for others, the end of that century heralded a new era. By then thinkers as diverse as John Stuart Mill, Durkheim and Freud had asserted that we had reached a new age in which places were no longer relevant. It was only a matter of time before even geographers joined in that chorus, and asserted that their traditional subject matter, the peasant communities of France or the agricultural villages of Africa, even small-town America, had been irremediably altered by the coming of the modern age. It isn't difficult to understand why this was believed; for students of society the evidence of changes in transport and communications technology, the rise of industrial capitalism, the development of vast nation states, all provided images in which tradition seemed increasingly enfeebled. For many this was a watershed; in the past people fundamentally belonged to groups, but now some did and some did not. As ways of life disappeared we needed new ways to talk about society, culture and the human occupation of the earth.

What accounts for this change? A thorough analysis even of the reasons for the changes in the use of the 'way of life' concept is well beyond the scope of this work. But it seems fair to say that there are both internal and external reasons. In the nineteenth century students of human ways of life seemed always to be working towards but never to be attaining a synthesis that could be characterized as constituting real progress. To the extent that geography was concerned with the development of accounts of the concrete it was doomed never to be finished. Similarly, those who took a neo-Wittgensteinian approach to philosophy seemed always to be analysing but never to be reaching closure.

Further, traditional approaches to the social sciences appeared in some way to undermine the possibility of the existence of real rigour, real standards of success, and modernism idealized nothing if not rigour. Rather, those approaches appeared to leave one doomed to a career of work in the intellectual salt mines. But they equally appeared to preclude the establishment of rigorous standards for the measurement of relative academic performance. Modernism seemed finally to provide a way of depoliticizing academics; it appeared that it might finally allow the truly good scholars to raise to the top of the profession, regardless of their political or personal idiosyncrasies. From the point of view of a society increasingly cowed by the complexities of modern technology, a way of life 'index' seemed far more useful – and scientific – than yet another copy of the *National Geographic.*

At the same time, it appears that there are external reasons, at least in geographical work, for the redefinition of the notion of forms of life. This recasting needs to be seen against the background of changes in the lives of those who write about ways of life. It should be no surprise that this thinking in some way resonates with the experience of those doing the work. And in contemporary society academics more than many other groups are removed from more overtly 'traditional' forms of sociability, particularly from place and the family. Further, they tend to be more reflective, if not always more accurate, about their experiences, and to see that reflection as something itself fundamental to those experiences. The result is that their analyses of the experiences of others tend, however slightly, to reflect their own experiences, and they come to see place and family as less important, social status as more so. But such modern interpretations seem little able to comprehend the depth and intensity of current reactions either those who attempt to undermine all manner of – sometimes non-place-specific – ways of life or to those who in recent years have begun – while invoking a conception of 'ways of life' – to reject the modernist understanding of progress in philosophy and the social sciences.

By contrast, to see these scientists themselves as in some sense partaking in a way of life is to suggest that the demise of the concept as a workable one may have been more than anything else the result of their inability in the modern world to reach a consensus about the rules to be applied in the creation of social-scientific explanations. Indeed, it seems reasonable, on this view, to imagine as some Wittgensteinians have that those ways of life have been there all along; that they are there, waiting to be studied, if only social scientists will give up their search for the ultimate social atoms. From the implication of the concept of a way of life is that those atoms are not to be found.

In support of this view, it should be observed that the term even now continues to be used. It seems likely that even its most rabid opponents, people like Hilary Putnam, or its most ardent abstractionists like Fischer use the term in everyday speech – or at least understand others when they use it. And this strikes me as saying something important about the nature of social life, about the nature of cultures, and about the persistence of places in human life. For were it not for the persistence of those and other forms of human sociability it seems quite certain that we would not understand others who use the term. And we certainly do, whether we're reading *Doonesbury*, or reading Raymond Williams, or even reading the work of a number of geographers, like Derek Gregory, into whose work the term has begun to impose itself (Billinge *et al.*, 1983). If there is a basic fact in social life, it is that when people talk about ways of life we largely understand them.

TOWARDS A CRITICAL SOCIAL SCIENCE

Of course, to appeal to one's experiences as a guide to what is human and what is not is to court danger: it is to lay oneself open to the charge of being unable to be critical. And certainly it is true that many ordinary-language philosophers have been none too critical of their own ethnocentrism. But there are those who have addressed the issue. Their work is best represented in philosophical circles by that group interested in the social relations of science. David Bloor, for example, has engaged in the extension of Wittgenstein's work in a more reflexive way.[13] Bloor and others have argued that it makes only sense, on Wittgenstein's view, to see the practice of philosophy – or of science – as itself constituting a form or way of life. Hence the implications of this way of thinking became 'critical'; it called into question the groundwork under those who had been talking about ways of life all along.

And here become most apparent the ways in which accounts of human life based on an image of forms of life and those based on an image of possible worlds differ. We recall that the basis of Wittgenstein's view was a rejection of essentialism and reductivism, and that implicit in these was the notion that any account of the human world must be in some sense reflexive. If we return to the earlier examples – to the geography of the post-Vidalians, or the urban sociology that has risen in the wake of Wirth, or the contemporary geography based on Q analysis – we note that each has in its way involved a turn away from the belief that ways of life are concrete entities, and at the same time an attempt to replace older versions of the concept with newer, more rigorous ones. Now it has seemed most clear in the case of Gould and Couclelis, although it ought to be equally clear in the case of the others, that the 'development' has been on a matter of the replacement of a social science based on ways of life with one based on possible worlds. The change has sometimes been subtle and gradual, sometimes not, but in no case have those involved recognized the magnitude of what was involved. One result, of course, is that today we can find works like Dear's that have the trappings of an anti-essentialist and anti-reductionist position but which are more fundamentally within the more contemporary possible-worlds school. Here anti-essentialism was replaced by essentialism; anti-reductivism was replaced by reductivism. And, perhaps most strikingly, the very mechanism for the recognition of the problems that attend those changes was removed.

Where in Gould or Couclelis or Atkin do we find a *sustained* consideration of the role of the social scientist in society? Nowhere. Where do we find a consideration of the relation between the social scientist and the object of the social sciences? Nowhere. Where do we find discussions of reflexivity, or power, or progress, of the assumed rationality of science? Nowhere. We merely find Atkin rhapsodizing:

> [W]hat was it like to be around in the fourteenth century? ... Well, I suggest that you need only be in this century and that you move yourself into a modern soft science and make the statement that 'education is the

> realization of a body's (person's) potential'. That makes modern sense? It is just as soft as medieval dynamics – so where is the Merton School which will demand to know where the kinematics for such a science is to be found? Where is the geometry which will allow us to describe that thing 'education'? Where are the measurements and the observations which will make this statement of Social Science into a hard statement? Where are the mathematics?
>
> (Atkin 1981: 39)

In the replacement of a social science based on forms of life with one based on possible worlds it has, remarkably, become possible at once to argue that we should 'let the data speak for themselves' and that we should refuse to listen to the data unless they are speaking the right language, nature's own language, as Galileo put it, mathematics.

What was Wittgenstein's reaction to this – to set theory, and to the attempt by Whitehead and Russell to reduce mathematics to logic? He referred to them as the 'sickness of a time' that:

> is cured by an alteration in the mode of life of human beings, and it was possible for the sickness of philosophical problems to get cured only through a changed mode of thought and of life, not through a medicine invented by an individual.
>
> (1983, II: 23)

The development of a Wittgensteinian view of the social sciences, as in its various and sometimes uneasy allies – the historically oriented sociology of science of Bloor (1983), the history and philosophy of science (Kuhn 1970 and others), the interpretive anthropology of Geertz (1983) or the philosophy of Rorty (1982) and MacIntyre (1984) – seems to undercut the philosophical arguments against the existence of ways of life. At the same time the empirical evidence in those works provides an argument different only in its eloquence from that which we find every day in the newspaper, on television and especially in our interactions with others, that there are indeed ways of life, and that those ways of life are fundamental to our understanding of the world.

Those who believe that the geographical concepts that we offer must fundamentally rest on (or at least appeal to) our understanding of the world need surely to take those facts into account. And, at the same time, those who deny the relevance of our understanding of the world to its explanation ought to consider that the accounts given by people like Bloor and Wittgenstein of this non-homocentric social science make it evident that that social science is merely a modernist venture, guided implicitly or explicitly by an ideology about life in the twentieth century that in no sense lets the data speak for themselves.

NOTES

1 I shall here refer to the terms 'way of life' and 'form of life', or *genre de vie* and *Lebensformen*, as though there is nothing to distinguish between them; indeed, many authors (see, for example, Sharrock and Anderson 1985) fail to make such a distinction. At the same time, it ought to be noted that these are only two of a family of terms, including 'ways of living' at one extreme and 'life forms' at the other; common parlance suggests, among other things, that these terms imply varying answers to the question of whether these ways or forms exist in a hierarchical relationship with one another (as seems to be implied in the term 'life form') or, by contrast, are on a par with one another (as in the term 'ways of living'). These questions involve the equivalent of the introduction of the concept of 'moral relativism' into discussions of 'cognitive relativism'; to put the matter another way, they involve a consideration of the moral ontology that underlies our rankings of the relative value of animals and humans. As such they are, however interesting and important, beyond the scope of this chapter.

2 I use the term 'ontology' here in a colloquial sense, to refer to an account of the kinds of 'objects' that are thought to exist within a social, cultural or geographical system.

3 Here, as above, I refer not to Kant's works on geography, but rather to his more general approach to ontological and epistemological questions, as outlined in his *Critique of Pure Reason* (1965).

4 It should be noted that Mumford's work was strongly influenced by that of Geddes (1915), who in turn was influenced by the French social scientist Le Play. Although Vidal's source of the term *genre de vie* is unclear – in the French style, he fails to cite his sources – Buttimer (1971: 24) argues that 'Le Play ... suggested that occupational groupings had distinctive family and life-styles, religious practices, beliefs, and attitudes toward nature. Vidal de la Blache no doubt learned from this how significant life-style was in promoting and maintaining regional differences, especially in agricultural regions where livelihood and milieu were so closely related.'

5 The introductory portion of the *Tableau de la Géoraphie de la France* (1903) was translated into English and reprinted as *The Personality of France* (1928).

6 For additional views of the nature of geographical 'personality' see Barnes and Curry (1983); they argue that Vidal *et al.* have been much too stringent in their establishment of limitations on the application of the notion of 'personality', and that judgements about the personality of places are constantly, and with no little justification, made.

7 This actually is more than a mere analogy; Sapir (1924) had put forward much the same argument in his 'Cultures, genuine and spurious', and for many of the same reasons. And Sapir's, of course, was just an echo of the argument in Matthew Arnold's famous *Culture and Anarchy* (1969). (It might be noted that this argument has not been without its detractors, the most eloquent of whom has probably been George Steiner in his *In Bluebeard's Castle: some Notes towards the Redefinition of Culture*, 1971.) Similarly, much the same can be seen in the geographical work of some self-styled phenomenologists; Relph, for example, has argued that if in the past the world was fundamentally a collection of places, today one has to search for places (1976). A similar position can be found in Lowenthal's work, especially in his article 'The American scene' (1968).

8 The ideal type, as explanatory tool, can of course be traced to the work of Weber (1949) and Simmel (1971, orig. 1903), and from there back to the neo-Kantians of the nineteenth century as well as forward to Wirth, Park, Burgess and the foundations of contemporary urban geography. For Wirth an ideal type was a construct, laid out with all the particularity of 'real' concrete reality, but only accidentally

referring to any existing social reality. Hence, although its aim was not unlike that which we associate with concepts in the physical sciences, both its means of formation and its means of use were very different from those of concepts in the physical sciences.

9 In fact we find the same sort of process in other areas – in anthropology, for example, where after World War II there was an attempt among developers of the Human Relations Area Files to create a 'way of life index'.

10 As where Wittgenstein, in the *Tractatus*, asserted that 'The world is the totality of facts, not of things' (s. 1.1).

11 See, for example, High (1972), Keightley (1976).

12 In the case of *Philosophical Investigations*, references are preceded by a part number; references to Part I (and to all of Z, and OC) are to sections; references to Part II, *Remarks on the Foundations of Mathematics* and to *Remarks on Frazer's 'Golden Bough'* are to page numbers.

13 Bloor (1983) refers to Mary Douglas's work on grid–group analysis, where grid defines the strength of structural arrangements and group defines the internal cohesiveness (against the outside) of a group; he suggests that this may be an appropriate way to order one's analysis of differences among ways of life.

ACKNOWLEDGEMENTS

The author would like to thank Mischa Penn, Fred Lukermann, Geraldine Pratt and J. Nicholas Entrikin for comments on earlier versions of this piece. The research was supported, in part, by the University Professors Program of Boston University, the Andrew W. Mellon Foundation and the Academic Senate of the University of California, Los Angeles.

8

READING THE TEXTS OF THEORETICAL ECONOMIC GEOGRAPHY

The role of physical and biological metaphors

Trevor J. Barnes

INTRODUCTION

The literary critic Stanley Fish (1980: 305) tells the story of a student who asks his instructor, 'Is there a text in this class?' The professor answers, 'The *Norton Anthology of Literature*.' This reply does not satisfy the student, though. For the student is not asking the name of the textbook, but rather how the text will be read. Will the class assume that the text admits only one meaning – students will learn to find 'the text' – or will it assume that the text is constructed through the very interpretations that the students themselves provide – students will create 'a text'?

The issue that Stanley Fish raises in this anecdote is as important for theoretical economic geographers as it is for theoretically-minded literary critics. Both fields of study must address the question of whether interpretive schemes or theories simply mirror reality, or whether in part they create the reality that they seek to interpret. This chapter argues the latter position with regard to theoretical economic geography. Specifically, the chapter suggests that the theoretical landscapes constructed by economic geographers – von Thunen's rings, Weber's triangles, Christaller's hexagons – are not necessarily reflections, not even distorted reflections, of an underlying, unchanging and pre-interpreted world. For if, as Fish (1980: 355) maintains, 'interpretation is the only game in town', our theoretical interpretations can reflect only other interpretations and not a bedrock reality. But if theoretical economic landscapes are created by, rather than corresponding to, our interpretive schemes and theories, then we must adopt a different strategy in dealing with theory. We must treat theory itself as a text to be read and interrogated, rather than according it a privileged position.

There are two main theoretical schemes found in economic geography: one drawn from neoclassical economics, the other from Marxism. Both assume that there is a single, albeit different, story to be told about the geographical landscape of economic activity; a story, each claims, that must be true in virtue of the correspondence between its theoretical category and an underlying geographical reality. In contrast, this chapter argues that there cannot be such a one-to-one correspondence because the text of landscape partly emerges from the very act of theorizing about it. In making this argument, the chapter focuses on the assumption of economic rationality in both schemes; for it is this assumption, more than any other, that gives shape to the theoretical landscapes created. Specifically, the argument is that the two notions of rationality employed by common versions of neoclassicism and Marxism[1] are based upon two different metaphors, one originating in physics and the other in biology. As a consequence the theoretical economic landscapes constructed do not necessarily represent an 'external reality', but primarily reflect the internal logic of the metaphor employed. In other words, theoretical economic landscapes are not necessarily testimony to the empirical robustness of the theories employed, but to the power that these two metaphors have in the economic geographer's imagination.

TEXTS, LUMPS AND METAPHORS

Let us begin by defining text. Following LaCapra (1983: 26), any written text is defined:

> as a situated use of language marked by a tense interaction between mutually implicated yet at times contestatory tendencies. On this view, the very opposition between what is inside and what is outside texts is rendered problematic, and nothing is seen as being purely and simply inside or outside texts. [For this reason] ... textual processes cannot be confined within the bindings of the book.

The implication of this admittedly dense definition is that the texts of economic geography cannot be seen as hermetically sealed from the broader milieu in which they originate. Text and context, inside and outside, are inextricably intertwined. This means that unless we interpret the texts of economic geographers, and the theories discussed therein, in terms of the interests, audience and social context that prevailed when such texts were written we will fail to understand their full implications. Furthermore, it also means that we must abandon the view that texts somehow mirror a brute reality. As LaCapra (1983: 27) again writes, 'the very construction of a "context" or a "reality" takes place on the basis of "textualized" remainders of the past'.

In elaborating upon this textual approach this section focuses on three issues: first, the nature of 'facts' within such an approach; second, the methodological position of relativism that it implies; and, third, the potential role of metaphor that ensues from adopting such a textualist position.

First, in arguing that interpretive schemes and theories create texts rather than necessarily corresponding to them, one is not denying the existence of brute facts or, as Richard Rorty (1985) calls them, lumps. For example, the *Norton Anthology of Literature* as a physical object is a brute fact in just the same way that, say, the pressure of light waves on Kepler's retina is a brute fact about looking through a telescope and observing heavenly bodies. The more important point is not whether we respect brute facts – we have no choice – but what we make of them. Thus in reading the *Norton Anthology of Literature* the new critics find complexity while Marxist-inspired readers discover ideology, while in astronomy the pressure of light waves represents for Ptolomey a geocentric celestial system but for Kepler a heliocentric one. More generally, recognizing a non-linguistic physical brutality in no way commits us to the view that 'truth' is somehow captured in the specialized vocabulary of scientists and social scientists. Brute physical resistance is one thing, interpreting it is something different. It might be objected that if this is really the case theories and interpretations can never be proved wrong; texts or lumps would be unable to resist the interpretations made of them. But it is not texts or lumps themselves that offer resistance, only other interpretations. As many philosophers of science argue, facts are not neutral but already come theory-laden (Hesse 1980a: chapter 3); that is, right from the beginning facts emerge only through the very act of interpretation, and do not exist independently of theory, thereby providing some kind of external resistance. As an illustration, in his play *Jumpers* Tom Stoppard has the Cambridge philosopher Wittgenstein assailing a colleague in the corridor and asking him why people believed that the sun went round the earth. The colleague replies that it was because it looked that way. To which Wittgenstein asks, but how would it appear if the reverse were true? The point is that in terms of the brute physical resistance of light waves the ancient Greeks had the same experience as Kepler. But it was not brute facts that overthrew the Greeks' astronomical scheme, but Kepler's interpretation of them, along with the changing social context in which he lived and which gave force to his new heliocentric vision.

This last point leads to the second issue of relativism. For how are we to choose between Kepler's scheme and the one presented by Ptolomey if we cannot appeal to the 'facts'? Are we in a world where one person's opinion is as good as the next? Such a subjectivist view, however, is unwarranted. Personal beliefs do not constitute our interpretations and theories. Rather, what we have to say is informed by knowledge that is both public and agreed upon by others. Our interpretations and theories come from what Fish (1980) calls 'interpretive communities'. Such institutions are internally bound by a common set of rules, methods and approaches in dealing with some facet of the world. Within such communities there is consensus on which interpretive scheme or theory is best. That said, such communities are never legitimated in absolute terms because they cannot justify a claim to privileged access to the 'truth'. As a result, these interpretive communities are only temporary shelters from relativism, albeit necessary ones. They are temporary in that usually sooner rather than later one interpretive community is supplanted by another, and necessary because they enable us to block off one set of questions while dealing with another set. But, having answered such questions, one cannot then claim to have made 'Progress', discovered 'Truth' or uncovered 'Rationality'. To establish these qualities one requires an absolute benchmark, but that is the very thing that the concept of an interpretive community denies (although members *within* a particular interpretive community may well believe that they have established such an absolute benchmark).

But if the concept of an interpretive community denies absolute benchmarks, on what basis do we understand the rise and fall of interpretive communities? Here, following our definition of texts, we must look to the broader context in which academics work. It is the academic debate found in journals, professional meetings, lecture halls and the seminar room that establishes consensus and division; it is there that interpretive communities are born and sometimes die. In addition, we must also look at the external relationships in which academics find themselves. What academics consider to be the good, the true and the beautiful must in part reflect what society as a whole thinks (for a superb example set within the history of science see Bloor 1982). More generally, because no interpretive community ever has a foolproof method, the success of any such community depends upon its rhetoric. Theories and interpretations, as Rorty (1985: 2) writes, succeed or fail not on the basis of 'antecedently plausible principles', but through the process of 'prais[ing] our heroes and damn[ing] our villains by making invidious comparisons'.

Finally, once the tie is cut between the form of theory and the form of 'brute' reality, and we see academic debate in terms of competing language communities, the nature of theorizing is also loosened up. In particular, an increasing number of philosophers argue that what shape the nature of the language community, and thereby the 'facts' it emphasizes, are the metaphors each employs. Although such a view is implicit in Thomas Kuhn's work (see the postscript to the second edition of *The Structure of Scientific Revolutions*, 1970), it is Mary Hesse (1963, 1980a) and more recently David Bloor (1982) who have worked out the details. In particular, Hesse (1980a), drawing upon what she calls a 'network' model, argues that new theories emerge out of an inherently local process. By drawing upon a set of local beliefs of a given time and place academics develop new theories by applying them to an extended range of circumstances (Hesse 1980a; Arib and Hesse 1986: chapter 8). In the sense that the application of old beliefs to new circumstances is an attempt to identify similarities, the acquisition of new theoretical knowledge is an inherently metaphorical process. For the defining feature of metaphors is their claim that one thing is similar to something else: sound is like a wave; culture is like a text; spatial interaction among people is like the interaction among heavenly bodies. In each of these cases, metaphors are not mere ornamentation or decoration (which is the view of some theories of language) but are central in formulating the problem and finding a solution. For Hesse, then, scientific revolutions are nothing more nor less than metaphorical revolutions. As Arib and Hesse (1986: 156) write, 'to make explicit the ramifications of metaphor is to engage in critique, evaluation, and perhaps replacement. Metaphor is potentially revolutionary.'

If we admit that different metaphors give shape to different language communities, and thereby the texts that emerge, are all metaphors as good as one another? Given my remarks about relativism, clearly we cannot provide any absolute assessment of the metaphor employed. None the less, we can enquire about the internal consistency of the metaphor as a metaphor. In other words, accepting the goal of the interpretive community, we can and should ask whether the metaphor really does identify similarities between the thing investigated (for example, sound, culture and spatial interaction) and the thing to which it is metaphorically compared (respectively waves, texts and planetary relationships). If, because of the very internal structure of the metaphor, no significant similarities are identified, then in a real sense the metaphor fails as a metaphor, and therefore must be judged inappropriate. Note that in using this criterion for evaluating metaphors we are not using any absolute benchmark. Rather, assessment is on internal grounds, that is, on the basis of the very definition of what it means to use a metaphor.

To help us evaluate metaphors on internal grounds we can once again draw upon Hesse's work. For she provides a set of criteria to judge the efficacy of metaphor use. Hesse's basic tenet is that the basis of scientific explanation is, as she dubs it, metaphorical redescription. That process consists of transferring words normally associated with one system (which she calls the secondary one) to redescribe the explanandum of another system (which she calls the primary one). For example, with the work of Issac Newton, sound (the primary system and explanandum) was metaphorically redescribed in terms of waves (part of a secondary system). Metaphoric redescription is successful, according to Hesse, because of the interaction between the primary and secondary systems. (See Black's pioneering 1962 essay on the 'interaction' view of metaphor.) In particular, such interaction transfers:

> the associated ideas and implications of the secondary to the primary system. These [then] select, emphasize, or suppress features of the primary; [with the result that] new slants on the primary are illuminated; [and that] the primary is seen 'through' the frame of the secondary.
>
> (Hesse 1980b: 114)

As a result, the thing explained – sound – is seen in a new light; it is seen to share features with other phenomena – waves – that are not usually associated with it. Recognition of such similarities is then the jolt to new forms of theoretical explanation, albeit tied to the original metaphor.

Within this framework, Hesse outlines three criteria by which we can assess the suitability of the metaphor in identifying similarity. To do this, Hesse first distinguishes among positive, negative and neutral metaphors. In applying, say, Newton's theory of gravity to spatial interaction there are some characteristics that are similar in both processes (e.g. objects – people or planetary masses – are separated from one another by geographical distance), the case of positive metaphor. Other characteristics, though, are dissimilar (e.g. that humans are conscious and sentient beings whereas planetary masses are not), the case of negative metaphor. Finally, there are some characteristics about whose relationships we are not clear, the case of neutral metaphor. Given these definitions, Hesse argues that for a metaphor to be appropriate three conditions must be met. First, that there is at least some positive metaphor. Second, the essential properties and relationships of the primary are not part of the negative metaphor between the primary and secondary. Note that in using the term 'essential properties' we are not implying a definitively defined set of characteristics. Rather, what is deemed essential is determined by the interpretive community itself. And finally, that the relationship among characteristics in the secondary system is carried over to the primary one. If one of these three features does not hold, we need at least to question the appropriateness of the metaphor. Note that we are not suggesting that metaphors are inappropriate because they do not correspond with the world; they are inappropriate because they violate the canons of good metaphor use. For if the hallmark of a metaphor is its claim to establish similarities, then on those occasions when similarity is not established (that is, if one of the three conditions that Hesse provides is violated) the metaphor is inappropriate. It is in this sense that the internal consistency of the metaphor is judged.

All this said, Hesse's scheme, useful as it is, remains a formal one. Its formality necessarily abstracts from the actual practice of metaphor use in the sciences and social sciences. In particular, lacking in Hesse's formal approach is any discussion of relations of power, that is, the authority to determine which metaphors, and which parts of metaphors, are accepted and used. Such an omission will be partly, albeit inadequately, addressed in the two specific studies that follow.

In sum, the implication of adopting a textual approach is the rejection of any correspondence theory of truth, whereby independent facts are the arbiters of theoretical validity. Instead, facts only emerge once the text is already written from within the canons of a given language community. Those canons are often metaphorical, a result of applying a set of local beliefs to new material. Although there are no absolute benchmarks by which to compare one language community with another, we can explore the internal suitability of different metaphors by using criteria suggested by Hesse. This theoretical argument is now put to work by examining the use of the rationality postulate in the substantive texts of economic geography. We begin with neoclassical economics and its use of a physical metaphor.

NEOCLASSICAL ECONOMIC LANDSCAPES AND THE METAPHOR OF PHYSICS

There is little doubt of the dominance of neoclassical thought in economic geography. In fact some argue that to all intents and purposes theoretical economic geography *is* neoclassical economics. A pivotal assumption in the neoclassical scheme is economic rationality, represented by the *homo economicus* postulate. This assumption is central for two reasons: first, it is intimately bound up with the definition of the economic problem that the neoclassical economists have set themselves, one that is conceived as matching a set of limited economic resources with a set of economic ends. Because resources are never sufficient to meet all the ends, choices must be made. Economic rationality's role is then one of a conceptual bridge that allows the right means to be connected with the right ends. Second, economic rationality plays a methodological role. In effect, it allows the neoclassical theorist to theorize. As Rozen (1985: 664–5) argues, the assumption of economic rationality allows economists to make 'claims to precision and exact inference ... [but] ... once the anchor of strict optimizing behavior is cast away, economic science can only aimlessly drift, incapable of making useful predictions and with no way of ensuring internal consistency'. Neoclassical economics is able to realize the goals of 'precision', 'predictability' and 'exact inference' by using the mathematical technique of constrained maximization to represent economic rationality. By defining certain mathematical conditions that must be met for maximization to occur, neoclassical economics is able to deduce theoretically the economic consequences of rational economic behaviour.

Because *homo economicus* represents a 'key code' in the neoclassical scheme, economic geographers who follow that school necessarily make use of the rationality postulate. As a brief example, take Alonso's model of household location in the city. In the introduction to *Location and Land Use* Alonso (1964: 1) writes that:

> The approach that will be followed in this study will be that of economics, and from this wealth of subject matter only a pallid skeleton will emerge. Both the Puerto Rican and Madison Avenue advertising man will be reduced to that uninteresting individual, economic man.

Uninteresting or not, economic man is crucial to Alonso's scheme, for it is individual rational choice-making that gives shape to the city. In particular, the city is conceived as a repository of means and ends. Thus, on the one hand, the urban landscape is composed of scarce resources (the economic means of land and location) while, on the other, it is full of householders who bring to the landscape the desire to maximize utility (the economic end). The resulting urban landscape is then the outcome of rational choice-making, a process that ensures that means and ends are brought together in the best possible way. Specifically, through the bid rent process, rational householders determine where they should live, and on how much land in order to maximize utility. The subsequent urban landscape that emerges out of the dictates of economic rationality is one where poorer groups economize on land and live close to the central business district, while richer groups use more land and live in the suburbs.

What lies behind *homo economicus*, and more generally behind neoclassical rational economic landscapes, is a physical metaphor drawn from nineteenth-century physics. Although many have recognized the influence of physics on the methodology of neoclassical economics, Mirowski (1984a, 1984b) has recently argued that the rationality principle has its origins in just one area of the subject, the conservation principle of energy and its associated mathematical technique of constrained maximization. Within such a physical metaphor, early neoclassical economists substituted utility for energy in the constrained maximization equation of the physicists, and made the rationality postulate the equivalent of the physicist's principle of least effort. Thus, just as the technique of constrained maximization reveals the path of least effort of any particle's movement, so the same technique is used to discover the most efficient (rational) actions of producers and consumers.

The consequence is that explanations based upon rational choice-making implicitly presume a correspondence between processes in the physical world and those of the social world. As a result, the neoclassical landscape text conforms not necessarily to what is really there, but to the dictates of the physical metaphor used. What are the dictates of this metaphor, and are they appropriate to the task neoclassical economics sets itself? Using Hesse's three maxims for appropriate metaphor use, let us now evaluate the internal consistency of the physical metaphor.

The first of the maxims is that there is some positive metaphor between the characteristics of the movement of particles and the characteristics of actions of individuals. This is met in that both particles and humans have mass, both experience movement, and so on.

The second requirement that nothing essential is lost in the negative metaphor is more problematic. Of course, this begs the question of what is essential. There can be no cast iron answers, but in this case there are two features of human behaviour that are denied by applying the physical metaphor, features that even some neoclassical economists consider important in discussing human action (that is, essential is defined by the interpretive community of neoclassicism itself).

First, the metaphor denies any deliberation on the part of the actor; humans are merely a conduit. This is because in the physical world particles move without any consciousness on their part; they simply behave in conformity with the restrictions of the principle of least effort. When this is carried over to the human world, the implication is that agents do not need to think about how to act: once certain parameter values are specified, human action is immediately determined. As Georgescu-Roegen (1971: 343) writes, in the neoclassical scheme 'man is not an economic agent because there is no economic process. There is only the jigsaw puzzle of fitting given means to given ends, which requires a computer and not an agent.'

The second feature lost is the context-dependent nature of human action. This is because the metaphor implies that rationality explains all human behaviour at every time and in every place; rationality is a universal. Again, such a view arises out of the very logic of the physical metaphor employed, in this case the extremal mathematical technique used to represent economic rationality, and the account of global maximization that it provides. Specifically, Rosenberg (1979: 523) argues that all theories based upon the constrained maximization technique:

> are committed *to explain everything in their domains.* In virtue of the claim that systems in their domains always behave in a way which maximizes or minimizes some quantity, the theory *ipso facto* provides an explanation of all of its subject's states. There is no scope for treating such a theory as only a partial account of the behavior of objects in its domain ...

Immediately one employs an extremal theory of human behaviour that presumes global maximization one has no choice but to apply it to every facet of that behaviour. If one did not, global maximization might not occur, thereby producing sub-optimality and an irrational outcome. Thus, by its very logic, rational choice theory must hold for all human behaviour, be it Trobriand Islanders exchanging kula or Wall Street dealers exchanging currency swaps.

Hesse's third requirement for appropriate metaphor use, for our case, is that the same type of relationships among particles carries over to the relationship among rational agents. Again this claim is problematic, for two reasons.

First, both the economic rationality postulate of economics and the principle of least effort on which it is based are, as already suggested above, extremal theories in that both attempt to maximize some value on the basis of a given set of constraints. For an extremal theory to work there must be some independent specification of the explanatory variable, in this case, respectively, the principle of least effort and the rationality postulate. Mirowski (1984b) argues that in the physicists' world the principle of least effort is justified by an appeal to the first law of thermodynamics, that is, the conservation principle of energy and matter. In economics, however, it is not clear that there is any consistent counterpart to the conservation principle, and therefore there is no comparable justification for the rationality postulate. It is true that parallels to the conservation principle have been proposed in economics – Say's law (and Samuelson's recent modification of it) suggesting that the value of output is conserved through the exchange process is one example, and the equating of *ex ante* desire with *ex post* satisfaction in utility theory is another (Mirowski 1984b: 469–71). But in both cases, as Mirowski (1984b) argues, neither holds up. Keynes forcefully demonstrated the inadequacies of Say's law, and Dobb, as early as the 1930s, recognized that if desire is the same as satisfaction consumers can never regret their purchases, an absurd conclusion. More generally, because the conservation principle of energy in physics has no obvious counterpart in the social world, the implicit teleology of *homo economicus* – the purpose of humans is to maximize utility – is severely undermined. In short, there is no natural order that ensures that things always work out for the best (Mirowski 1984b: 472).

The rationality postulate is also faced with the similar necessity of providing an independent specification of utility maximization. No theory, however, has ever been worked out on that basis, nor is one likely to be. The difficulty is that to account for utility maximization one presumably must make some reference to one or more characteristics of humans. Such human characteristics, however, can never account for utility maximization, because, since extremal theories explain everything within their domain, such characteristics must already be maximized. Thus we *are* left with a circular argument: utility maximization accounts for maximizing human behaviour, and human behaviour, because it is subject to maximization, accounts for people maximizing utilities. More generally, the problem is that the set of relationships postulated in the physical theory – one where an independent grounding is given – do not carry over to the human world. This is a clear case of a negative metaphor between the relationships of the secondary system and the relationships of the primary one.

The second area where the relationship among particles does not correspond to the relationship among rational agents is in respect of 'motivation'. In the physical world an impulse of energy *causes* a particle to move in accordance with the principle of least effort. The parallel in neoclassical theory is the consumer, who is motivated by his/her desire to maximize utility.

The problem with this parallel is that a number of philosophers argue that desires are not causes. The argument is that cause in the Humean sense refers to the constant conjunction of two independent events: if event *a*, then event *b*. The conception of cause further implies that the relationship between events *a* and *b* is a contingent one. This is because in our observations we ascertain merely a regular empirical association. Because we do not know whether such a constant conjunction will hold in the future, the relationship is by definition contingent. In the case of the relationship between desires and action, however, we are dealing with one that is necessary rather than contingent. Specifically, the word 'desire' as conventionally used necessarily implies, *ceteris paribus*, that people will take actions to realize their desideratum. These actions stem from the very meaning of desire. In fact, if people do not take action to realize their desire, then one must conclude that they never had the desire in the first place. It is in this sense that the relationship between desire and action is a necessary one: there is something about the internal relations of the concept 'desire' that demands that it be acted upon in a specific way.

If we apply this to the neoclassical scheme, all individuals there desire to maximize utility. And this desire translates into the purchase of a particular bundle of commodities. From what was argued above, *ceteris paribus*, the relationship between a consumer's desire and the action of purchasing those goods is a necessary one. But if that is true, then the relationship cannot also be causal, because in the Humean definition of cause events are contingently related. It should be noted that, in arguing that desires are not causes of action, one is not denying that some aspects of human behaviour may be described causally. The point, however, is that desires and causes cannot be conflated, because in so doing one confuses two distinct types of relationship.

In sum, the physical metaphor employed by neoclassical economic geographers fails because the relationships and features of the primary metaphor are not carried over to the secondary one. In particular, the conservation principle of energy that justifies the principle of least effort has no counterpart in the social world, thereby undermining any parallel argument used to legimitate the rationality postulate. We should also note that although framing this critique of neoclassical economic geography in terms of metaphor is novel, many of the substantive criticisms that were actually made are not. Over the last two decades economic geographers, whether they knew it or not, have in effect critically explored the limits of the physical metaphor found in economic geography (see the pioneering critiques of Pred 1967 and Sayer 1976). In fact such critiques have already created a theoretical revolution in the way Hesse suggests. As economic geography enters the 1990s it is clear that it is no longer dominated by neoclassical metaphors, but rather by the metaphors of classical political economy. It is to these that we now turn.

MARXIST ECONOMIC LANDSCAPES AND THE METAPHOR OF BIOLOGY

Although Marxist economics is clearly a very different interpretive community from its neoclassical counterpart (this difference explains why adherents of one community often view the work of the other with incomprehension and incredulity), it shares with it an emphasis on rationality. However, the radical view of rationality, as seen in the work of David Harvey (1982, 1985a, 1985b) and Neil Smith (1984) at least, is very different from the neoclassical one. This difference, we will argue, is a result of both authors using a biological metaphor rather than a physical one.

Economic rationality as it emerges in radical geography is couched not at the level of the individual but in terms of the structure of the economy as a whole. The economy is rational provided that there is crisis-free accumulation. Such a condition is pivotal to Marx's scheme because of the circular view of production that he adopts. Such circularity is readily seen in Marx's well known diagram tracing the reproduction of capital:

$$M - C\begin{matrix} LP \\ MP \end{matrix} \ldots P \ldots C' - M'$$

where M is financial capital; C is production capital; LP is labour power; MP is the means of production; P is the production process itself; and $'$ represents the (increased) value of the variable after the production process. Such a scheme depicts the circulation of capital around the production process from the moment necessary inputs are purchased to the conclusion when commodities are sold. But, once they have been sold, the revenue is used to purchase more inputs and the cycle continues once more.

The general hallmark of this circular view, originating with the Physiocrats, and later extended by Ricardo, is that economic outputs from the past production period are inputs for the present one. The result is that, if accumulation does not occur, then the economy, by its very definition as a circular system, is unable to reproduce itself. It is precisely here, however, that Marx's analysis of capitalism begins. For Marx's objective was not to celebrate capitalism's rationality (its ability to maintain the circular flow), but to suggest that that rationality constantly fails. For capitalism by its very nature is unable to maintain smooth reproduction; it is jolted by crises that interrupt the circular flow. Because the circle of reproduction is not completed, the analytical focus is then necessarily on capitalism's irrationalities. To discuss crisis, Marx none the less requires the benchmark conditions for smooth reproduction, represented in his theory by the schemes of simple and extended accumulation (simply more elaborate versions of the figure above). With these conditions identified, Marx then shows that they are very difficult to realize because of, for example, the nature of technical change, lack of effective demand, the nature of business organization, and so on.

In describing the landscapes of capitalism David Harvey and Neil Smith follow Marx in first setting up a rational landscape as a benchmark – one that allows crisis-free reproduction. (Smith 1984 says that such a landscape represents 'equilibrium', and Harvey 1985a: 190 writes that 'It is impossible to imagine [the capitalist] ... material process without the production of some kind of ... 'rational landscape' within which the accumulation of capital can proceed.') This rational landscape is one that allows reproduction of both the wider capitalist system and the specific spatial configurations associated with it. But it is also a landscape that is fleeting. Because 'capital ... must represent itself in the form of a physical landscape created in its own image' (Harvey 1985a: 43), the contradictions of capitalism necessarily make such a landscape short-lived. Specifically, Harvey brilliantly shows that under capitalism the very act of making a rational landscape necessarily creates conditions that produce irrationality. His argument is that in order to increase surplus value (profit) capitalists are impelled to 'annihilate space by time', which in practical terms means that spatial patterns of production (including the associated infrastructure) are continually reconfigured. But that very reconfiguring of the landscape helps create a crisis that the reconfiguration was in part designed to avoid. For in the process of locational change large amounts of fixed capital are scrapped (devalued) and large numbers of labourers are made unemployed. Places such as Youngstown, Hamilton and Manchester are left to die as capital uproots and creates a new geographical landscape, one that matches the new conditions of accumulation better. But sooner rather than later the new locational configuration will itself become a spatial barrier to accumulation, thereby eventually requiring 'annihilation' like its predecessors. As such the equilibrium rational landscape is at best ephemeral and is always likely 'to burst asunder' at the next crisis. As Harvey (1985a: 44) puts it:

> Capitalist development has to negotiate a knife-edge path between preserving the values of past capitalist investments embodied in the land and destroying them in order to open up fresh geographical space for accumulation. A perpetual struggle ensues in which physical landscapes appropriate to capitalism's requirements are produced at a particular

moment in time only to be disrupted and destroyed ... at a subsequent point in time.

More generally, the theoretical landscape that Harvey and Smith construct mirrors the reproduction view of Marx. Both set up benchmark conditions defining rational reproduction, one that is crisis-free. This means for Marx that the same mix of commodities is reproduced, while for Harvey and Smith it is the same combination of social and physical landscapes of capitalism. They then both argue that rationality begets irrationality, with the result that the cycle of reproduction is broken, thereby precipitating crisis. For Marx crisis continues until there are fundamental changes in production, while for Harvey it is until the physical and social landscape is transformed. But in both cases such changes are never fundamental enough to prevent future crisis, and the processes associated with restoring 'rationality' are ones that will also undermine it in the future (e.g. in Marx it is often technological change, while for Harvey it is locational change).

The metaphor of reproduction that underlies this landscape, and the notion of economic rationality associated with it, are clearly drawn from biology. The Physiocratic circular system of production was in fact explicitly based upon Harvey's seventeenth-century discoveries about the pulmonary circulation system, and when Ricardo made use of the reproduction metaphor he set it within the context of his 'corn model'. Some have also suggested that Marx's biological metaphor of reproduction may have come from Darwin, although the evidence here is weak (for a review of the literature see Warren 1987). Note that, in saying this, we are claiming neither that Marx drew only upon biological metaphors (although they certainly figured large), nor that he rests his arguments exclusively on the reproduction metaphor. We do claim, however, that a central metaphor was reproduction, especially with regard to his theory of crisis and social transformation.

Perhaps more important than the specifics of how the metaphor of reproduction was imported into Marx's scheme is the question of its methodological import. We argue that in taking the idea of reproduction from biology and applying it metaphorically to society one is compelled to employ some kind of functionalist argument. As Jon Elster (1979: 28) writes, 'functionalist explanation ... [in the social sciences] rests upon an ill-conceived analogy from biology'. Incidentally, it is here that we can see the link between Marx and Darwin. What Marx and Darwin share is not their survival-of-the-fittest vision of respectively capitalism and nature, but the general type of explanation that they employ, functionalism. And they share this methodology because both make the idea of reproduction central. Once again, it should be made clear that we are claiming neither that Marx rests all his arguments exclusively on a functionalist explanation, nor that all subsequent Marxists have followed him in using this methodology. Rather, the more limited argument is that Harvey and Smith, by drawing upon Marx's theory of crisis, make use of that part of his work that is most susceptible to functionalism because it rests upon the reproduction idea.

Let us begin by noting the hallmark of a functionalist explanation, which is that the 'character of what is explained is determined by its effect on what explains it' (Cohen 1978: 278). For example, in Darwin's theory of evolution, species have the characteristics they do because they are best fitted for reproduction in the environment in which they find themselves. Here ability to survive and reproduce, the effect, explains the character of the species. As such, the functionalist argument reverses our normal notions of cause and effect. Usually we assume causes explain effects, but here we are saying, 'the cause occurred because of its propensity to have that effect' (Cohen 1978: 281). The reason functionalist arguments are so common in biology is the prevalence of what Elster (1979: 33) calls the 'regulative idea'. That is, 'the structure of behavior of organisms [is explained] through the benefits for reproduction' (Elster 1982: 463; see also Rosenberg 1985: chapter 3). In this light, Marx's scheme of reproduction represents a recasting into the economic and social world of that regulative idea.

In its geographical form the argument is that the landscape of economic activity possesses those characteristics that are most conducive to continued capitalist accumulation. Thus, using a phrase that he repeats in a number of different places, Harvey (1985b: 36) writes, 'capitalist society must of necessity create a physical landscape ... in its own image, broadly appropriate to the purpose of production and reproduction'. In so doing, Harvey is clearly making the effect, the reproduction of capitalism, explain the cause, the form of the physical landscape created. One might object at this point and argue that Marx's and Harvey's concern is with crisis, which is surely dysfunctional. But, although crisis is disruptive, it is in fact functional to capitalism. Thus Smith (1984: 127) writes, 'for no matter how disruptive and dysfunctional, crises can also be acutely functional for capital. The mergers, takeovers and bankruptcies ... that accompany crisis also prepare the ground for a new phase of capitalist development.' The broader point is that, although Harvey and Smith allow crisis into their theoretical landscapes, it is done within a functionalist argument.

If the metaphor underlying the landscape constructed by Harvey and Smith comes from biology, is it an appropriate one? Once again using Hesse's threefold criteria we now address this question.

First, it is clear using the first of Hesse's criteria that there is at least some positive metaphor between reproduction in the biological world and reproduction in the social and economic one. For example, new commodities like individual members of species are reproduced.

Hesse's second criterion, in our case that the essential characteristics of the social world are retained once metaphorically redescribed, is more problematic on two grounds.

First, it is not clear that there is any place for referring to individuals once the metaphorical redescription is made. For functionalism, as Elster (1982: 459) writes, is a theory of 'action in search of an actor'. The difficulty is that once one identifies a function individuals drop out of the analysis because their role is simply that of 'bearers' of the function. This is clearly seen in a well known passage taken from Althusser's (Althusser and Balibar 1970: 180) *Reading Capital*:

> the structure of the relations of production determine the *places* and *functions* occupied and adopted by the agents of production, who are never anything more than occupants of these places, insofar as these are 'supports' (Träger) of these functions. The true 'subjects' ... are not ... 'concrete individuals' – 'real men' – but *the definition and distribution of these places and functions.*

If we cannot refer to individuals in the social world, many would argue that something central is missing from our analysis; the metaphor occludes rather than illuminates a vital characteristic of the thing that is investigated. From his writings on Paris that highlight individuals Harvey (1985b) himself must be aware of this problem.

Second, use of the reproduction metaphor also derogates human reflexivity. That is, the ability to modify action in the light of its consequences. To understand why this is a necessary outcome of the logic of the metaphor it is helpful to make a distinction between antecedent causes of an event – usually labelled proximate causes – and its functional consequences – usually called ultimate causes. For example:

> If one were to ask, 'Why is a polar bear's coat white?' an answer in terms of ultimate causes would look at its consequences for the bear's ability to hunt prey in a snowy environment and functionally relate this to reproductive viability. An answer in terms of proximate consequences might look at discrete causal processes in killing prey, at the physiological connections between nourishment and reproduction, at the biochemistry and genetics of hair colour, and so on.
>
> (Warren 1987: 264)

For polar bears and the like the relationship between proximate and ultimate causes, following Darwinian theory, is due to random variations. Because of a chance variation in genetic make-up (the proximate cause) bears whose coats are white and who live at the North Pole are able to survive and reproduce (ultimate cause), whereas those whose coats are of a different colour do not. In the case of humans, however, the relationship between proximate and ultimate causes is very different. For the proximate causes of human behaviour are people's intentions, desires, and the like. But the relationship between these proximate causes and ultimate ones cannot be one of chance variation because 'humans can intentionally modify their (proximate) behaviours in the light of ultimate consequences' (Warren 1987: 265). For example, if capitalists realize that the ultimate cause of capitalism's downfall is their desire to exploit workers unmercifully, then they may well curb their predatory urge; that is, they modify proximate causes (their desire to exploit unmercifully) in the light of ultimate consequences (the death knell of capitalism). As a result, if we wish to maintain the characteristic of human reflexivity in our analysis the distinction between ultimate and proximate causes cannot be sustained in the human world (they are interdependent, not independent). Furthermore, functionalist explanation cannot be sustained either, because it rests precisely on just such a distinction. For this reason the only way in which the reproductive metaphor can be applied to human behaviour is to deny its reflexivity – a denial that many would consider unwarrantable.

Finally, the third of Hesse's criteria in our case is the carrying over to the social world of the relationships found in the biological one. Once again, there are at least two problem areas.

First, an issue that we have already touched upon, the mechanism relating the beneficial consequences of reproduction to the specifics of how that reproduction occurs seems very different in the biological world compared to the social one. This problem, in fact, has been a particularly contentious issue in the debate between Marxist functionalists (for example, Cohen 1978) and the so-called rational-choice Marxists (for example, Roemer 1982, Elster 1982). For Cohen (1982: 491) traditional functionalist Marxism 'is *at best* in a position like that occupied by natural history before Darwin transformed the subject'. By this he means that although writers before Darwin believed that a species' characteristics are explained by the beneficial consequences of those characteristics for reproduction, they were unable to provide a theory that linked the two. It was only through Darwin's mechanism of chance variation and natural selection that an explanatory feedback loop was provided. In this sense the landscapes of capitalism discussed above are also often pre-Darwinian. There is usually no rigorous mechanism that connects the nature of the landscape with the beneficial consequences of reproduction. Furthermore, as Elster (1982) argues, this is not because Marxists have not had sufficient time to develop such mechanisms; rather, to supply such a mechanism in the social sciences, at least, analytical Marxists argue that we must begin with individuals and their intentions – a requirement that functionalism by its very definition is unable to meet.

Second, and related, the relationship implied by a functionalist explanation drawn from the reproduction metaphor is a teleological one. By making the claim that an organism or, in our case, the economy has the purpose of reproduction we are appealing to teleology to explain the phenomenon. Such a view has been criticized even in biology (Rosenberg 1985: 44), but it is even more problematic in accounting for human action. The teleological view reverses the accepted relationship between human purpose and action. Normally motives, desires and intentions are taken to precede action. But in the teleological view purposes lie in the future, and somehow reach back and direct present action. What allows future purposes and present actions to be linked is never made clear.

In summary, as with the case of the physical metaphor, the Marxist view of rationality based upon a reproduction metaphor drawn from biology fails on two of the three criteria suggested by Hesse. To a more limited extent than for neoclassical economic geography, some of the criticisms of the reproduction metaphor raised here are also foreshadowed by previous critiques of radical geography (see Duncan and Ley 1982). Unlike the case of users of the neoclassical rationality assumption, however, there has been an attempt by at least some on the left in geography to respond constructively to such a critique – although it should be noted that the reproduction metaphor still remains powerful even among this group. For example, in the collection edited by Wolch and Dear (1989) there is a strong editorial statement in favour of the reproduction metaphor.

CONCLUSION AND IMPLICATIONS

In summary, the principal argument of this chapter is that the activity of theorizing as practised by neoclassical and Marxist economic geographers is a constitutive one; theories create worlds rather than corresponding to them. This was shown by arguing that the theoretical landscapes described by neoclassical and Marxist economic geographers correspond not to what is 'really' there, but to the logic of one of two metaphors drawn from physics and biology respectively. In making this argument we are not denigrating the importance or necessity of using metaphors. Metaphors are both enabling (all research is done from some point of view) and constraining (some questions are necessarily bracketed out by the metaphor in order to answer others). However, the specific purpose here was to raise questions about the appropriateness of using some of the metaphors found in economic geography in fulfilling the tasks that are set them. Because we take the metaphors within which we work for granted, we rarely question whether the metaphor is congruent with all the things that we expect of it. By using Hesse's threefold criteria the chapter drew attention to the need to recognize the nature of the metaphors we use, as well the need to be self-critical of them. More generally, in following Arib and Hesse (1986: 156) the chapter attempted 'to make explicit the ramifications of metaphor', for it is of this that revolutions are made.

In this light, it is appropriate to make one final comment about the recent interest in 'place' as it emerges in economic geography, both in Massey's work on locality studies and in Scott and Storper's work on industrial districts. If we take seriously the claim that 'place matters', meaning that we must allow for the richness and texture of different locales, then we must also scrutinize the metaphors that we employ. From examining the neoclassical and Marxist landscapes, represented respectively by Alonso's, and Harvey's and Smith's work, it is clear that place in those two schemes often does not matter. Place is reduced to the single metaphor that underlies each of the two analyses – either an inflection point on an indifference curve, or as a deposit of capital secreted from the circular flow. There is no room for the diversity of individual places because only one metaphor is used to account for all places. In looking at Massey's and Scott and Storper's work we find that their accounts too rest upon single, albeit different, metaphors. Thus Warde (1985) argues that Massey employs a geological metaphor, while from Amin and Robins's (1990) commentary it is clear that Scott and Storper's work relies on the metaphor of flexibility. But if this is so, can Massey and Scott and Storper fulfil the charter of making place central? Are not places reduced to either the logic of layers of capital accretion (Massey's view) or the logic of agglomeration economies (Scott and Storper's view). And it is precisely here that the respective critics of Massey, and Scott and Storper, are most compelling. For the criticism that Warde (1985) makes of Massey, and Amin and Robins (1989) make of Scott and Storper, are precisely about all the things that get left out of such single metaphors. We should emphasize that this is not to trivialize the work of Massey or Scott and Storper – it is the most exciting and innovative research being done in economic geography. Furthermore, there is acknowledgement on all their parts of some of the things that are excluded from the metaphors they use. None the less, such acknowledgement is partial, thereby pointing to the need both to scrutinize systematically the metaphors we use, and to inspect critically the congruence of our metaphors with our broader goals. More generally, it seems clear that if the new interpretive community that values place in economic geography is to succeed, then it must self-consciously make use of the panoply of metaphors available in our culture and language; to confine ourselves to only one metaphor is to engage in synecdoche – to mistake the part for the whole. The consequence is not only the denial of the richness of our own language, but for geographers it is also a denial of the richness of place that they seek to describe.

NOTE

1 I am excluding from my discussion the so-called analytical Marxists who follow John Roemer. Their view of rationality is identical to the neoclassical one, and therefore a physical and not a biological metaphor underlies it.

ACKNOWLEDGEMENTS

I would like to acknowledge the very helpful comments of Les Hepple, Nigel Thrift, and especially Ian Gordon that greatly improved this paper. Any errors remaining, however, are my responsibility alone.

9

METAPHOR, GEOPOLITICAL DISCOURSE AND THE MILITARY IN SOUTH AMERICA

Leslie W. Hepple

The metaphors we construct, adopt and employ in our academic geographical texts sometimes escape from these confines and affect a wider scene, with social and political consequences that can be dramatic and even tragic. This chapter sets out to explore one such metaphor: the metaphor of the state as a spatial organism first developed by Ratzel in his paper in *Petermanns Mitteilungen* in 1896 (Ratzel 1896).

The organic metaphor of the state, in the specific form used by Ratzel, was rapidly taken up by Rudolf Kjellen (1917) and became the basis of much of classical geopolitics in Germany and elsewhere. Geopolitics is a discourse that developed at the beginning of this century in Europe and which gives a geographical perspective on the development and history of states. It emphasizes the importance of space and environment, and from this draws historical lessons, generalizations and 'laws' about the growth of states, to advise politicians and design policies, usually expansionist and nationalistic in tone. From its original European source, the organic metaphor also became an important part of the large geopolitical literature that grew in South America, and it is the impact of the metaphor within South America that this essay examines. Yves Lacoste, in *Le Géographie, ça sert, d'abord, à faire la guerre* (1985), has argued that there exists a close historical connection between the military establishment and its interests and the institutional and intellectual development of geographical knowledge, and this connection is very important in the case of South American geopolitics. Geopolitical discourse, and the organic metaphor, have been very influential in the military academies and in military thought about the state. Moreover, the state-as-organism metaphor has not only been adopted and used, but has been extended and transformed to apply to the internal as well as external security of the state. These concepts have had a major political impact in providing the ideology of the military regimes that have dominated the major South American states over the last twenty-five years. Leading geopoliticians (like Pinochet in Chile, or Golbery in Brazil) have become major government figures, and the impact of geopolitical ideas can be seen both in major programmes of development in Amazonia, and in the repressive apparatus of the security police state. It is unfortunately the case that a direct line can be traced from the original geopolitical metaphor to the ideology of the Chilean junta and the Argentine 'Dirty War', and it can be argued that, to some extent, the original metaphor carried within it the seeds of such political repression.

The chapter first of all briefly reviews the history and scope of South American geopolitics and its military and political significance, before examining why it has appealed especially to military groups in South American society. It then turns to the use of the organic metaphor within South American geopolitics, and looks at the way in which metaphors work, highlighting some aspects of their subject but at the same time obscuring and neglecting others. The chapter then outlines the major extension and transformation of the state-as-organism that took place in Brazil, turning the metaphor inward to the internal health of the organism and leading to the National Security Doctrine. The political impact of these ideas is discussed, and then the final section looks at the likely death of the organic metaphor in contemporary Latin America and how this affects the future of geopolitical thought there.

SOUTH AMERICAN GEOPOLITICS

In the Spanish and Portugese-speaking states of South America there is a long history of geopolitical writing and theorizing, stretching back to the early years of the century. Until Child's review (1979) this literature was largely unknown to European and North American scholars, but during the 1980s it has become accessible to the English-speaking (or reading) world through Child's studies and those of other writers (Child 1985; Hepple 1986a; Kelly and Child 1988), and also to French-reading groups (Foucher 1986). Because these studies survey the literature and history of South American geopolitics in detail, only essential background features will be outlined in this essay.

The first of these is the sheer scale of the geopolitical literature. It originated in the early twentieth century, linked with classical geopolitical theorizing. Thus a Chilean military geography of 1905 referred to Ratzel's ideas, the Argentinian admiral Storni in 1915 made use of Mahan's analysis of sea power, and the Brazilian Backheuser drew on both Raztzel's and Kjellen's work, publishing in the German *Zeitschrift für Geopolitik* (1926). This writing continued through the decades, but expanded in the late 1950s and 1960s, with literally hundreds of books, published in all the Latin American states but especially in the three major southern cone states of Argentina, Brazil and Chile. In addition, there are numerous geopolitical journals such as *Geopolítica, Geosur, Estrategia, Segurança e Desenvolvimento* and the *Revista Chilena de Geopolítica.*

The second feature is the close association of geopolitics with the military in South America. Many, if not most, writers are either active or semi-retired military or naval officers. This contrasts with the European and North American case, where, despite the title of Weigert's *Generals and Geographers: the Twilight of Geopolitics* (1942), few military officers other than General Haushofer himself were involved in classical geopolitics, and German geopolitics was essentially a civilian literature. In South America geopolitics is well established (indeed, an essential core subject) in the major military academies, such as the ESG (Escola Superior de Guerra) in Brazil since the 1950s, and in other states. These academies have played a major role in the education not only of military elites but also of business groups through courses and seminars (Stepan 1971). In Chile geopolitics is taught in the National Senior Security Academy and its associated Institute of Geopolitics, and military-approved courses are taught in the universities. In addition, semi-official, part-funded institutes of geopolitics and strategy exist in states such as Argentina, with the Instituto Argentino de Estudios Estratégicos y de las Relaciones Internacionales (INSAR) and the Instituto de Estudios Geopolíticos, and such institutes, largely staffed by former military officers, have been influential.

The writings are of especial interest because they are not just the speculations of uninfluential groups or individuals. In South America military regimes have exercised power for various periods in most of the countries since the 1960s, and military regimes have controlled the southern cone states for most of the last quarter of a century: in Brazil from 1964 to the late 1980s, in Argentina after 1966 and again from 1976 to 1982, and Chile from 1973 to 1990. Geopolitical writings are significant within these regimes for two reasons. Individual and leading geopoliticians from the military have become major political figures in the military governments. Thus General Pinochet, formerly Professor of Geopolitics at the Chilean War Academy and author of the textbook *Geopolítica* (English translation 1981), became military dictator and President of Chile. In Brazil the leading geopolitician General Golbery do Couto e Silva became head of the Security and Intelligence Service SNI, *chef de cabinet* and a central figure in several military governments, a key role acknowledged in all the political histories of Brazil. In Argentina the personal impact has been less focused, but several geopoliticians have served in ministerial posts, and General Osiris Villegas, a prominent writer, was Interior Minister, ambassador to Brazil and a key negotiator in the Beagle Channel dispute with Chile.

The second reason is even more significant. Geopolitical thought has played an important intellectual and legitimating role in the military regimes. It has had a key role in shaping the ideology of the regimes, in constructing a theory of the military-dominated state (the National Security Doctrine), in articulating national rivalries between states (such as Argentina–Brazil, Argentina–Chile, Argentina–Britain), and in constructing major regional development strategies such as the speculative 'Southern Project' in Argentina and especially in the policies for Amazonia's development – policies which major studies such as Hecht and Cockburn (1989) regard as geopolitically driven by the generals.

These several factors all mean that geopolitics is not only a well developed discourse in South America but has also had real political impact and has itself become a facet of the political life of the countries, or, as Foucher neatly expresses it, 'The importance of "geopolitical" discourse in Latin America is a fact which has, in itself, a geopolitical significance' (1986: 269, translation by present author).

GEOPOLITICAL DISCOURSE

The South American geopolitical literature has considerable internal diversity, especially between different nationalistic interpretations, but also sufficient common direction and internal coherence to be seen as a discourse in Foucault's sense of the term. A discourse is a set of rules or perspectives for the acquisition and organization of knowledge, with its own dominant metaphors that facilitate further knowledge and insights, but simultaneously limit it. Further, again following Foucault, the texts of geopolitical discourse are not free-floating, innocent contributions to an 'objective' knowledge, but are rooted in what he calls 'power/knowledge', serving the interests of particular groups in society and helping to sustain and legitimate certain perspectives and interpretations.

Lacoste, himself directly influenced by Foucault's writings, has emphasized the basic historical connections between the military and geographical knowledge (Lacoste 1985), stemming from the fact that the state is, in Michael Mann's terms, a 'territorial arena' and a 'power-container'. The military's role of defending the state requires territorial control and knowledge about space in the form of cartography, survey, frontier definition and the geography of communications and resources. Geopolitics goes beyond this basic connection and provides a more conceptual and comprehensive foundation for an ambitious military–political vision or theory of the state. Hence its strong appeal to military groups in South America, serving to enhance and legitimate their role in the state, as against other groups and interpretations. At times this role of geopolitical discourse gets formally institutionalized, as in the Brazilian ESG's *Manual básico da ESG* (1976).

The nature and appeal of geopolitical discourse in South America can perhaps best be seen as a series of three layers or stages: first, as a geographical perspective on the state and its history; second, as a geographical interpretation of history, with policy lessons; and finally as a theoretically structured geographical interpretation.

The first layer, and the one that gives much of the broad appeal of geopolitical writing, is simply to provide a spatial or geographical perspective on the state and its history – a political 'geohistory'. In the Brazilian Golbery's work there is explicit reference to Braudel's geohistory and to the French *Annales* school. Conventional political, legal and constitutional studies of the state and its history usually neglect this perspective, or treat it very dryly, yet geographical change and territorial development are central to the history of all the South American states. The geopolitical perspective brings this dimension to life, and it was precisely this element of breathing life into the dry bones of legal and constitutional political history that so appealed to the Swedish founder of geopolitics,

Rudolf Kjellen (Kristof 1960). Thus in the case of Brazil the geopolitical interpretation of the state's history as a 'march to the west' of settlement and political control from the original coastal possessions, with territorial struggle between the Portugese and Spanish empires, is legitimate (in that it brings to the fore aspects of Brazilian history often neglected), and has been found appealing and refreshing by a broad readership from a wide range of social and political groups.

Injecting a geographical dimension into historical analysis, where appropriate, is itself hardly controversial, but in practice all historical writing requires interpretation, the selection (and omission) of perspectives, materials and modes of analysis: where is the geographical dimension appropriate, how important is it, and how should it be analysed and presented? In remedying the traditional neglect of space, geopolitics and geohistory not only push geography into the historical play but push it into centre stage. As such, environment and space tend to become the key actors in the narrative as nations play out their 'manifest destinies', thereby constructing a geographical interpretation of history, usually nationalistic in tone, from which the geopolitician discerns 'permanent national objectives' and draws historical lessons and directions for future policy. Thus, in Kjellen's original system of political analysis, geopolitics was only one of five political science fields (the other four were demopolitics, ecopolitics, sociopolitics and cratopolitics, covering demographic, economic, social and governmental perspectives), but it soon became the dominant perspective, and is the only one of Kjellen's terms to have survived and flourished. Moreover, within this overall spatial perspective, the geopolitical literature emphasizes certain themes and neglects others. Issues of political conquest, territorial acquisition and control, and the occupation of frontier areas are given prominence, but the literature can be very thin on the economic aspects of regional development or the social dimensions of frontier societies. Even the more recent South American geopolitical literature contains few references to the economic literature on core–periphery models by Friedmann, Myrdal and others, let alone more formal economic studies.

Because of these selections and emphases, geopolitics has had a strong appeal to the military establishment, more so than to any other group. It gives an historical prominence and political priority to military control and security issues, and at the same time plays down and neglects other aspects of the state and society. This initial appeal has been self-reinforced as military writers have contributed to the literature and kept these concerns to the fore.

There is, however, a third layer to the appeal and success of geopolitics in South America: the interpretation has more force and coherence, especially as an organized discourse sustained through institutions, journals and a large literature, if it has theoretical structure, a dominant metaphor or framework for the interpretation. Such a framework is provided for much South American geopolitics by the organic metaphor of the state, and we now turn to this central focus of the essay.

ORGANIC METAPHORS AND THE STATE

Classical geopolitical thought, as developed by the German geographer Ratzel and the Swedish political scientist Kjellen, is strongly rooted in the metaphor of the state as a spatial organism. Ratzel's pioneering 1896 paper 'Laws of the spatial growth of states', developed in his books on political geography and *Lebensraum* ('living space'), views the growth of states in biological terms (Ratzel 1896, 1901). This perspective was taken much further by Kjellen in his writings, notably *Der staat als lebensform* ('The state as life form', 1917). States are seen as two-dimensional spatial organisms – usually of limited internal complexity, so the amoeba rather than the human body is the model – which compete with each other for living space, survival and growth. The frontier regions are the territorial skin of the state, and strong states will expand and grow at the expense of weak states in the struggle for political existence. For success the strong state needs development of its core region (the heart of the organism), with good communications along the transport arteries between the heart and the frontier regions.

Views of the state, or of society, in terms of the body or an organism were nothing new. Organic metaphors for the state or society go back at least to Aristotle's *Politics* and his comparison of the city state to the human body and its constituent organs, the 'body politic'. There is no one organic or organicist metaphor of the state or government but many, each developing the perspective in different ways. The organic metaphor is one of the 'big metaphors' of social and political theory throughout history, repeatedly rearticulated in new ways. Indeed, part of the appeal of organic metaphors (and part of their danger also!) is the facility to slide from one form of the metaphor to another. All the variations, however, tend to share an emphasis on the 'natural order' or construction of some perspective on society, and also on the central role of harmony and common purpose within the state, with each social group fulfilling its appropriate role.

Within geography as a discipline organicist metaphors have been widely employed, and Berdoulay (1982) has reviewed many of these uses, extending from Davisian geomorphology and French studies of the personality of regions to the German geopolitical concepts. As Berdoulay notes, together with Stoddart (1967), geographical use of the organic metaphor tends to draw on a specific version, in contrast to the much more varied tradition in social and political theory. This version draws on Darwinian evolutionary theory, with its strongly biological form of the organic metaphor. These ideas were especially fashionable, through the work of Spencer and others, in social theory in the late nineteenth century, a very formative period for geographical theory (Berdoulay 1982). By drawing on Darwinian theory the geopolitical version of the metaphor places very strong emphasis on struggle for survival and growth between states, legitimating a power politics perspective, and one that gives very limited consideration to the internal complexity of the organism in terms of the structure of society or government, a marked contrast to some other organic metaphors in political theory.

Discussion of the organic theory of the state in geopolitics has been bedevilled by arguments about how literally the state was seen as a super-personality in its own right. The writings of Ratzel, Kjellen and later geopoliticians are full of incautious statements, but, as Kristof (1960) has shown, they wrote in metaphor. Metaphor always provides a partial vision, a particular perspective, but even careful recognition of this – and much geopolitical writing is not cautious in this way – does not avoid the risk of being captured by one's metaphors, imprisoned in their visions and the limitations of those visions.

Metaphors 'see' or explain one system of interest (here the geopolitics of the state) in terms of the characteristics and structure of another, using language from one sphere in discussion of a second. As Hesse (1963) has noted in a general discussion of the nature of metaphor in science, metaphorical explanation will always, and necessarily, emphasize some aspects at the expense of others. A particular metaphor illuminates and highlights some aspects of the object of study, but of necessity at the very same time it shadows and hides other aspects. The strategic silences of the metaphors we use are as important as the aspects that are thrust centre-stage in the language and vision. Nor are these statements and silences innocent: they help sustain and legitimate particular social and political orders, and are adopted by social groups in their own interests, and the spatial organic metaphor of the state is a good illustration of this.

Within social science and social theory, successful and fruitful metaphors are those which are capable of sustained extension and considerable development, not purely literary devices or unexpected comparisons designed to surprise or shock (Bicchieri 1988; Hesse 1963). They have to support research programmes, with their language and vocabulary becoming established as 'normal', so that they become 'dead metaphors', part of the taken-for-granted world. There is an extensive contemporary debate between philosophers such as Davidson, Hesse and Rorty on the nature of metaphor and the differences (or lack of them) between scientific and literary uses, with social theory somewhere between the two extremes. The present context can provide ample material for this debate, but there is no need to take sides here. Successful social metaphors need to be developed, yet many of their real social dangers, their Kafkaesque qualities, emerge only when they become dead metaphors, accepted as part of the literal or 'real' world, unproblematic and so imprisoning. Equally, many successful literary metaphors are also successful precisely because they are sustained and developed, at least within the individual text itself: one thinks of Donne's use of the geographical metaphor in his love poetry, seeing the female body and physical love in terms of the geographical discovery and exploration of the globe.

When metaphors in social and political theory 'run out of steam', when their creative extension and deployment falters and stalls, then one of two things happens. Either they become fossilized, no longer a source of creative thought but still capable of influencing our intellectual visions and our social lives – dead as metaphors but still very operative as mental prisons – or they are replaced by new metaphors and perspectives. Such overthrowing of dominant metaphors is often as much a matter of the political and social rejection of the particular vision of the metaphor as of scientific or intellectual decision. In the case of the 'big' metaphors, however (such as the organic state metaphor), there is always the potential of renewal and transformation of the metaphor itself. Both systems in the metaphor (here 'state' and 'organism') are complex and rich enough for only selected linkages to have been exploited: there always exists the possibility of new, original insights into different linkages, or of cross-fertilization from alternative forms of the metaphor in other areas of social theory. The transformation may be the result of intellectual advances (for example, the society-as-physical-system metaphor was originally seen in terms of Newtonian gravitational laws, then of gaseous laws, and more recently of the physics of chaotic systems) or of pure serendipity (as Richard Rorty would emphasize), or of shifts in social and political context that encourage the development of the metaphor in new directions previously ignored or unattractive. The organic metaphor of the state provides a good example of such a transformation, and all these stimuli are involved to different degrees.

The organic metaphor of the state pioneered in geopolitics by Ratzel and Kjellen provided impetus and a theoretical framework for the development of German geopolitics and other inter-war versions in Japan, Italy and Spain, giving a language and a perspective that specific regional studies could employ explicitly or draw on selectively. The same pattern is true in South American geopolitics, where the Brazilian geopolitician Backheuser drew on Kjellen's ideas and extended them in his concept of 'living frontiers', applying the ideas to the development of Brazil and the need effectively to occupy frontier areas before other states penetrated them. Within the South American literature there is a continuous development, not hindered by the demise of European geopolitics after 1945 (Hepple 1986b). Child comments:

> Geopolitics in the sense used by contemporary Latin American writers ... generally accepts the basic concept of the state as a living organism that responds to geographic, political, military, economic, demographic and psychological pressures in its struggle to survive in competition with other states.
>
> (Child 1979: 89)

Most South American textbooks make use of the organic state framework, and discuss Kjellen's ideas (e.g. in Argentina Atencio 1965, in Brazil Golbery 1967 and Meira Mattos 1975, and in Chile Pinochet 1981). Frequently the original German sources are filtered through intermediate Spanish-language versions of English or German texts. Most texts are favourable to the metaphor, though rejecting a literalist or deterministic view and dissociating themselves from the Nazi connection and excesses.

Not only has the organic metaphor provided a more formal framework for applied geopolitical studies in South America, but the very form of the metaphor has also helped reinforce some of the military appeal of geopolitics. The simple biological model of competition for survival between organisms allows little logical space within the metaphor for internal diversity between social groups in the nation, or for internal debate and discussion about the goals of the state organism, still less for the value of toleration of conflicting views and goals. The internal complexity within the state organism in geopolitics is virtually all spatial (core region, frontier, etc.) rather than social or political. The strongly nationalistic form of the metaphor (the struggle between nation states as organisms) also leaves little scope for class-based analysis or for class solidarity across nations. The result is that the metaphor leaves itself very little space for most of the insights of social science or political theory, and so geopolitics has had very little appeal for liberal or Marxist writers, or for most professional social scientists in Latin America. On the other hand, the metaphor does bring to the fore the key concerns of the military, and has had considerable appeal in institutions like ESG as a model of the state and the central role of military defence and security. The very neglect of internal diversity and complexity that so repels the liberal mind is part of its attraction in military authoritarian circles.

It is important, however, not to give the wrong impression of the South American literature. Most of the vast number of geopolitical studies are not explicitly concerned with discussion and development of the organic metaphor. They are regional studies of historical territorial development or current problem issues, empirical and specific. However, the context within which they address the issues of control of the national space, external threats, the need to occupy and develop frontier zones, their use of vivid and simplified cartography of circles and threatening arrows of penetration, and many details of the language, all draw on the *controlling* metaphor of the state-as-organism.

Writers in Brazil and Chile have made particular use of the metaphor, and their writings have been very influential. In Brazil Backheuser's concept of 'living frontiers' has been continued by later writers such as Golbery and Meira Mattos to justify the need for the rapid occupation and development of Amazonia, and the linking together of Brazil's 'archipelago' of regions by new roads. Meira Mattos writes of borders resembling 'a skin of a growing organism ... An actual frontier of a state is always the result of a phase of its evolution' (1975: 39–40), later arguing for occupation of the living space of the Amazon basin before the overpopulated parts of the world start to covet it.

In Chile, Child notes, 'geopolitical writers place special emphasis on Ratzel's organic view of the state as a living organism struggling to survive in a world where might makes right' (Child 1979: 102), and this is best seen in two important texts produced in Chile in the late 1960s: the textbook on geopolitics by General (then Colonel) Pinochet and a monograph by his War Academy colleague Von Chrismar (Pinochet 1981; Von Chrismar 1968). Pinochet's study is a general text on geopolitical theory, and is built around the organic metaphor. Because of Pinochet's subsequent political role as military dictator and President of Chile the text has been reprinted both within and outside Chile, and there is also a largely unknown English-language version published in 1981.

Pinochet views 'the state as an organism, acquiring in its composition a structure similar to that of an amoeba', and then structures his text around the elements of borders ('the enveloping layer'), the hinterland ('feeder space of the Vital Nucleus'), the Vital Nucleus or heartland, communications ('the nerves that connect the different zones with one another'), and the life cycle of the state: how states are born, expand and then, in the long run of history, eventually die, as do organisms (Pinochet 1981: 29).

> Geopolitics tries to give a scientific and reasoned explanation of the life of these super-beings who, with unrelenting activity on Earth, are born, develop, and die, a cycle during which they show all kinds of appetites and a powerful instinct of conservation. They are as sensible and rational beings as men.
>
> (Pinochet 1981: 65)

The development and growth of states is governed by a number of laws or 'tendencies', and Von Chrismar's (1968) monograph takes up his theme in detail. He combs the geopolitical literature for elaborations and extension of Ratzel's original laws, lengthening the list of geopolitical generalizations considerably, aiming at providing a general set of guides for state policy. Both texts are conservative and politically cautious. There are few locally politically contentious statements, and most of Pinochet's examples are drawn from outside Latin America (for example, the spatial growth of France, Russia and the USA). One notable exception is Pinochet's support for the Chilean position on the War of the Pacific of 1879, when Chile expanded and cut Bolivia off from the sea. Pinochet sees this as an example of the need for living space of a strong, vital state. But the overall tone is that of the lecture notes of the professional, non-political military officer, an important aspect in the light of politics and geopolitics in Chile after 1973.

South America's international relations are riddled with frontier and territorial disputes, many a legacy from the collapse of the Spanish American empire and disagreements about what each successor state had legitimate title to, and geopolitics has helped in the extensive sabre-rattling and identification of 'threats'. In fact few of these disputes have got as far as the outbreak of war (most notably the 1879 War of the Pacific, the 1941 Ecuador–Peru dispute and the 1982 Falklands/Malvinas conflict) but geopolitics has helped to keep such issues to the fore, sustaining the case for a strong role for the military establishment, both to repel such 'threats' and to help secure and develop frontier zones. In this the organic metaphor assists by presenting a very partial perspective on the state, emphasizing the struggle for survival and growth and the central role of the military. However, one can argue that by the time of Pinochet and Von Chrismar's texts the metaphor was becoming rather tired. Although South

America had provided a fruitful context for its application, the metaphor itself had undergone only limited development from the early ideas of Ratzel and Kjellen, and extensions such as Backheuser's 'living frontiers' were themselves now old. Ideas of half a century ago were being reiterated and redeployed, in part because the geopolitical literature had become rather isolated. With the post-1945 decline of geopolitics outside Latin America, there were limited external sources to draw upon, and the geopolitical literature has always been weak on links with developments in the other social sciences, so that as time went by it became more inbred and suffered from the deficiencies of inbreeding. Writers like Pinochet set out and codified the organic view, but there was little sign of creative extension in much of the literature. Yet developments in Brazil were to show that this decay was not inevitable and that renewal of the metaphor was indeed possible.

TRANSFORMING THE METAPHOR

If the use of the organic metaphor became somewhat hackneyed, it was revitalized and transformed in Brazil during the 1950s and 1960s through the work of General Golbery do Couto e Silva and his colleagues at the ESG, creating new aspects of the metaphor that were to have a profound effect on the countries of the southern cone and the lives of their inhabitants. Although developed in Brazil in the 1950s and 1960s, these extensions became widely diffused to Spanish-speaking South America only in the 1970s.

In a series of studies from 1952 onwards General Golbery reviewed traditional geopolitics, both Brazilian and international, integrating the 'standard' geopolitical literature with wider references from international relations, strategic planning and political thought (Golbery 1955, 1967). His writings are difficult, often opaque, and sometimes mystical in their references to Western civilization and the nature of the state and society. His work deserves, but has not yet received, detailed exposition and critique. Through his work, and that of others at ESG (e.g. ESG 1976), the National Security Doctrine emerged that was to provide the ideology of the Brazilian military regime after 1964.

Golbery draws on a number of different metaphors, but in terms of the organic metaphor he develops it in a new direction. All previous work views the state organism as facing an external threat across its borders from a neighbouring state. Intensive development of the Vital Nucleus, control of transport arteries and the occupation and security of frontier areas, together with the overall 'cultural strength' of the nation, are necessary for the state organism to survive such external threats and exploit international opportunities. Golbery also sees these threats but, influenced by the Cold War of the 1950s and American anti-communist ideology, he sees a more widespread threat: not only external threat but internal subversion. In this perspective, initiated by General Golbery, the internal health of the state organism becomes central, with the threat of subversive, cancerous cells within the body of the organism. As his colleague General

Meira Mattos expresses it, "The enemy is now within, not a threat of direct attack across our borders.... The real international threat is revolutionary war' (Meira Mattos, cited by Kelly 1984). In Golbery's view, 'What is certain is that the greater probability today is limited warfare, localised conflict, and above all indirect Communist aggression, which capitalises on local discontents, the frustration of misery and hunger' (1967: 198).

In this perspective, internal territorial development of the state is necessary to secure areas against subversion by guerilla and terrorist groups, not just against external forces. For development to take place, the region must be made secure by the state. Equally, permanent security requires development, both of the infrastructure and of the economy, but also to end the frustrations of misery and hunger that subversive movements exploit. Security and development became the twin watchwords of Brazil's national objectives, pointing the way to the Brazilian model of rapid economic development under overall military control. Such growth must involve Golbery's plans for roads to link together disparate regions of development, and in particular his plan to 'flood the Amazon with civilisation', creating the strategy of Amazonian development.

Golbery's work, and that of his colleagues at the ESG, also drew on American (and to some extent French) counter-insurgency doctrine, and some writers have seen the National Security Doctrine that emerged as no more than the South American expression of North American military concepts (Markoff and Baretta 1985). However, the evidence is that Golbery and the ESG's work antedates the major rise in US counter-insurgency warfare, which took place after the Cuban revolution of 1959 (Hepple 1986a). In addition, the South American version differs markedly: it is not a limited military and operational view of counter-insurgency strategy (as was most of the North American version) but a more comprehensive theory of the state. It is precisely this theoretical frame that geopolitics provides through the organic metaphor. In the Golbery–ESG view, the external threat is downgraded – corresponding with the reality of the Brazilian situation – but the perceived internal threat enhances the need for security and hence the role of the military. Protecting the state organism from internal subversion, and the linking of security with development, puts the military in a central political role, defining 'permanent national objectives' and ensuring total security by eliminating subversion wherever it occurs or is perceived to exist.

This grouping of ideas became known as the National Security Doctrine (NSD), and has been reviewed and criticised by a number of scholars (e.g. Alves 1985; Pion-Berlin 1989). Alves summarizes it:

> the planning and running of the national security state involves the development of government directives for determining policies and structures for the control of every area of political and civil society. To carry out the program, it has been necessary to take full control of state power, centralise it in the executive branch, and place those closest to the

> information network and programming of internal security policy in key positions in government.
>
> (Alves 1985: 23)

The concepts of the NSD have been taken in different directions by various groups, with distinctions between hard-line and soft-line versions (Hepple 1986a; Pion-Berlin 1989). Golbery himself always emphasized the need for democratic institutions and individual liberty, and opposed the excesses of the military regime, in the end resigning over government failure to investigate abuses. He himself created the State Security Service, SNI, but later commented, 'I have created a monster,' a statement that can also stand for the doctrine at large, as it led to total military control, the suppression of all political expression and dissent, torture and other abuses. The National Security Doctrine certainly developed in ways unintended by its main originator, another example of a text escaping its author!

The 'internal' development of the state organism metaphor took a particular form, and these specific aspects need to be spelt out. The most crucial role has been to provide a general theory of the military state, a bridge that allows the military a legitimate and commanding place in all aspects of the political and economic life of the state, all in the name of the security of the state organism from both internal and external threat. The specifically spatial aspect, which is all-important in the traditional version of the metaphor, remains significant but is secondary to the more general role (Hepple 1986a). The NSD literature can, and quite often does, ignore the spatial aspect, but the organic view of the state is central to its justification, and the centrality of this link is accepted in the literature (Alves 1985; Pion-Berlin 1989). The spatial dimension remains important, especially in terms of Amazonia's development and regional integration, but a notable gap in the geopolitical literature in Brazil and elsewhere is detailed study of spatial strategies for countering guerilla warfare and terrorism. Yet, in the abstract, one would have expected such studies to be a central aspect of internal 'cancer cell' geopolitics, and the lack is indicative of the general, theory-of-the-state NSD role of the metaphor, rather than its professional, counter-insurgency role. This gap is still somewhat surprising, and may be partly a result of the 'real' guerilla threat being very limited in Brazil. (In practice most of the real terrorist threats in the southern cone states have been urban-based rather than rural peasant-based 'foco' on the Cuban and Che Guevara model, and the counter-insurgency programme in the Araguaia–Tocantins region of eastern Amazonia was a grand-scale example of NSD overkill and military repression, with 20,000 troops deployed after a group of sixty-nine core militants.)

These ideas, developed in Brazil in the 1950s and 1960s, were also to have a major political impact on the other southern cone states of Chile and Argentina during the 1970s. The Chilean case is interesting because the geopolitical writings of Pinochet and Von Chrismar were particularly strong expositions of the organic state metaphor in geopolitical theory. Yet their overall content and tone are almost entirely in the traditional mould, with little hint of the geopolitics of internal subversion or the NSD. This is especially true of Pinochet's text, and it is framed within the constitutional and professional tradition adopted by the Chilean armed forces prior to 1973 (see Nunn 1976 for a full discussion of this tradition). Pinochet's (1968) *Geopolitica* aims to provide principles and guidelines for the statesman, but he himself advocates few explicit policies, and there is little emphasis on internal security (despite Pinochet's personal involvement in such security operations in the 1960s) or on the communist threat so much at the centre of Golbery's writings and the NSD. Von Chrismar's study is also largely traditional in this way, through his concluding section contains many Golbery-esque phrases on 'permanent objectives', 'development and security', and:

> our Geopolitical school could consider as one of its basic principles the concept of the development and integral security of the state, ensured by an adequate balance and close relation between both concepts in order to maintain the health and strength of the nation, preventing accidents or infirmities, in order to ensure a natural and organic growth which is the only beneficial and lasting one.
>
> (Von Chrismar 1968: 239; English version from Pittman 1981)

Von Chrismar, however, leaves discussion of security and development at this very general level, and there is no explicit development of the NSD theme.

This traditional perspective is all the more remarkable given the extreme and hard-line form of NSD rapidly adopted by the Chilean military regime after the overthrow of Allende in 1973. The 'Brazilianization' of Chile was quick and dramatic, both in the direct military implementation of the NSD and in the actions of the internal security service, DINA, and also in the adoption of the NSD and the new form of the organic metaphor of the state. As Pion-Berlin (1989: 411–12) comments:

> The organic metaphor is saved by visualizing 'uncooperative' individuals as cancerous cells, which must be excised by the state so that the entire organism may endure. This view provides a strong sanction and justification for dictatorship. The Chilean junta took this metaphor seriously in eliminating some 10,000 sympathizers of the Salvador Allende government in the first year after the 1973 coup.

The military regime made strong and explicit use of the cancer metaphor, and geopolitical writing rapidly became Brazilian in form. In 1974 the Academia Superior de la Seguridad Nacional was established, with geopolitics as an integral aspect, and the new journal *Seguridad Nacional* has been a major vehicle of the new geopolitics, with a strong emphasis on internal security. In 1981 an Institute of Geopolitics was created, attached to the academy, with its journal *Revista Chilena de Geopolítica*, and this journal has also carried papers on the development of the organic metaphor (e.g. Stack 1985). Chile's strong traditional emphasis on the organic metaphor has made it fertile ground for these developments, and in addition provides a good illustration of how different forms of a 'big' metaphor can be conjoined. In Chile there is also a strong tradition of Catholic, Christian Democrat ideas of the organic state, with its corporatist emphasis and the need for order, discipline, morality and the role of the family, and in post-1973 writing, and in political speeches by Pinochet and others, one gets the different forms of organic metaphor joined together, even producing papers on 'National security and matrimony' (cited by Child 1985).

In Argentina geopolitics also became linked with the diffusion of the NSD, especially after the 1976 military government. It is best seen in General Goyret's *Geopolítica y subversión* (Goyret 1980) and in his journal *Armas y Geoestrategia*. However, although there is widespread appeal to 'security and development' language, the Argentinian literature has remained more tied to territorially based studies of disputed zones (perhaps because Argentina faces several perceived threats), southern region development and the potential of the south-west Atlantic and Antarctica. At the same time the NSD itself has had a major impact, with a strong influence of French ideas on counter-terrorism, and found violent expression in the 'Dirty War' and the disappearance of more than 15,000 individuals.

In summary, the organic state metaphor underwent a major extension and transformation in Golbery's Brazil, with a new emphasis on internal health and the need for security against subversive 'cancer cells' within the state organism. This not only revitalized the metaphor and turned it in a new direction, but was also the major element in legitimating a military view of the state, helping to turn narrowly military-professional counter-insurgency into a full-blown ideology for the military authoritarian state of the last quarter century.

In retrospect the new direction of the organic metaphor can appear very obvious and even trivial. Yet it was not. One has only to look at Pinochet's text to see that it was not obvious, even to someone with a mind and the will to implement later a very hard-line version of the NSD, even if the direction was latent within the earlier use. Its role in the construction of the NSD and its political impact were anything but trivial. Tragic, maybe, but certainly not trivial.

To some extent this still leaves unanswered the important question of why the metaphor was transformed when it was and not before. Why was the internal element not always present in the geopolitical literature, and why did Golbery develop it and Pinochet ignore it? The reasons are implicit in what has already been discussed, but it is worth bringing them out explicitly. In part it is a matter of General Golbery's serendipity – his cleverness and his luck – in seeing what others had not seen, but, true though this is, more lies behind it. First, there were reasons why it was neglected earlier. Classical geopolitics, and the form of its organic metaphor of the state, developed at the turn of the century in a particular historical and political context, and the use of the metaphor reflected this. Geopolitics grew in a world where newly emergent nation states like Germany and Italy were in potential conflict with other states competing for empires and dominance. The classic, locked military conflict of 1914–18 trench warfare lay only a few years ahead. The Darwinian view of competition between state organisms for survival and growth was the most appropriate form of the spatial metaphor. Internal subversion and especially guerilla warfare were not seen as key military or political issues (though the Russian revolution and civil war, and events in Germany itself, should have given more pause for thought after 1920). Nor was the pathology of cancers and tumours in such medical or common currency as it is today, when all of us are used to seeing photographs of malignant cells inside organisms, and so the internal, more medical view of the metaphor was less available for use. South American geopoliticians took over this classical form, and, until Golbery, did little to develop it, although they made great use of it, and it helped legitimate and sustain the traditional role of the military in national defence against external threats.

In Brazil in the 1950s General Golbery was prescient enough to recognize and respond to a new context. His books show he read widely and was much more open to outside ideas than most geopoliticians, and this is a major factor in his originality. Issues of subversion – and communist subversion especially – were becoming current in the military and strategic literature at the time. At a very early stage, before this literature became the torrent it was later, Golbery took these elements and synthesized them with the traditional themes of Brazilian geopolitics in the way described earlier. With the success of Castro's guerilla forces in Cuba, and the possibility of similar revolts in other Latin American countries, Golbery was a man writing to his times, identifying current fears and concerns. Twenty, or even ten, years earlier the discussion of internal subversion and cancers would have seemed largely irrelevant. But this shift towards an internal security view, and its elevation into a new organic theory of the state, also involved a rethinking and redefinition of the whole role of the military in the state. Despite coups and interventions in political life, in the 1950s most military doctrine and thinking still justified the military in terms of external defence. Internal intervention was seen as exceptional, justified as an occasional act of constitutional guardianship. The new internal view that became the National Security Doctrine challenged all this, as Stepan (1971) has emphasized. The new view gave the military a central and continuing role across a very wide range of the state's activities, all in the name of 'security and development'. Acceptance of this perspective within the military required a major change in military culture. In Brazil, Golbery and the ESG began such a change in the 1950s, in response to what they identified as communist-led subversion threats to the peaceful and orderly development of the state, but in other states the older military tradition lasted longer. In Chile, as Nunn (1976) has shown, it lasted until 1973, when the military and political culture altered in a catastrophic shift. Pinochet's text of 1968 belongs with the older culture.

So whilst one can reasonably argue that the internal aspects of the organic metaphor were always latent within it, one must not underestimate the scale of the change of vision that had to go with it. In Golbery's work the organic metaphor of geopolitics formed the basis of the new National Security Doctrine as a theory of the state. Adoption of the internal view had, as its corollary, acceptance of this new, more comprehensive role for the military, and this itself was a major step for military thinking and culture.

DEATH OF A METAPHOR?

The last decade has seen the fall of the military regimes of South America and a return, sometimes somewhat uncertain, to democracy. In Brazil there has been a gradual liberalization and democratization superintended by the military (among whom General Golbery himself was a principal advocate), leading to the civilian presidency of Sarney in 1986 and the free elections of 1989. Argentina saw the overthrow of the military after the Falklands/Malvinas war of 1982, and Chile has seen the end of Pinochet's presidency in 1990 after the 1989 elections had voted him out.

The ending of military rule inevitably diminishes the role and political importance of military geopolitical theorizing and geopolitical perspectives. But it should not be taken as ending the significance of geopolitics in the southern cone states or elsewhere in South America. First, the military still remain important political actors in those states, as political pressures and revolts in Argentina since 1983 have shown. In Brazil (as of early 1990) the military retain a strong hold on the influential National Security Council, and under the civilian President Sarney in 1986 the Calha Norte project, developed by the NSC under General Bayma Denys, zoned a huge northern frontier area of Amazonia for development under military and security control, with major implications for the large Indian population (Hecht and Cockburn 1989). More recently the politically important Amazonian environmental programme Nozza Natureza was also largely shaped and defined by General Denys. In Chile, Pinochet remains head of the army, and the future role of the military remains uncertain. So in all three states the continued impact of military thinking (and therefore of geopolitics) should not be underestimated.

A second factor is that geopolitical thinking is now well rooted in some civilian political circles, and this has always been part of the tradition, with journals such as *Geosur* and the semi-official and independent geopolitical institutes. Many of the ideas are well embedded in popular thinking, reinforcing the long-standing territorial nationalism of the countries. In post-military Argentina the geopolitical priority given to the southern regions of Patagonia, the south-west Atlantic and the possibilities of Antarctica remain important in Argentinian political life, diverting attention from other priorities, as is witnessed by President Alfonsín's proposal for a new Argentinian capital at Viedma in northern Patagonia. In Brazil the priority and impetus given to the development of Amazonia – 'The generals' blueprint', in Hecht and Cockburn's words (1989) – will be very difficult if not impossible for any civilian government to reverse.

So the military and their geopolitics, together with the excessive attention to territorial issues and speculative regional development plans that geopolitics helps foster, will remain significant. But one crucial element in geopolitics has changed with the return to civilian rule: the National Security Doctrine and its associated organic metaphor of the state. In the civilian regimes of both Argentina and Brazil – the situation in Chile has yet to be established but will probably be the same – the military continue to have potential political power, but in defining a new role for themselves they cannot use the NSD for their legitimacy. That legitimacy, and all the language and metaphors involved, died with the regimes it helped to sustain. An important legacy of the Brazilian security service, SNI, the activities of Pinochet's security service, DINA, and the events of the 'Dirty War' in Argentina is that no writer trying to define a new role for the military would attempt to use the NSD or with it the internal security version of the organic metaphor. In Argentina military writers after the 1982 war soon attempted to define a new role for both the military and geopolitics. Thus Ballester *et al.* (1983) argue for a legitimate military role in a civilian, democratic society, with the military subordinate to the civilian power and the constitution, and playing an important external role, but not having an internal role in suppressing dissent. Similarly for Brazil, Foucher (1986) has cited the work of Cavagnari, defining a new military geopolitics but drawing on new ideas from international relations and politics, with names like Aron and Castoriadis replacing Ratzel and Kjellen.

The organic metaphor of geopolitics, launched at the beginning of the century by Ratzel and Kjellen, and revitalized by the Brazilian General Golbery in the 1950s, has now run its course. The drastic implications of the vision of the state constructed through it, and the political implementation of that vision, have destroyed its usefulness today. As a metaphor it cannot be dismissed as epistemologically 'wrong'. On the contrary, all metaphors are partial, and this metaphor does usefully highlight some important aspects of the spatial development of the state. But the way it was developed and became a dominant vision now makes the metaphor morally wrong, an unacceptable view of the state. Yet this epitaph does not mean the metaphor is killed off for ever. Like the vampire Dracula, major metaphors only sleep, available for rebirth as new uses for the perspective are discovered or constructed by later generations.

Many academic and leftist political writers in the liberalized states of South America would like to see this killing-off of a major character in the geopolitical play associated with an end to the play altogether – a rejection of geopolitics itself (e.g., in Brazil, Chiavenatto 1981; in Argentina, Reboratti 1983). For reasons outlined earlier, I think this unlikely, but also undesirable. Geopolitics certainly does give an exaggerated and partial interpretation of the role of space and geography in the state. The texts of South American geopolitics are certainly ideological and do not mirror a geographical reality. But neither do they present an entirely invented or mythical picture. Territorial issues, both of economic development and of international tensions, are politically important in the South American states, as are legitimate questions of military security. Engagement with geopolitics, and its enlargement into a more critical geopolitics, is what is needed, not its dismissal. Unfortunately alternative theoretical perspectives on the state, such as Marxism and neoclassical liberalism, deal very inadequately, if at all, with these spatial and territorial issues.

Wider debate of geopolitical issues, with a new range of metaphors linked into other contemporary debates on the relation of space and society, is needed. The form will, however, have to change. Perhaps General Golbery himself identified the crucial issue. In his *Geopolítica do Brasil* (1967) he cites Isaiah Berlin's (1957) essay on interpretations of history. In this essay Berlin uses the aphorism of Archilochus, 'The fox knows many things – the hedgehog one big one,' to compare limited, localized historical interpretations with overarching grand theories. Applying this to geopolitics, Golbery writes:

> Geopolitical truth is like a porcupine. [Golbery refers to porcupine rather than hedgehog, presumably to aid familiarity to the Brazilian reader.] It doesn't know much, but it knows one big thing. And here is the power of geopolitics properly applied. It is robust in perspective, admittedly partial, always incomplete, schematic even, and at times fanatic. In the end it unifies and clarifies, and imposes on complex reality its imperative, to plan and to act.
>
> (Golbery 1967; as translated by Hecht and Cockburn 1989)

It is a perceptive observation, revealing many of the strengths of geopolitics, but also qualifying those strengths, qualifications rarely observed in practice. Today, however, perhaps what South America needs are geopolitical truths and analyses like the fox rather than the hedgehog. Political discussions of frontier issues, territorial development, the major objectives of the state, and the form of both internal and external security policy, are all legitimate and necessary, but to try to link them all together into one overarching framework, as the state organism metaphor has done, does a disservice to political debate. We need more limited, fox-like geopolitical studies, and in this the use of various metaphors will remain unavoidable.

10

FOREIGN POLICY AND THE HYPERREAL

The Reagan administration and the scripting of 'South Africa'

Gearóid Ó Tuathail

> Many Americans, understandably, ask: given the racial violence, the hatred, why not wash our hands and walk away from that tragic continent and bleeding country? Well, the answer is: we cannot.
>
> In southern Africa, our national ideals and strategic interests come together. South Africa matters because we believe that all men are created equal and are endowed by their creator with unalienable rights. South Africa matters because of who we are. One in eight Americans can trace his ancestry to Africa.
>
> Strategically, this is one of the most vital regions in the world. Around the Cape of Good Hope passes the oil of the Persian Gulf, which is indispensable to the industrial economies of Western Europe. Southern Africa and South Africa are repositories of many of the vital minerals – vanadium, manganese, chromium, platinum – for which the West has no other secure source of supply.
>
> The Soviet Union is not unaware of the stakes.... If this rising hostility in southern Africa – between Pretoria and the front-line states – explodes, the Soviet Union will be the main beneficiary. And the critical ocean corridor of South Africa and the strategic minerals of the region would be at risk. Thus, it would be a historic act of folly for the United States and the West – out of anguish and frustration and anger – to write off South Africa.
>
> (President Reagan, 'Ending Apartheid in South Africa', address before members of the World Affairs Council and Foreign Policy Association, 22 July 1986)

The formulation and practice of United States foreign policy towards South Africa has, it is generally agreed, been characterized by greater degrees of continuity than discontinuity during the last two decades (Noer 1985; Bender *et al.* 1985; Minter 1986). Perhaps the most significant contributing factor to that continuity derives from the remarkably durable representation of 'South Africa' as a particular type of place in US foreign policy discourse. United States foreign policy is characterized by a structured way of seeing South Africa which involves, as the quotation from Reagan makes clear, two essential moves. The first is an obligatory act of disavowal. South Africa is a morally repugnant place. The very name 'South Africa' offends the moral sensibility of Western society. Few foreign-policy issues demand such a grand gesture of moral rebuke on the part of public leaders – using standard locutions such as 'abhorrence' and 'repugnance' – as that of apartheid, a situation pervasively represented, in both political and civil society in the West, as a 'system' of 'legalized discrimination' and 'racial segregation' found uniquely in South Africa. Such a set of representations, complete with standardized descriptions and often colourful imagery such as the 'tragic continent and bleeding country' impels policy in the direction of moral clarity and unambiguous *non-involvement* with the government of the Republic of South Africa.

The second move that is characteristic of US foreign policy discourse is one of reinscription. Whereas the first move represents South Africa as a morally repugnant place the second represents the same place as 'one of the most vital regions in the world'. Within this second set of representations South Africa is a place of 'strategic interests', 'vital minerals' and some vulnerability to a savvy, preying enemy. Accompanying this set of representations is the scenario that abandoning South Africa would put the 'critical ocean corridor' and 'the strategic minerals of the region at risk'. The United States, by such logic, must be unambiguously *involved* in safeguarding the security of South Africa. Even though dealing with South Africa involves 'anguish, frustration and anger' it would be 'folly' on the part of the United States to 'write off' that state.

This chapter is an investigation of the structured way of seeing and writing about 'South Africa' in US foreign-policy discourse. It seeks to make a case for the following four arguments.

1. United States foreign policy towards the government of the republic has been framed or scoped by two coexistent but often conflictual scripts of South Africa as a place. The term 'script', as used here, is meant to describe a set of representations, a collection of descriptions, scenarios and attributes which are deemed relevant and appropriate to defining a place in foreign policy. The first of these scripts represents 'South Africa' as tragedy, whereas the second represents the place as a valuable prize in a global strategic game.

2. These scripts structure the very reality of South Africa for US foreign policy and for Western political and civil society more generally. It is by means of these dominant scripts that the complexities of South Africa – its histories, peoples, places and struggles – are disciplined and rendered broadly meaningful to Western society. Scripts participate in the very constitution of the real, and one can argue, as Baudrillard (1987: 22) does, that events no longer have any meaning because they are preceded by scripts or 'models with which their own

process can only coincide'. A riot in a township or a mass strike by coal miners, for example, is immediately read as an element in the drama of apartheid and the tragedy of South Africa. A similar strike in a place such as Poland, for example, becomes a scene in a different script, the popular contemporary script concerning the 'historical failure' of communism. In both cases one has a script which precedes actual events and appropriates those events as part of itself. The consequence is the marginalization of alternative meanings and alternative scripts (the struggle between capital and labour, for example), even to the extent of ignoring the meanings of the participants themselves. Reality is made by scripts, not by raw events, which are never immanently and univocally meaningful.

3. The operation of scripts in the practice of foreign policy is a useful example of what Jean Baudrillard has described as the 'hyperreal'. For Baudrillard (1983, 1987) hyperreality is a condition where reality has lost its referent, and models, simulations or scripts of the real become more real than the real itself. Usually identified as a characteristic of late capitalist development or the contemporary era of postmodernity, the condition is closely associated with the development of media technologies, particularly film and television. The blurring of fact and fiction or the real and the imaginary that is said to be characteristic of the hyperreal, however, is hardly a new phenomenon. The very shape of the post-war world was determined, in large part, on the basis of a fantastic reality organized around an immanent Soviet 'threat' to the West and an idealized vision of 'modernization' in what became known as the 'Third World'.[1] The principle of hyperreality merely reached a grand apotheosis in the Reagan years with, on one hand, hyperreal *threats* from Nicaragua (a nation of 3 million), the window of vulnerability (a non-existent nuclear Achilles heel) and terrorism (which claimed fewer US deaths than lightning in 1985; Pringle 1986: 60) and, following on from these, the hyperreal *solutions* of the *contra* 'freedom-fighting founding fathers', the MX 'peace keeper', 'Star Wars' and spectacular 'surgical strikes' (as if bombing were clean). The case of 'constructive engagement', the name given to the Reagan administration's policy of accommodation with the white racist government in southern Africa, is, I wish to argue, a lesser appreciated example of hyperrealism in US foreign policy.

4. The means by which one is given a 'South Africa' to be seen in the contemporary period should be placed within a global political economy context. Scripts, and the metaphors and genres that organize them, are never politically innocent. The dominance of one script over another – why 'South Africa' is a 'tragedy', for example, and not 'racial capitalism' – and the appropriateness of a certain set of predicates and not others to particular situations have tremendous political significance. In helping to constitute a 'reality' scripts structure ways of seeing and admit only certain political possibilities as ways of responding to that 'reality'. They are propagandistic not because they manipulate some supposedly non-discursive real but because they constitute it and attempt to tie it into a persuasive story designed to explain the messy complexity of events in a simple

fashion. Scripts are not screens behind which omnipotent elites work to preserve racism or reproduce capitalism, but they do have a materiality and participate complexly in economic, political and ideological relations of power. One can understand scripts, in a Gramscian fashion, as particular productions of 'common sense' upon which a consensual political mythology can be constructed. Elements of that consensual political mythology such as the conjuring up of the Soviet Union as an 'Other' and omnipresent threat, the affirmation of the United States as world guardian and exemplary non-racist state, and the pedagogy of capitalism and *laissez-faire* economics as liberative forces – 'capitalism will destroy apartheid' – have been crucial to the functioning of the post-war global political economy, a political economy constructed around the now fading political, economic and ideological hegemony of the United States.

In order to make the investigation manageable and relatively concise this chapter concentrates largely on US foreign policy during the Reagan years from 1981 to 1986, although other material is used. It is divided into two parts, the first documenting the fantasies involved in the strategic representation of South Africa and the second documenting and commenting on the attempt to portray South Africa as a tragedy. The source material is drawn largely from the *US Department of State Bulletin* (hereafter *SDB*), a former weekly and now a monthly publication of the US Department of State which provides a record of the public speeches and policies of the US government towards various regions of the world. All the policy statements and speeches recorded in the journal dealing with South and southern Africa from 1981 to 1988 were examined. These include statements by the President, the Secretary of State, the Assistant Secretary for African Affairs, the Under-Secretary for Political Affairs and the members of the US delegation to the United Nations. Because such statements are public and designed for wide circulation it is sometimes implied that they are of superficial and not substantive value in understanding how foreign policy really works. Such a naive view relies on a popular and rather simplistic prejudice against rhetoric. The public exposition of foreign policy is never simply a screen, tool or ornament to the actual practice of foreign policy. The public exposition of foreign policy *is* the practice of foreign policy, for it gives meaning to the variety of concrete practices involved in carrying out a policy. A rhetoric that ignores or refuses to participate in the prevailing norms of political discourse, particularly those concerning what constitutes inconsistency and misrepresentation (i.e. the popular public criteria for defining something as "lies") will marginalize itself and never be credible and persuasive. It is absolutely crucial to the power of political leaders that their rhetoric is seen to make common sense and appears reasonable to political and civil society. It need not, of course, be an accurate or adequate representation of its object, as is the case with our subject of investigation.

STRATEGIC FANTASIES

In volume one of *Male Fantasies* Klaus Theweleit explores the multitude of fantasies clustering around women, floods, bodies and history in the writings and practices of the men of the Freikorps, the proto-fascist volunteer armies in the early Weimar Republic that were later to make up the core of Hitler's SA (Sturm Abteilung, Storm Troops). The fantasies Theweleit explores were not specific to the Freikorps nor to Weimar but part of what can be described as the collective unconscious of Western civilization for at least 200 years. Much of the basic form of the fantasies concerning women, the body and floods were crucial in giving shape and character to the post-World War I political landscape in Europe and elsewhere. George Kennan, one of the key policy-makers involved in the creation of the post-war world, represented the Soviet Union in his famous X article (first published anonymously in *Foreign Affairs* in July 1947) in the following manner:

> Its political action is a fluid stream which moves constantly, wherever it is permitted to move, towards a given goal. Its main concern is to make sure that it has filled every nook and cranny available to it in the basin of world power. But if it finds unassailable barriers in its path, it accepts these philosophically and accommodates itself to them. The main thing is that there should always be pressure, unceasing constant pressure, towards the desired goal.

The image of the Red tide being appealed to here was central to earlier Nazi ideology ('the Red flood'). The flood was a threat because it represented the transgression of boundaries: it was soft, gushing, unrestrained, anarchic and dangerous. In tandem with this image was the representation of the Soviet Union as a desiring 'Other' ('unceasing constant pressure, towards the desired goal'), a role previously historically occupied by the Turks (Said 1979). The predicate field or set of representations constructed around the Soviet Union was that appropriate to the potential rapist. The Soviet Union was aggressive; it was determinedly thrusting outwards; it had long-standing designs on certain vulnerable areas and it cleverly adopted a number of different policies (disguises, e.g. peaceful coexistence) to further its base desire. Given such an historically preordained object – the Soviet Union was this way and there was nothing one could do about it – serious dialogue and diplomacy with the Soviet Union was deemed impossible. The policy of the United States, therefore, must be 'that of a long-term, patient but firm and vigilent containment of Russian expansive tendencies' (Kennan 1947: 575). The consequences of this strategy of containment, as we know, were the unprecedented militarization of the global political economy and the pervasive disciplining of international politics and regional conflicts in terms of its logic. Such a disciplining, in the case of Western Europe, could still be found in the early Bush administration, wherein increasingly anachronistic geopoliticians insisted on representing the Soviet Union and

Mikhail Gorbachev's new thinking in foreign policy as a 'charm offensive' aimed at the 'seduction' of Western Europe.

The hyperreal fantasies of the Cold War were projected on to South and southern Africa like many other regions in the post-World War II period. Appraisals of the strategic significance of the region involved, according to Bowman (1982), 'four central arguments'. Each is briefly examined in turn.

The Cape route scenario

The commonsense basis of the Cape route fantasy is that there are certain features of the physical geography of the world which, because of their configuration and location, are of great military significance yet extremely vulnerable to enemy attack. Such places are written as 'choke points' or, in contemporary strategic parlance, 'strategic lines of communication' (SLOCs). The scenario simulated around these choke points involves the premise that it is important for the West to maintain control of the sea lanes around the Cape of Good Hope because this route has become a 'vital lifeline' of commercial (freight and oil) and military shipping. In June 1981 Chester Crocker argued that 'Southern African states form the littoral to one of the most vital lifelines of the industrial democracies' (*SDB*, August 1981). The 'strategic Cape sea route' was a 'lifeline of Western commerce' (Crocker, *SDB*, June 1982). This fetish with 'lifelines', a consequence of geopolitical discourse's historical representation of space as organic, leads policy planning and scenario construction into sado-masochistic themes. What if a sadistic Other began to choke this 'lifeline' of the West or succeeded in slashing this exposed 'jugular vein', as the retired General Walker (1980) graphically portrays it on a minimalist map in his book? The use of 'lifeline' maps, which represent the flow of oil from the Middle East to Western Europe and North America in terms of a thick line, lends such reasoning *qua* fantasies a cachet of objectivity, for maps are held to be touchstone documents of objectivity and clarity.[2] Maps are considered uncluttered forms of representation which supposedly provide magical powers of insight (an exhortation to 'look at the map' is a common triumphal move in strategic discourse), for 'reality' is supposedly mapped on to the page without the distortion of words. Maps, however, as every good cartographer knows, are arbitrary interpretive documents which are not outside or external to discursive reasoning: rather they are part of it (see chapter 13 of this volume).

The 'South Africa as bulwark' scenario

Perhaps the most common paranoia of intellectuals of statecraft concerns disease, contamination and the penetration of vital areas by the Other. The paranoia is spatial, for diseases (cancer is a favourite trope of the Cold War because its origins are mysterious and inexplicable; see Sontag 1978) 'spread', 'proliferate' and 'diffuse'. Space is also a container which either penetrates or is penetrated, is dominated or contaminated. The self-proclaimed task of the superpower to protect the region threatened by the aggressive Other necessitates the use of, on the one hand, buffers, bulwarks and a *cordon sanitaire* and, on the other hand, techniques for 'stripping' the aggressor and showing his or her true designs. In southern Africa, it is argued, both the Soviet Union and Cuba have made important strategic gains in the last decade by penetrating Angola and Mozambique. Richard Bissell (1979: 220), a former CIA operative, argues that:

> the Soviet Union has been in the role of the aggressor, taking up power positions in areas vacated by the West . . . to the extent that South Africa is seen as detrimental to more extensive Soviet influence in Africa, it has a strategic importance.

The Soviet Union has a 'desire to disrupt the political and economic system established by the European powers in Africa in the eighteenth and nineteenth centuries' (Bissell 1980: 200). '[T]he main thrust,' according to Roy (1980: 193), 'of Soviet efforts to attain political influence in ex-colonial territories beyond the borders of the USSR has been in Africa.' Soviet objectives in southern Africa, according to Crocker in March 1982 (*SDB*, June 1982), are 'objectives which would push the people of that area deeper into an environment of chaos, violence, and disorder'. The US, by working for evolutionary change in Namibia and South Africa, will 'strip the Soviet Union and its surrogates of any excuse they have to continue to fuel violence in southern Africa' and

> strip the Soviet Union not only of any justification that it may put forth to justify its efforts to fan tensions within South Africa itself into a racial war, but we also make it very clear to the people of other African nations and to the world the gravity with which we view developments in southern Africa and the strength of our own policy.
>
> (*SDB*, June 1982)

The hard, firm strength of US policy has an equivalent in the hard, firm brutality of the South African Defence Force's (SADF) murderous raids into the front-line states in the 1980s. Reagan, in July 1986, condemns the South African raids but notes that 'In defending their society and people, the South African Government has a right and responsibility to maintain order in the face of terrorists.' The terrorists, for Reagan, were not the SADF but the 'Soviet-armed guerrillas of the African National Congress'. Such a script of Soviet aggression, terrorism and strength is assembled from the same material as Prime Minister (later President) Botha's 'total onslaught' scenario (address of 21 March 1980). It had the forces of international communism 'encircling the Republic of South Africa' under 'the guidance of the planners in the Kremlin', whose goal it is to 'overthrow' the republic and 'create chaos in its stead, so that the Kremlin can establish hegemony here' (Botha, quoted in Bowman 1982: 164).[3] The similarity (some would argue, collusion) of scripts was further evidenced in the United States' championing of the original South African argument that Namibian independence must be linked with the withdrawal of Cuban troops from Angola. Such a formal linkage between the two issues put, as Minter (1986: 316) correctly notes, Namibian independence – a cause with virtually universal international and legal legitimacy – on the same level as Angola's sovereign decisions on self-defence against South Africa.

Arguments supporting the value of South Africa point to the fact that it can monitor Soviet traffic around the Cape of Good Hope, its military strength deters the 'spread' of conventional Soviet forces in the southern African region and it co-operates with a number of US allies around the world, in particular Taiwan and Israel. For Bissell (1979: 227) it was 'an outpost of orderliness and well-managed enterprises in a region of disorder and highly uncertain political futures'. For Reagan, in response to a question from Walter Cronkite in 1981, it was a country that 'has stood by us in every war we've ever fought, a country that is strategically essential to the free world' (Minter 1986: 315).

The 'South Africa as a regional power' scenario

The construction of certain actors as regionally dominant powers has been an important element in post-war US defence strategy. Dulles's ring of regional alliances in the 1950s was given a new codification in the 1970s in what became known as the 'Nixon doctrine'. This envisaged a number of regional powers who would be responsible for their own and Western security interests in their respective areas. Instead of direct US military intervention the US government would supply certain chosen 'local Leviathans' or regional policemen with adequate military hardware.[4] South Africa was conceptualized as one such regional policeman in National Security Study memorandum No. 39, written in 1969 and leaked to the press in 1974. Option two, the favoured track of the Nixon administration and regarded by Coker (1986) as the first formal articulation of 'constructive engagement', was premised on the assumption that 'The whites are here to stay [in power] and the only way the constructive change can come about is through them' (El-Khawas and Cohen 1976: 105). This assumption later proved to be disastrously flawed in the cases of Angola, Mozambique and Rhodesia, but the Reagan administration continued the pursuit of 'constructive change' through a white racist minority in the case of the republic.

The position of the Reagan administration with regard to regional powers is replete with tensions, paradoxes and ironies. It has been the stated position of US administrations that Africa should not be a theatre of East–West conflict or competition (see, for example, Crocker, in *SDB*, January 1984). This itself presents a contradiction, for it is the US, the leading Western power, that has determined it. Crocker adds that 'our strategic goal in Africa is to limit and thwart the application of outside force in African conflicts and thereby to permit Africans to shape their own futures'. In October 1981 (*SDB* January 1982) and in 1982 (*SDB*, December 1982) he states that it is the United States' policy to establish the 'rules of the game among the most powerful nations that will limit the application of external force' (*SDB*, January 1982) in African conflicts. The US has the strategic goal of not letting anyone have strategic goals; it wants to establish the rules so that Africans can establish the rules; it engages in intervention so that there can be no interventionism! Once again one finds disavowal and reinscription.

The criteria for determining outsiders and interventionism from genuine Africans and mutual co-operation is even more ironic. Libyans and people who carry Soviet weapons are outsiders, aggressors and adventurists, as are the Cubans. Violence for these outsiders is not a means to achieve national self-determination or to end apartheid but 'an end in itself, a political vehicle to enhance external influence and permit the political and ideological subjugation of independent Africa' (Lawrence Eagleburger, Under-Secretary for Political Affairs, 23 June 1983, *SDB*, August 1983). The purpose of this violence is to turn the landscape of southern Africa into 'an enlarged version of Lebanon'. The people trained by the US military assistance personnel in Zaire, Kenya, Somalia, Morocco, the Sudan and Egypt are, by contrast, Africans who are dedicated to a non-aligned and independent (i.e. westernized and anti-communist) Africa. There are two types of involvement in southern Africa: legitimate 'aid' and mutual co-operation, and interventionism and opportunism. United States military aid, Western capital and transnational profit-seeking are non-interventionist, constructive, mutually beneficial and enlightened. Soviet and Cuban activities are exploitative, cynical and a distortion of the natural order:

> In this environment [of economic crisis], outside powers are tempted to exploit instability. There is no excuse for some 35,000 Cuban troops in Africa – trained, equipped, financed, and transported by the Soviet Union – inserting themselves into local conflicts, and thereby internationalizing local problems. This Soviet/Cuban meddling has no precedent; it distorts African nonalignment; it injects an East–West dimension where none should be, making fair solutions harder to achieve.
>
> (George Shultz, February 1984, *SDB*, April 1984)

The power of such an example of scripting is in its scoping of a world wherein US values and principles are the horizon from which all else is judged. Order, moderation, balance, communication and the constructive are, by definition, equivalent with the United States. These are self-evident truths.

The 'strategic minerals' scenario

The 'strategic minerals' argument in relation to South and southern Africa is premised on the rather dubious assumption that Western industrialized consumers, not underdeveloped Third World producers, are becoming increasingly dependent on the international trade in certain key 'strategic' minerals. Energy and strategic minerals are the equivalent of food in the language of Western strategic discourse. A ready and plentiful supply of these minerals is necessary, it is argued, for the maintenance of national defence capacity and the smooth functioning of industrial economies. The pervasive concern of strategic planners and other intellectuals of statecraft is with the geographically defined concentration of these minerals in only a few places. Southern Africa, it is claimed, is a Persian Gulf of minerals. The Republic of South Africa is the world's leading producer of chromium, vanadium and antimony. It possesses a major share of the known world reserves of these minerals and produces significant proportions of the world's supply. The only major reserves of these minerals outside southern Africa are to be found in the USSR and China. The scenario simulated from these 'realities' is that any radical transformation in southern Africa will jeopardize the West's access to essential minerals (*SDB*, January and December 1982). The West cannot let the Soviets take over control of these resources. A long-term rise in the price of minerals 'would bleed the West of its economic vitality and raise defence costs prohibitively' (Bissell 1979: 218). Ortona (1980: 206) argues that the 'master plan' of the Soviets in this 'era of the resource war' 'is the gradual cutting of the jugular vein of the flux of energy supplies to the Western side'.

A number of commentators in the United States have taken issue with the scenarios simulated from the given facts of mineral demand and mineral production. Price (1978, 1981, 1982) argues quite persuasively – it is not difficult, given the hyperrealism – that the notion of a resource war in southern Africa is 'a fantasy' and 'any government in power [in South Africa], whatever its ideological slant, would be locked into selling its industrial raw materials to the West just as the West is locked into buying them' (Price 1982: 64). What such arguments do not question is the very scoping of the issue from the outset, for the supposed 'realities' or 'facts' concerning 'strategic minerals' are themselves questionable. Many so-called 'strategic minerals' are arbitrarily such, with greater significance for private capital accumulation than for national security. A significant proportion of the United States' imported chrome is used in the automobile industry. Cobalt is an important raw material for the aero-space industry. It is ironic, given the concern with concentration, that the economic concentration of extraction and distribution in the hands of a few transnationals is never a matter of concern. Even more ironic is the adherence of most strategic thinkers to neoclassical economic philosophies wherein scarcity is not conceivable because of the 'magic of the market place'. The fact that South Africa relies on Western markets for 89 per cent of its mineral production (Coker 1986: 22) would seem to indicate that it is South Africa, not the West, which is the dependent party. The 'facts of geography', whether they concern the Cape route or the location of certain minerals, have more to do with choices made by industry than with constraints imposed by the earth.[5]

The 'realities' of strategy in the post-war period have a peculiar unreality to them. The proud 'realism' of the Western strategic community is a realism which in its very operation has produced some of the wildest, most exaggerated strategic fantasies of the post-war period. The strategic real, in the post-war period, has consistently been a hyperreal, a condition wherein the model of the real – in this case, strategic scenarios – has become more real than the real itself. The boundary between fact and fiction, realism and fantasy, blurs into indistinguishability, and places like Vietnam, El Salvador and South Africa become the scenes of fantastic hyperreal threats to Western civilization. The proliferation of threats is an integral part of the post-war global political economy and a very effective means of disciplining people and politics. The hyperreal threat from communism, both within and outside the Western community, provided a justification for the permanent development of an enormous peacetime military establishment supplied, supported and serviced by a burgeoning military industrial complex. Whether it be 'South Africa' or 'Star Wars', the fantasies of strategy are what keeps this configuration of power and political economy in business.

THE TRAGEDY OF 'SOUTH AFRICA'

> We Americans are witness to a mounting tragedy in South Africa that stirs our emotions and prompts us to ask ourselves those very American questions: what can we do about apartheid? What can we do about the violence and destruction it generates and about the spill-over effects of South Africa's trauma on its many neighbors? . . . We have concluded that, despite narrowing odds, we should be doing all we can to reverse an impending tragedy. . . .
>
> ('The US Approach to South Africa', statement by Secretary of State George Shultz before the Senate Foreign Relations Committee on 23 July 1986, reprinted in *SDB*, September 1986)

The appeal to 'realism' and 'strategic interests' has, historically, proved to be a standard effective means of generating attention, support and legitimacy for the conduct of US foreign policy. An even more effective means of generating support and spontaneous consent for foreign policy, historically, has been the use of what the strategic community understands as 'idealist' themes. Within the United States at least, a foreign policy that is explained and understood in universalist terms (e.g. safeguarding freedom and promoting democracy throughout the world) has a greater positive force and constituency than a foreign policy principally articulated and understood in narrow strategic terms, terms that can be popularly stigmatized as amoral, detached and even un-American. United States foreign policy is typically a mix of both 'idealism' and 'realism', with strategic simulations and fantasies (e.g. tumbling dominoes in South East Asia or Central America) and idealist simulations and fantasies (e.g. South Vietnam or El Salvador as 'democratic') existing side by side. In the case of the Republic of South Africa the idealist simulations and fantasies that coexist with the strategic ones revolve around what Shultz and others demarcate as the 'impending tragedy' of South Africa and the heroic role ('despite narrowing odds') of the United States.

Tragedy is a genre with its own particular set of structures and rules. While there are many different types of tragedy one can, nevertheless, identify how a certain set of literary conventions work to produce that which is recognizably tragic. Tragedy involves the violation of a just and natural moral order. It usually concerns a sickness or madness that inflicts a person or place considered rich and full of promise. Unnatural conditions and abnormal affairs result. The leading protagonist is caught in the vortex of a tragic movement brought on by the existence of a fatal flaw. This flaw is a human flaw, something recognizably corrupt yet something to which all people, at one time or another, are vulnerable. Its presence provokes an anguishing conflict of emotions within the leading protagonist between the moral and the immoral, the light and the dark. Throughout the unfolding of tragedy one feels empathy – the key tragic emotion – for those caught up in its movement. One identifies with the torn protagonist while simultaneously recognizing the evil brought about by the fatal flaw. Inevitability is a key element in tragedy, for the fatal flaw not only brings abnormality and convulsion but carries its protagonists forward to what is a foreseen cataclysmic end. With the abnormality cleansed, the social order returns to its natural state.

The 'South Africa' scripted in US foreign policy discourse is not a complete but a 'mounting' or 'impending' tragedy. Its representation as such is bitterly ironic, given the strategic fantasies just examined. There conspiracy was the organizing narrative: every move of the Soviet Union, Cuba and their 'surrogates' was imbued with significance and malicious, cynical intent. Competitive gaming metaphors predominated – moves, stakes, blueprints, prizes, hands, pawns, – while exhortations to awaken from slumber (a favourite strategic metaphor) and remain vigilant framed policy discussions. The 400 year-old history of white domination, exploitation and control of the black community in southern Africa, by contrast, is a 'tragedy'. The consciously created and systematically constructed system of apartheid is presented as the consequence of a moral failing, not a cynical racial plot. Apartheid is a fatal flaw, not the latest in a long history of schemes violently to suppress black power and ensure white supremacy. The white minority government is a locus of empathy, not an implacable enemy. Its intentionality is ambivalent (it too wants to be rid of apartheid), not cynical and hard-boiled. The situation is a tragic flow of events, not a brutal and violent game of power. As a means of understanding some of the details of how such a script is written and held in place analysis of the US foreign policy discourse is organized under three different dramaturgic headings.

Background and immediate setting

There is a remarkable consistency in the way in which the setting of South and southern Africa is described in US foreign-policy documents. The State Department's periodic 'country profiles' script the background of southern Africa as similar to that of the United States. Southern Africa, like the United States, is a physically beautiful area with a tremendous resource base. It has also had a long frontier history and, like the US, a history of racial difficulties. Many of the motivations and ideals of the settlers were similar to American motivations and ideals, but unfortunately, unlike the United States, such ideals never reached fruition because of the 'human failure of racial strife'.

> Africa is a continent of hope, a modern frontier. The United States from the beginning has been a country of the frontier, built by men and women of hope. The American people know from their history the meaning of the struggle for independence, for racial equality, for economic progress, for human dignity. I am not here to give American prescriptions for Africa's problems. Your program must be African. The basic decisions and goals must be African. But we are prepared to help.... One is struck by the similarity of philosophy in the American Declaration of Independence and in the Lusaka Manifesto.... There can be no doubt that the United States remains committed to the principles of its own Declaration of Independence. It follows that we also adhere to the convictions of the Lusaka Manifesto.... Southern Africa has all the prerequisites for an exciting future. Richly endowed with minerals, agriculture and hydroelectric potential, a favorable climate, and most important, great human resources, it needs only to overcome the human failure of racial strife to achieve bright prospects for all its peoples. Let us all strive to speed the day when this vision becomes a reality.
>
> (Secretary of State Henry Kissinger at Lusaka, 27 April 1976, reprinted *SDB*, 31 May 1976)

Henry Kissinger, whose axis of significant space in world politics did not include the south (he told the Chilean Foreign Minister, Gabriel Valdes, in June 1969 that the south was 'of no importance': Minter 1986: 221) gives voice to a pious script which draws upon elements of American mythology and biblical parable. Southern Africa, site of some of the oldest civilizations on earth, is paternalistically addressed by an older Christian father. It is a 'modern frontier' which will have to struggle, as the American people did, for independence, racial equality, economic progress and human dignity. Both America and southern Africa are represented as sharing the same ideals encoded in documents 'similar in philosophy' (even though one document enabled its proclaimers to retain their African slaves). Both are lands of great potential and Africa can, through hard work and enterprise, become like America. However, their programme must be 'African' (as understood by the United States) because the US has no right to tell Africans what their programme should be (disavowal – reinscription)! The parable of the good Samaritan, himself a victim and sufferer, is frequently appealed to in US foreign policy. The United States, itself a victim of colonialism, racial strife and underdevelopment, can supposedly empathize with the sufferings of less developed peoples and help lead them to the promised land. Africa can become a variant of the American story, which itself is a variant of biblical themes. Africa, like America, is a chosen land: it has been richly endowed with talents and resources but it faces a challenge in the form of 'racial strife' that it must overcome before it reaches the light ('bright prospects for all its peoples').

The representation of Africa as 'a continent of great promise' (Shultz, *SDB*, December 1988) and southern Africa as a region of great potential contrasts sharply with the stark reality of the continent. The 1980s, Marcum (1989: 159) remarks, have been a period of ominous economic, social and environmental decline, with depressed world commodity prices, declining agricultural production, increasing desertification, crippling drought, rising external debt and endless civil and inter-state warfare. It is the mythological background of great promise, however, that helps drive the tragedy story line.

The immediate setting of the drama is in a troubled and controversial land. Shultz's portrait of 'South Africa' before the Senate Foreign Relations Committee on 23 July 1986 is of a country not at ease with itself. Certain actions beget others and the consequences accumulate until a series of interconnected deeds roll towards what seems an inevitable calamity:

> Politics in South Africa is increasingly polarized and shrill; suspicion and mistrust abound. The youth, black and white, are being schooled in a style of politics that sees violent retribution, rather than open debate, as the natural reaction to any expression of views different from their own. The rising violence provokes terrorism from extremists on all sides.... Soviet-armed ANC guerrillas have embarked upon expanded terrorist violence inside South Africa, dragging neighboring states inexorably into a cauldron of conflict with the South African Government increasingly eager to shift the blame for its internal woes to its neighbors. The MPLA regime in Angola, encouraged by massive Soviet arms shipments, has used South African attacks and internal problems as an excuse to suspend negotiations and pursue an illusory military option against UNITA.... The fundamental cause of all this damage is the system of apartheid and the mounting and inevitable reaction to it.
>
> (*SDB* September 1986)

South Africa, a place of 'extraordinary resources and talent', is not as it should be: politics is polarized and shrill, suspicion and mistrust abound and the young are learning the opposite of that which they should be naturally taught. Like Denmark in *Hamlet* or Scotland after the murder of the king in *Macbeth*, the country is seized by an unnatural sickness: one has tremendous social upheaval, riots, states of emergency, boycotts, disruptions of trade, sanctions and the subversion of the democratic process. Violence and destruction abound. The region is a 'cauldron of conflict', things are illusory and protagonists are eager to deceive. The root cause of this unnatural state of affairs is the 'system of apartheid', the great fatal flaw that has provoked a mounting and inevitable reaction.

The fatal flaw: 'apartheid'

> The United States condemns unequivocally the system of apartheid. The Administration of President Reagan holds apartheid directly responsible for the tragic events occurring at this time in South Africa.
>
> (Statement by Ambassador Vernon Walters, US Permanent Representative to the United Nations, to members of the Security Council, 25 July 1985, *SDB*, October 1985)

The factor that defines the drama of 'South Africa' more than any other is *apartheid.* The word, an Afrikaans one meaning 'apart-ness', has become an infamous, untranslatable idiom for a universally condemned system of racial oppression. The very definition of 'South Africa' by means of this singular infamous idiom and the set of descriptions by which apartheid is known are highly significant. Apartheid has its own script: it is held to have begun in 1948 after the electoral victory of the Nationalist Party, to be a unique phenomenon found only in the republic and Namibia, to be a system which has an institutionalized basis in law, and to be a mechanism of racial separation and discrimination. Among the many definitions one can find in the *Bulletin* are: 'a system of institutionalized discrimination' (Richard Moose, Carter's first Assistant Secretary for African Affairs, *SDB*, 19 December 1977), 'institutionalized racial domination' (Donald McHenry, US Ambassador to the United Nations, 13 June 1980, *SDB*, September 1980), a 'system of legalized racism' (Patricia Derian, Assistant Secretary for Human Rights and Humanitarian Affairs, 13 May 1980, *SDB*, October 1980) and a 'rigid system of racial segregation ...' (President Reagan, 22 July 1986, *SDB*, September 1986). 'Our own history,' Reagan states, 'teaches us that capitalism is the natural enemy of such feudal institutions as apartheid.' 'Apartheid, South Africa's system of legally enshrined racism,' a statement from the Working Group on South and Southern Africa (*SDB*, September 1986) declared, 'is contrary to the principles of liberty and equality of opportunity on which the United States was founded.'

The commonsense quality of these descriptions belies their significance. What they scope is a simplified world wherein (in contrast to earlier) the United States is the opposite of South Africa and a mythological world of capitalism the opposite of a mythological world of apartheid. In South Africa one has apartheid, feudalism, legalized racism, a rigid system and racial domination. In the United States one has no apartheid, capitalism, equality of opportunity, liberty, flexibility and freedom. The policy implications, crudely put, are that the solution to apartheid – an unnatural phenomenon – is the Americanization of South Africa. South Africa needs the revolutionizing power of capitalism and American industry. The creation of a supposed 'New South' in the southern states of the USA since the 1960s is sometimes evoked as a model for South Africa. Andrew Young, the Carter administration's first ambassador to the UN, played and personified this theme. The practices codified by the Sullivan principles (now renounced by Sullivan) were portrayed as a concrete example of how US industry (those parts of it that signed) could be a constructive and progressive force. United States firms were to be little islands of Americanism teaching by example, beacons of hope in a dark land. The simulated scenario was one where economic growth would erode apartheid and the enlightened practices of US branch plants would strengthen black economic power and leadership while improving the racial attitudes of white co-workers. What South Africa needed, in the long term, was a multi-racial democracy based on the principles of consent and participation in the political process, the rule of law and the protection of the rights of majorities, minorities and individuals by constitutional guarantees.

The problem with such a story-line is that it relies on a series of rather problematic segregations such as that between the United States and South Africa. The United States is hardly a model non-racist state when traditionally issues such as the education of minorities have never been a priority in the US. The landscape of many American cities is as segregated as one is likely to find in South Africa. Apartheid may not be codified in legal terms but it can still exist in practice. Similarly the problem *of* South Africa is not simply a problem *in* South Africa. The very existence of southern Africa as a region, and the Republic of South Africa as a state (which belongs to the white and not the black population) is a consequence of global developments in the history of European overseas expansionism, imperialism and the development of capitalism. One cannot separate the development of apartheid from the global demand for valuable minerals, such as gold and diamonds (Saul and Gelb 1986). Apartheid, both ideology and practice, is unthinkable without the European concepts and histories of state, race and religion (Derrida 1985, 1986b). Its historical origins, which are older than 1948, are intimately bound up with capitalist relations of production in both mining and agriculture, where cheap labour was a vital component of the accumulation process. It is, thus, hardly feudal and much more complex than simply 'legalized racism' (a small black 'homeland' elite is part of the operation of apartheid). A more appropriate definition is that it is a racially influenced form of capitalism found throughout the global political economy. Racially based exploitation, institutionalized discrimination and the rigid segregation of space are hardly unique to South Africa.

The pervasive conceptualization of apartheid as something discretely legalistic has shaped the policy agenda of South African and US politics. Apartheid has been declared 'dead' by elements of the South African government. The crass examples of 'petty apartheid' no longer exist and some of the laws governing 'grand apartheid' have been abolished. Its place has been taken by philosophies which have undergone several name changes: 'separate development', 'multi-nationalism', 'self-determination' and the American-sounding 'democratic federalism'. This argument has not been persuasive internationally – South Africa is still held to have apartheid – but the understanding of apartheid as discretely legalistic has given credence to the view that the South African government is 'reforming'. The key assumption here is that apartheid is a system, not a process, a static phenomenon and not a condition central to the functioning of which is continuing movement *qua* 'reform'. The South African government, has, on this point, been very successful in creating an international consensus. The 'indisputable fact which we must recognize', Lawrence Eagleburger argued in June 1983, 'is that the South African Government has taken the first step towards extending national political rights beyond the white minority'. 'In South Africa, unlike Nicaragua and Afghanistan, there is a government that is moving towards change and reform' (US Department of State, 'Misconceptions about US policy towards South Africa', *SDB*, September 1986). 'South Africa is not a totalitarian society,' both President Reagan (*SDB*, October 1985) and Ambassador Walters, (*SDB*, January 1986) declared, noting that it had a relatively free press and a degree of openness in its society.[6] Society, 'far from being frozen in a rigid status quo', was in transition and US policy was to 'both react to and encourage the real ferment in South African society, to capitalize upon the growing realization among all sectors there of the imperative of change' (Crocker, *SDB*, July 1986). This evocation of 'change', which is paternalistic (the US knew change was best: all sectors in South Africa are only now realizing this) and assumes change is equivalent with progress, is consistently routed, protestations about 'black empowerment' notwithstanding, through the white racist government. The whites are there to stay, and constructive change can come only through them. If US policy has any real conspiracy it is a conspiracy of co-operation between white elites to manage, as best they can, change to their advantage. 'The day has not yet come,' a former US ambassador declared, 'when white men will fight white men in Africa for the sake of black men' (quoted in Lelyveld 1985: 229).

The waste of the good

It is the white racist government that is the ultimate locus of empathy in US foreign-policy discourse. Chester Crocker, in the article which won him the post of Assistant Secretary for African Affairs in the Reagan administration (1980), portrays Botha as a potentially tragic figure torn between reform and the necessity for repression, between right and left, the *verkrampte* (or hard-line) and the *verligtes* (literally, 'enlightened ones').[7] The world of the white is, after all, much more familiar than that of the black community. It is an ordered world, a Christian world and a prosperous middle-class world characterized by a familiar religious and business sensibility. South Africa (the mythological white state) is part of the West, with recognizable features of a contemporary democracy: a free press, a loyal military subordinate to civilian rule, general elections and recognizable problems with terrorists. This world has a familiar cachet, and herein lies the tragedy, for South Africa, like the loyal soldier Macbeth, is a part of the moral order which it violates. South Africa is part of the West yet in its violent maintenance of apartheid it is untrue to its soul of goodness.[8] It thus suffers and lives uncomfortably. There is no tragedy in the destruction of evil: the tragedy is that it involves the 'waste of the good' (Bradley 1949: 37).

The potential waste of the good involves not only South African whites but the South African economy. The tremendous effort the Reagan administration put into the task of warding off substantive Congressional sanctions was revealing. Not only did it demonstrate a double standard – sanctions were a legitimate means of policy towards Nicaragua, Cuba and the Soviet Union but not South Africa – but the issue touched a real sensitivity (it was something worth fighting about). The terms used to describe those who sincerely held that sanctions (something leaders in the black community, including the Congress of South African Trade Unions requested) could be an effective instrument of policy were reminiscent of earlier descriptions of Soviet and Cuban activities in Africa. Those who supported sanctions were frustrated primitives who wanted hot emotional satisfaction. Speaking to the General Assembly of the United Nations, Vernon Walters declared that

> apartheid will not be undone by demagogic posturing and sloganeering. Exhortations to bloody revolution, calls for mandatory sanctions and hypocritical talk about liberation ... will not bring peace. Ending apartheid is a task that demands more than hot rhetoric, no matter how emotionally satisfying it may be.
>
> (*SDB*, January 1986)

Sanctions were supposedly self-evidently negative, 'a "cop-out," cloaked in a fastidious false piety' (Crocker, *SDB*, January 1984). They represented an '"ostrich" policy', a turning of one's back on South Africa, a washing of one's hands, a rejection of influence. Americans were builders, not destroyers. Indignation alone was not a policy. The frequent repetition of this stock set of descriptions was accompanied by the argument that critics of Reagan administration policy are 'misguided' and 'distort' that policy. In the wake of Nobel Peace Prize winner Bishop Tutu's criticism of 'constructive engagement' and call for sanctions Crocker argued that 'Bishop Tutu has a considerable degree of lack of information and misunderstanding as to what it is that we stand for.'[9] The statement, which is potentially racist and certainly paternalistic, is indicative of the fundamental failure of US policy to appreciate the black experience of apartheid. The world of US foreign policy and the world of the South African black community are not the same. If the Reagan administration was close to any South African world it was to that of the white population, for it spoke their language on sanctions and reproduced their paternalism – 'we know what is best' – in its foreign policy.

Despite the Reagan administration's protestations to the contrary the debates surrounding sanctions represented a case where the US was, for the first time in years, directly facing up to its participation in the maintenance of apartheid. The South African situation was being confronted, not avoided, and the exercise of influence was being considered, not rejected. In late 1986 the US Congress passed the Comprehensive Anti-apartheid Act of 1986 and overrode a subsequent presidential veto. It instituted a set of limited and inexpensive sanctions against the republic, including a ban on certain imports (excluding strategic minerals), a termination of South African Airways' landing rights in the USA and a ban on new investments. The phrase 'constructive engagement', which had been effectively rendered unacceptable in political and civil society (by virtue of the divestment movement and projects such as the *Sun City* record and music video). The passing of sanctions did not debunk but recast the heroic US role. Interviews with pro- and anti-sanctions advocates were said to reveal more concern for being on the 'right side' of the issue than for strategic or realistic thinking about how best to facilitate fundamental reform (Marcum 1989: 172). How US political and civil society sees and understands the issue, however, is not necessarily how it is seen and understood among the black community of southern Africa. United States foreign policy has not moved to aid such established liberation organizations as the African National Congress, nor has it even explicitly called for one person, one vote. Coexisting with a certain fascination with the black community in the social unconscious of Western society is a dark fear of black power and difference. Stories of communism and scenarios of anarchy and crazed violence still find their way into discussions of the ANC. The ANC, according to Shultz after his meeting with Oliver Tambo in January 1987, needed to 'think seriously' about its strategy ('violence' to Shultz but 'justice' to others). Much has been made of the practice of 'necklacing', whereby black collaborators with apartheid are draped with a petrol-filled tyre and set alight. Although the ANC has condemned the practice Senator Jesse Helms, President Reagan and others would frequently use it to stir the social unconscious of the white population (black violence as crazed, irrational and demonic).[10] The preferred solution of US foreign policy is probably 'a good multiracial oligarchy' (the phrase is Gavin Relly's, chairman of the Anglo-American Corporation: Lelyveld 1985: 234) where apartheid would survive, an apartheid with a human face, a face that could be brown or black. In this collective dream of transnational capital apartheid would still exist on the ground but be no more in reality, their reality.[11]

CONCLUSION

The contemporary practitioners of foreign policy, whether they be hardheaded strategic thinkers or teleprompted great communicators, like to think of themselves and their activity as 'realist'. By this they understand themselves to be incisive, unsentimental and non-ideological readers of world politics – the kind of people who resist giving into 'anguish and frustration and anger' and thereby committing acts of 'folly' like writing off South Africa. But isn't this move part of the ideology of foreign policy? The making of foreign policy is a cultural practice like any other, and proclamations of 'realism' are one of its rituals. So is the evocation of tribal enemies (the cynical or emotionally hot Other who glories in violence and destruction), the celebration of tribal mythology (America as frontier, progress, future), the reaffirmation of a moral centre (avoiding the 'extremism' on both sides) and the investment of faith in tribal remedies (enlightened business practices, the revolutionary power of capitalism). The proud realists are proud mythologists, the skilled story-tellers of their tribal community. Their scripts are hyperreal artefacts of genuine power, stories which, while illusory, are simultaneously real. They drive much of the accumulation process in the US and have, historically, enabled regularized contact and co-operation between Western governments, transnational capital and white South African racists while, at the same time, preserving – though at times with difficulty – the legitimacy and reasonableness of this contact.[12]

What, then, is the task of the critical intellectual, given this situation? One response, a common one, is to insist on an alternative script, a different real, a repressed reality that is systematically distorted or misrepresented by official policy discourse. The strategy is problematic not only because of the attribution of omniscient intentionality to official policy (people are being duped by scurrilous politicians) but because it too relies on the assumption of an immanently real South Africa, a truth uncontaminated by ideology, language or discourse. A different response is to play with the illusion of the real and the power it exerts. Irony and paradox become one's allies in a skirmish war against official conceptualizations of the real (stories, for example, of Cuban troops guarding US enterprise – Gulf Oil – in Cabinda). One issues a challenge, in Baudrillard's (1987: 46) terms, to the 'real' – the attempt is to put the real, quite simply, on the spot. To the extent that this latter practice refuses to grant an objectivity to the world or put a closure on meaning, such would seem to make it different from the former. But can it proceed without some notion of the former, without some conception of a real South Africa which is different from the scripts we are given? I think not. The strategy of critical intellectual practice should never involve abandoning the 'real': what it does involve is a refusal to let the powerful in our global political economy represent it without contestation.

NOTES

1 One of the best documentations and critiques of the fantasy world of development and modernization is Dorfman and Mattelart (1975). Representations of the Soviet threat too had a cartoon-like, 'larger than life' quality to them.

2 For examples of such 'lifeline' maps and such reasoning see Kemp (1977: 51, 1978), Lewis (1976), Hanks (1981: 21, 1983: 11) and the recent editions of *Goode's School Atlas*. It is worth noting that the original *Goode's School Atlas* (first edition 1932: 39) represents the production of petroleum by green circles, with areas in proportion to the output of various countries. 'Lifelines' of petroleum, given ample US domestic production, were not a concern then.

3 It is noteworthy that 'South Africa' for the white racist regime is a mythical predominantly white state which is periodically spoken of as if it were not even located in Africa! See Lelyveld (1985).

4 The Carter administration moved away from sole reliance on this strategy to develop a US-based mobile combat unit which could be flown to 'hot spots' in a matter of hours. The unit subsequently became known as the Rapid Deployment Force.
5 Why there was a dramatic increase in oil traffic around the Cape in the 1970s, for example, was due to a cost-minimization, profit-maximization decision by oil companies to build supertankers. Supertankers are too big for the Suez Canal.
6 The distinction between totalitarian (not open to reform) and authoritarian (potentially open to movement towards democracy) governments was popularized in the Reagan administration by Jeane Kirkpatrick, the Reagan administration's pugnacious first Ambassador to the UN.
7 The equally conflictual emotions of Nelson Mandela and Bishop Tutu – can the use of violence be justified to combat the daily structural violence of apartheid? – do not resonate with Western intellectuals of statecraft in the same way the governance dilemmas of the South African state do.
8 The *Guardian* greeted the drastic press clamp-down of December 1986 with the headline 'South Africa turns its back on the West.'
9 Senator Christoper Dodd, who questioned Crocker on his statement in Senate hearings, noted that Tutu was 'not just any black leader from South Africa' but to 'many people in this country, indeed, around the world he is the most prominent' ('US Policy towards South Africa', hearings before the Committee on Foreign Relations, United States Senate, 99th Congress, 1st Session, 24 April, 2 and 22 May 1985, p. 129).
10 Winnie Mandela's apparent glorification of revolutionary violence ('By our necklaces we will set ourself free') has similarities not only with Frantz Fanon but also Thomas Jefferson, who wished the tree of liberty to be periodically watered with blood.
11 Like a law, it would be 'abolished'. Like a machine, it would be 'dismantled'. Like the form of capitalism it is, it will restructure.
12 The extent to which capitalism runs on fantasies and illusions is not always appreciated. One cannot avoid confronting the question in the contemporary period, for global financial markets, state budgets, capital restructuring and defence spending all run on an increasingly fantastic economics.

ACKNOWLEDGEMENTS

Earlier versions of this paper have been presented at the annual meeting of the Association of American Geographers, Portland, Oregon, 22–5 April 1987, and to the Center for International Studies at the University of Southern California, Los Angeles, September 1987. I would like to thank the Roscow Martin Foundation and the Syracuse University Senate Committee on Research (Award No. 168) for supporting my research. I would also like to acknowledge the support of the Department of Geography, Syracuse University, the Center for International Studies, University of Southern California, and the Departments of Geography and Political Science, University of Minnesota, Minneapolis.

11

PORTLAND'S COMPREHENSIVE PLAN AS TEXT

The Fred Meyer case and the politics of reading

Judith Kenny

INTRODUCTION

Summarizing the dominant view of its citizens, the City of Portland Comprehensive Plan (1976) begins:

> Portland is more than a geographic area – it is a way of life. Many characteristics combine to provide the unique livability of the city: the physical setting of hills, trees and rivers, accented by snow-capped peaks on the horizon; a dynamic urban setting enhanced by the intense, yet human, character of Downtown; an active seaport a hundred miles from the ocean; thriving businesses and industries providing diversified employment; and a variety of neighborhoods, each unique in character, allowing for a broad range of lifestyles.

Associated with this proud description of the city and its quality of life, the plan expresses a concern that without care these attributes might be lost. Addressing this possibility in conservative tones, the document indicates that 'the task facing us is to retain the most important characteristics of our city in the face of change we cannot control'. In apparently simple terms, it is accepted that the city can be defined by its landscape, since it exhibits qualities that are 'Portland'. Beyond any 'sense of place' that might be attributed to this appreciation are values, attitudes and beliefs which define what is good and what should be reproduced in Portland's landscape.

In its broadest terms, the purpose of this chapter is to demonstrate the value of textual analysis as a means of interpreting the landscape. A planning document, possibly more than any other written text, articulates the ideology of dominant groups in the production of the built environment. Consequently an examination of a document such as the Portland Comprehensive Plan provides a means of analysing the social and political ideologies that are to be transformed into physical form. In the quotations cited above a liberal belief in the role of government and a commitment to certain qualities in the city's landscape are clearly indicated. The authors of the plan have created a vision of Portland's future informed by local values and an interpretation of national trends. The resulting document represents layers of meaning that are drawn from the multiple discourses within American liberal ideology.

Thus, more narrowly defined, I will evaluate the relationship between the Comprehensive Plan, liberal planning practice and the landscape of Portland. This is far from being a straightforward task. The work of several geographers warns against the assumption of a simple relationship between landscapes, ideology and social practice (Cosgrove 1984; Duncan and Duncan 1988). James and Nancy Duncan argue that while 'virtually any landscape can be analyzed as a text in which social relations are inscribed ... naive readings commit the error of "naturalizing" these social relations' (1988: 123). Too easily the landscape's political and social underpinnings are masked by a 'taken-for-granted' world.

Conflict over the plan and the landscape to which it refers, however, effectively denaturalizes the 'taken-for-granted'. In such disputes the political nature of the interpretation and the influence of economic and social conditions are emphasized. Although the planning process and the plan's legal status imply a single meaning for the document, conflict in land use cases highlights not only competing interests but also, at a more abstract level, tensions between various discourses of liberal thought. These tensions can allow competing material interests to produce opposing textual readings. This is demonstrated in what has been described as Portland's most controversial land use decision, the Fred Meyer Inc. Comprehensive Plan amendment request. The Fred Meyer Inc. case is noted not only for the level of controversy surrounding the decision, but also for the company's traditional hostility to planning (MacColl 1979). Despite this reputation Fred Meyer Inc.'s representatives were able to appropriate the text by focusing debate on contradictions inherent in it. A review of this case will demonstrate the impact that the politics of reading can have on the production of landscape.

Before interpreting the conflict, however, I will outline both the textual and the contextual bases of the study. First I will address the interpretive method used and the concepts of text, textual community, discourse and discursive field. Then I will discuss the liberal foundations of Oregon's and, more specifically, Portland's planning programme. Since Portland provides one of the few examples in which nationally accepted prescriptions for planning are being tested (Abbott 1983: 8), an effort will be made to relate this to national trends in liberal planning theory as well as to the specific values of Portland's textual communities.

TEXT AND DISCOURSE

To call the Portland Comprehensive Plan a text may evoke fewer questions than applying the same term to describe a landscape. The meaning of written texts is generally thought to be accessible to any member of the 'reading public'.

However, it is acknowledged that there may be cases where the meaning, being obscured by the language of planning, must be elucidated by various interpreters, such as city planners, planning consultants and lawyers employed to translate its message. In these terms the Comprehensive Plan would appear to fit the traditional definition of a text as an 'entity which always remains the same from one moment to the next' (Hirsch, quoted in Fish 1980: vii).

Challenging this view, however, literary critics and more recently social scientists have removed texts from the realm of objective meaning. As redefined, a text is 'the structure of meaning that is obvious and inescapable from the perspective of whatever interpretive assumptions happen to be in force' (Fish 1980: vii). This move away from understanding a text as a single stable meaning is a relatively recent development in poststructuralist literary criticism. In its most radical form, reception theory transfers authority over meaning in a text from the author to the readers. Authors' intentions are lost in the act of reading where meaning is created by a reader's beliefs and expectations. The literary critic Terry Eagleton (1983: 77) describes the reception theorist's view of the reader and text as follows:

> Reading is not a straight-forward linear movement, a merely cumulative affair: our initial speculations generate a frame of reference within which to interpret what comes next, but what comes next may retrospectively transform our original understanding, highlighting some features of it and backgrounding others.

Thus reading is a complicated activity with alternative layers of meaning which the reader interprets dialectically.

This theory can deteriorate into radical individualism unless we recognize the social and historical bases of the reader's belief system. I would argue that, without acknowledgement of the social construction of meaning, this theory becomes of questionable use to social scientists and presumably of limited value to literary critics as well. Eagleton (1983: 87) warns that a radical reader response theory is 'a simple fantasy bred in the minds of those who have spent too long in the classroom'. His argument proceeds with the reminder that:

> such texts belong to a language as a whole, have intricate relations to other linguistic practices, however much they might also subvert and violate them; and language is not in fact something we are free to do what we like with.

Although there may be multiple readings of a text, readings are not in any important sense unique to an individual. Meaning is contained within the limits of language, as Eagleton describes, but also has an element of stability based on the social and historical context of interpretation or discourse.

I will argue that the Comprehensive Plan text was produced reflecting discourses within the discursive field of liberal planning ideology. Inherent in definitions of discourse and discursive field is the politics of interpretation. As used in this essay, a discursive field is a range of competing discourses that are relevant to a particular realm of social practice. Discourses can be defined as social frameworks that enable and limit ways of thinking and acting. The definition of discourse followed in this essay is not the deterministic one of Michel Foucault (1967), which assumes that discourses are incommensurable or indisputable from the outside. Instead, following the interpretation of James Duncan (1990: 16), it is assumed that while some discourses are hegemonic others are contestatory. As Duncan (1990: 16) states:

> Whereas words may have different meanings within different discourses as Foucault has argued, I do not assume the impossibility of translation between discourses, nor do I reject the possibility of real, resolvable conflict between those subscribing to the terms of different discourses. Of course, I would acknowledge that the difference between the discourses may be based in real and irreconcilable material interests, and thus resolutions may often be the product of an unequal power struggle within which one group loses its voice.

Since a 'strong' definition of discourse is not employed, it is also possible to assume that a stable discursive order may exist in which competing discourses coexist in an uneasy alliance. I will argue that this is the case within the discursive field of liberal planning.

The final definition required is that of textual community. It is with the goal of analysing the production of social meaning that the historian Brian Stock introduced the concept of a textual community. Taking the asocial definition of poststructuralism's 'text', 'the structure of meaning that is obvious and inescapable from the perspective of whatever interpretative assumptions happen to be in force', the range of meaning is narrowed by recognizing the socio-historical limits placed on those interpretive assumptions. Stock's definition of a textual community organizes meaning around a particular understanding of text (1986: 299): 'What brings people together and makes them act is the articulation of a text within the group and the binding of a group's behavior to the rules set forth in the text.'

In the case of the Portland Comprehensive Plan the text is produced by a textual community, based on its interpretation of a larger set of texts both written and otherwise inscribed (in the landscape, for example). This interpretation is the product of social discourses reflecting the hegemonic value system of dominant groups within Oregon, and more specifically Portland. Environmental interests and control of land use are significant aspects of this value system. A rather unsympathetic interpretation of these views was expressed by an editorialist in a 1985 issue of *Sales and Marketing Management*:

> You remember Oregon. That's the state where the governor used to ask people to 'come visit – but don't stay.' The place where the locals sport bumper stickers that read 'Don't Californicate Oregon.' The state with

> some of the most restrictive land use laws in the nation. Friendly, yes; but friendly to people – not industry. Friendly to residents – but not to immigrants.
>
> (Kern 1985: 20)

While the impression made on those outside the state is significant, a closer inspection of the textual community's land use documents reveals a more expansive view of planning goals. The 'most restrictive land use laws in the nation' also represented a complex interpretation of liberal planning ideology, reflecting a concern for both the rights of individuals and the rational management of society.

DISCOURSES OF LIBERALISM

Social order and individual choice

Within liberal ideology there are multiple strands of meaning which have varied in significance historically. Contrasting American and British liberalism, Gordon Clark and Michael Dear have identified two primary strands: the concern for social order and the concern for an individual's natural rights (1984: 10, 169). In the one, government's role is to act as a social engineer to redress socio-economic imbalances and maintain fairness for disadvantaged groups; the other is concerned with individualism and the guarantee of the potential for individual choice.

The American tradition, primarily, is one of Lockean liberalism. The conception of natural rights is central to Locke's notion of individualism and justice. Throughout the Revolutionary period an individual's natural rights of life, liberty and property were proclaimed. During most of the nineteenth century liberals continued to assert that government's exercise of power diminished the liberty of individual Americans (Volkmer 1969: 3). Decentralizing power to the local level was viewed as the most advantageous means of governing, since local control was associated with a greater responsiveness to the will of the people.

The other major strand of liberalism is associated with Bentham and the utilitarians, who defined the state in terms of the need to respond to the 'greatest happiness of the greatest number' (Clark and Dear 1984: 10). This was to be accomplished with scientific rationality and state involvement in managing the problems of society. Although Clark and Dear identify these beliefs with the British tradition of liberalism, twentieth-century conditions resulted in revisions of American liberalism which increased government's role based on a similar rationale. Replacing the older fear of government, liberalism accepted the need for a relatively strong government which would protect the welfare of weaker members of society and thus create greater equality. As summarized by the political scientist Walter Volkmer, this transformation of liberal thought was based on the belief that 'True liberty required the ability of a man to use his talents and energies in a constructive fashion – it meant the positive freedom to achieve and accomplish' (1969: 4). The basic philosophy of twentieth-century liberalism is pragmatic and secular, placing reliance on humans and their capacity to find solutions to political, social and economic problems. 'Positive freedom' for each member of society is viewed as the goal of a liberal government.

While the view of government's role has evolved, concern for the welfare of the individual has persisted as the central concept of liberalism. The traditional liberal belief in the individual's right to life, liberty and property stresses the individual. This belief distinguishes the liberal from the conservative, who is likely to emphasize the fundamental nature of property rights over the needs of society. Such a fine distinction can result in misconception of liberal philosophy. There is no hostility to property rights within liberalism, nor is there opposition to capitalism. On this matter, Abraham Lincoln has been cited as providing the best summary of liberal beliefs: liberals 'are both for *man* and the *dollar*; but in cases of conflict, the *man* before the dollar' (Volkmer 1969: 12).

As Richard Walker and Douglas Greenburg argue, a common error is to view liberal reform movements as diametrically opposed to business interests. Instead, liberalism 'tends, even when serious problems have been identified, to maintain a faith in the possibility of expert repair of all maladjustments *within* the existing social order' (1982: 26). Liberal ideology does not so much question the capitalist social order as suggest rational social management as a means of obtaining more equitable 'positive freedom' for society's members. Walker and Greenburg describe this rationalist ideology of scientific management as the basis of the contemporary environmental movement as well as of much planning theory (1982: 23). This assessment is particularly appropriate in the case of Oregon's land use laws.

Liberal planning theory

Oregon's land use laws were introduced in the late 1960s when confidence in the economic climate and planning solutions was strong. Describing the decade as a time of remarkable ferment in planning theory, Michael Dear has labelled the 1960s the period of 'new scientism and the rise of popular planning' (1986: 377). The first philosophy stressed technical solutions and rational management for the public's good, while the second emphasized citizen participation in the definition of the public good. The emphasis on popular planning here reflects the more individualistic discourse deriving from classical liberalism, whereas the emphasis on scientific management reflects the 'public good' discourse within contemporary liberal ideology. Although these two different discourses fit within the larger discursive field of liberal planning ideology, they interact uneasily. There are tensions between them which are sometimes glossed over by planners anxious to serve as many different interests as possible, but which can quickly erupt into obvious controversies. Planners attempting to juggle these two competing discourses are known to use the phrase 'a win–win situation', which summarizes the optimism of liberal planning. Conflict in land use decisions could be eliminated, it was assumed, through the technical skills of the professional planner and the guidance of an active citizenry. The implied assumptions are that goals are uniform among members of a community and that there are no contradictions between various goals.

Oregon would serve as one of the first laboratories for these new planning prescriptions. When in 1969 Governor Tom McCall stated that 'Oregon has wanted industry only when that industry was willing to want what Oregon is,' liberal state-wide planning efforts were being initiated. McCall (1977: 192) explained his concern for 'balanced' state growth in these terms:

> We are being realistic. We know we cannot, at this time, support a human tidal wave of migration. We haven't the jobs for that kind of onrush – we haven't the facilities – and we are determined to maintain our magnificent environment . . . when we say, 'Visit often but don't come to live,' we aren't being hostile or provincial. We are being prudent.

Sprawled development and its pressure on agricultural land was a concern shared by many in the rural areas of western Oregon. The impulse for state-wide planning came as much from these areas as it did from the new environmentalists associated with the cities.

Ten thousand citizens participated in public meetings where state goals were developed, linking urban and rural concern. With an unusual combination of liberalism and conservative language, the goals were written to direct growth while maintaining 'what Oregon is'. At this time the phrase the 'Oregon tradition' was introduced and cited frequently. That tradition was described in terms of Oregon's history as a 'pioneer' rather than a 'boom' state. Various historical references were made, citing the efforts of nineteenth-century residents and their 'pains taken to build up an orderly and enlightened society' (in contrast to California, where 'the land systems and titles were in a state of chaos', Lyman 1903: 38). Its twentieth-century land use and environmental efforts were felt to be a modern pioneering effort, and Oregon's landscape became synonomous with the Oregon tradition. An example is provided in a book of scenic photographs. The brief introduction provided by the editor instructs: 'If you move there be prepared to love it and protect it. That's become an Oregon tradition' (*Oregon*, 1985).

Public choice theory, with its emphasis on citizen participation in planning, was translated easily into another Oregon tradition, that of local control. By creating a plan through the efforts of the community, it was assumed, the interests of members of that community would best be met. At the same time, planners at the state level were responsible for reviewing local plans for consistency with overall state requirements. This two-level process was intended to acknowledge both the interests of local control and the necessary collective action required for scientific and rational planning in the interests of the state population. For example, the Portland plan's housing policy statement indicated that the city's goal was to:

> Provide for a diversity in the type, density and location of housing within the city consistent with the adopted city housing policy at price and rent levels appropriate to the varied financial capabilities of city residents.
> (*Comprehensive Plan* 1976)

An analysis of those needs, based on current and projected population figures, was conducted to determine the appropriate supply of land for the necessary housing stock. Planners, with public review, would then identify suitable locations for future housing stock and zone accordingly. Confirming that the work was completed in a manner consistent with city and state goals, the state land use planning staff would approve the goal statement and the plan.

Within the text of the plan, individual property owners are assumed to share group interests which will be negotiated and represented in the adopted planning document. Where their interests are not adequately addressed, the plan can be modified through a plan amendment process. Such a modification, however, requires that the change conform with the public's interest. The individual applicant bears the burden of developing that argument and proving that the amendment is in conformity with community goals.

While the collective need is used as the standard of decision-making, there is a lengthy amendment appeal process that focuses on the rights of the individual. The amendment request can be heard by up to six official 'interpreters' of the text before the process is exhausted. First, the city's Hearings Officer listens to and decides the case in a public meeting. Then the following bodies may hear an appeal as requested by the applicant or anyone opposed to an approved plan amendment: the City Planning Commission, the City Council, the State Land Use Board of Appeals, the State Court of Appeals and the State Supreme Court. This quasi-judicial process can be quite time-consuming and expensive, but it is viewed as planning's system of checks and balances.

As a legal document with the necessary support to maintain its enforcibility, the Comprehensive Plan is a specific written text with a perceivable impact on the landscape of Portland. The basis of this planning apparatus is found in the 1960s debates of planning theory, inspired by the strands of liberal philosophy. Other reform groups of the 1960s and 1970s clearly shared similar influences and expressed similar goals (Ley 1980). Portland's plan and its implementation, however, must be 'read' as a local interpretation of culturally and historically specific discourses. The 'Oregon tradition' legitimizes certain values and beliefs, while at the same time the phrase is related to, if not created by, contemporary academic and popular views of American society.

Those who have authored the plan have relied on the texts of planning theory and the larger 'text' they call 'tradition'. The future landscape of Portland, however, will not simply be structured by authors' intentions. Various competing interpretations of the plan, influenced by economic and social conditions, will be based in different political discourses within the larger discursive field of liberal planning ideology, and it will be this competition which will determine what aspects of the larger ideology predominate in the 'text' on the ground. Ultimately the relationship between plan and landscape is an issue of power – whose interpretation will prevail in the 'politics of reading'?

POLITICS OF READING

Authors' intentions

During the major battles that ensued over the construction of a large Fred Meyer shopping facility in inner north-east Portland, various residents repeatedly voiced belief in the planner's ability to resolve issues of public interest. One resident expressed these views by saying:

> Sullivan's Gulch has a record of working in a positive fashion in land use issues. We are proud of the work done on the Comprehensive Plan. Many volunteer hours were put into developing a strong, viable, workable plan to take us through the next twenty years of growth. That is evident in the final Comprehensive Plan for our neighborhood. But the growth must be with a deep concern for our neighborhood, the other neighborhoods and the City at large.
>
> (Public hearing, November 1985)

In effect, neighbourhood representatives claimed authorship of the plan and assumed an interpretation necessarily consistent with those who shared in its creation:

> I believe the City's decision regarding this proposal will affect the future integrity of the Comprehensive Plan and that the future of the Northeast neighborhoods lies in the balance ... The validity of the Comprehensive Land Use Plan as a set of development guiding policies would be surrendered. I think that the people working on the development of this plan did not intend to have it thrown away so easily. The City and State are responsible for upholding these policies just as the police are responsible for upholding our laws ...

This knowledge of, and commitment to, the planning process on the part of neighbourhood activists contributed to the length and significance of the Fred Meyer land use battle. Portland's mayor, Bud Clark, described the Fred Meyer Comprehensive Plan amendment case as the most difficult decision he had faced as mayor. Given the other issues dealt with by Portland's leaders, including racial tension between the police and Portland's black community and a threatened police strike, one might wonder at such a sense of priorities. An explanation might be provided in the mayor's background as a former neighbourhood activist and a small businessman. From the first of the hearings the mayor was identified as the 'swing vote' in the land use case. The controversy surrounding the plan amendment was undeniably a challenge to notions of a singular meaning for the text.

Plate 11.1 The Hyster facility – proposed Fred Meyer site

Plate 11.2 House in Grant Park neighbourhood with anti-Fred Meyer sign

The controversy began in 1985 when Fred Meyer Inc. presented a request to the City of Portland for an amendment to the Comprehensive Plan map and zone change on a 13.4 acre parcel of property in north-east Portland. Formerly the site of a manufacturing operation, it was necessary to amend the property's industrial use designation before construction of the proposed 153,000 square foot shopping facility would be allowed. As the applicant for the amendment, the Fred Meyer company was required to demonstrate that the plan map revision was not only appropriate but necessary. The argument provided was that the new shopping centre would fill unmet market demand in the area. Citing the Comprehensive Plan's economic development goal, Fred Meyer asked the city to demonstrate that it truly was 'open for business'. Given the poor economic climate and the city's campaign to let firms know that Portland was hoping to grow, this posed a more significant challenge than it might have when the plan was adopted in the 1970s. At that time, population and jobs were increasing. In the early 1980s the impact of recession on Oregon's timber economy resulted in a decline in Portland's and the state's population.

The proposed development site is located in the midst of several older neighbourhoods. It has been said that these older residential sections within five miles of the downtown give Portland its recognizable character. Although there has been residential change in Portland's inner-city neighbourhoods, this change is not characterized as gentrification so much as a transition from an older to a younger population. An older housing stock that belongs to working-class and upper middle-income occupants has been maintained in these residential areas. In each, a strong neighbourhood commitment has been demonstrated in planning and civic activities. The eight organized neighbourhood associations that represented the inner north-east voiced a strong negative reaction when the Fred Meyer proposal was made public. Objections were based on the proposed development's traffic impact on the neighbourhood and the anticipated adverse impact on the existing businesses in the area. The neighbourhood groups argued that the manufacturing designation should be retained in order to maintain the area's existing character and encouraged the city to search for an appropriate manufacturing firm to occupy the vacant facility. Citing its violation of the neighbourhood, transport and economic development goals of the Comprehensive Plan, the neighbourhood organizations launched a co-ordinated opposition to the plan change and the proposed development.

The organized protest against the proposed revision demonstrated the well articulated belief system of the active residents. Consistently, residents cited their commitment to the urban character of their neighbourhood and explained this on the basis of both their interest in the particular environment of the residential area and their concern for a 'healthy' Portland. For example, one resident stated that:

> We need the City's protection. We need the City to protect the history we've been a part of and want to preserve. We need the City to protect the

> values that attract and keep families in the inner city . . . to articulate the balance that every city has to achieve and by which a city makes a statement about itself.

Consistent with the positive belief in planning's ability to resolve different issues, the neighbourhood association called on the city to work with Fred Meyer to consolidate property within the existing Hollywood commercial district in order for the facility to be built in compliance with the plan's goals and guidelines:

> I think the tax-payers of this city deserve and can get a 'win–win situation'. We can get economic development that will not destroy the fabric of our neighbourhood and we can get economic development within the planning here at present.

It was the effort of the neighbourhoods that mobilized the area's small businesses against the development. Initially inactive, the businesses joined in the protest when an economic consultant hired by the residents indicated the negative impact competition from a new commercial centre would have on the older commercial area. As summarized by one resident, the area's businesses expressed a concern for the neighbourhood's economic viability, saying:

> It is my understanding that the City Plan is for the protection of older, established districts as well as quality environments for neighbourhoods. The Hollywood District already has many vacancies it hopes to fill following the renovation and Light Rail construction happening there now. A new Fred Meyer complex housing virtually every service and product now being offered by independent businesses in Hollywood would simply be the death knell for this fine old district.

The liberal rhetoric in opposition to the project caused one businessman to sympathize with the Fred Meyer management, saying that they were being faced by 'a cabal of out-of-work lawyers, over-their-heads activists, and just plain meddlers' (Winningstad letter). The land use issue, however, only served to reinforce the feeling of shared neighbourhood values. One resident described it as follows:

> You can understand why I feel it's important not to lose this. These people read two books a week and don't eat sugar. I don't read two books a week and do eat sugar, but you can see why I want my daughter to grow up with their children.
>
> (Phone conversation, February 1986)

The neighbourhood associations and individual residents who were opposed to the proposed land use amendment supported a reading of the plan that they felt was consistent with its liberal text. Rather than calling for no growth, they argued for 'controlled growth'. Economic development and neighbourhood goals were weighed equally in their arguments. The proposed shopping centre, with its potentially negative impact on both residential and business areas, was viewed as a proposal that 'simply does not fit the place or time' (public hearing, November 1985). City planning staff and the Hearings Officer also recommended denial of the requested plan map amendment.

Contested readings

Fred Meyer Inc. appealed against this negative decision to the City Council, while at the same time a Fred Meyer official appealed to public opinion by describing frustration with the planning process (*Oregonian*, November 1985): 'In Alaska people are great. They welcome you. In Oregon and Portland, people talk of economic development, but fight plans to spend $10 million to develop vacant land.' In a more private forum, Fred Meyer management called for the help of Portland business people:

> The business climate of Oregon is *the* topic of discussion – in board rooms, over lunch, on street corners, in every political utterance before any gathering and even during football games. We strive daily to improve it for our companies, ourselves and all Oregonians ... Now we must ask for your support, help and suggestions. Our request for this land use change has been heralded as the 'acid test' of Oregon's land use laws; as the land use decision of the 'century', the 'true' test of Oregon's quest for economic development; and the test of the cities [*sic*] responsiveness to Neighborhood Associations.
>
> (Fred Meyer letter, 30 October 1985)

The founder of Fred Meyer Inc., Fred Meyer, had been known for his opposition to public planning and his belief that 'people will act in their own best interests and society will benefit accordingly' (MacColl 1979: 630). Seven years after his death, Fred Meyer's successors were presenting similar arguments to the business community and the public at large. In the land use hearings, however, a different sentiment was required for a successful campaign.

Using the language of the plan text, the Fred Meyer consultants, lawyers and management staff attempted to demonstrate that there was an interest other than their own to be served by constructing the proposed facility. The public would benefit because the facility would serve an unmet market demand and create new jobs. Those opposing the development accused the Fred Meyer consultants of 'magical thinking' in their efforts to develop a case for a new land use that would not create any change in the surrounding area. Minimizing any potential negative impact, the Fred Meyer consultants argued that the new facility would neither bring in significant amounts of new traffic nor drain business away from the existing commercial centres. Community organizations, according to the Fred Meyer representatives, were overreacting to the potential negative impact of the facility based on miscalculations. Hours of hearing time were spent debating the findings of the applicant's technical experts as opposed to city planning staff's and the neighbourhood organizations' consultants' reports. The proposal was argued to be consistent with the goals of the city as expressed in the Comprehensive Plan. In effect, it was said, this proposal offered a 'win–win' decision for the city even if it was not recognized as such by the neighbourhood groups that would most directly 'benefit' from it.

In addition to the testimony of experts, Fred Meyer built its case on the interest expressed by individuals within the area, conducting several large surveys of residents. Significant support for the store was documented among area residents. While Fred Meyer was reminded by members of the neighbourhood association and the City Council that planning was not a democratic process, their figures did challenge the neighbourhood organizations' ability to represent the position of area residents. With the Comprehensive Plan's liberal emphasis on citizen involvement, certain City Council members questioned the role of the neighbourhood organization versus the individual residents who responded in support (City Council Record, January 1986). The neighbourhood organizations, it was implied, were elitist and not capable of addressing the issue of public need. Arguing in this vein, a Fred Meyer representative referred to a letter in support of the proposed development. The complaint expressed in the letter cited a previous land use case in which the neighbourhood organization had supported the construction of a methadone drug treatment clinic within its boundaries. Clearly, the author of the letter stated, this was an obvious case of misplaced priorities and suggested irresponsibility on the part of the activists (City Council Record, January 1986).

The definition of citizen participation in the planning process became problematic. Although the neighbourhood organizations argued that their position represented informed opinion, as a result of their participation in the development of the Comprehensive Plan, a cloud was cast on the issue of public choice. A resident in opposition to the development responded to this debate by saying:

> Land use planning means never having to say you blew it. No amount of public opinion will make our streets safe again after several hundred to several thousand vehicles a day are added to every block in my immediate neighborhood. No amount of postcard returns will bring back the ambiance to this inner-city neighborhood after it's been rendered commonplace.

Fred Meyer's arguments, however, implied that the neighbourhood group represented self-interest rather than informed opinion. From their position close to the proposed development site, it was stated, the neighbourhood activists had lost sight of the public good.

Increasingly the arguments in the land use hearings focused on the contradictions between the two discourses of liberal planning. Although the textual community did not agree with the terms of debate, the neighbourhood activists were assigned the discourse of public choice planning. Highlighting the economic development goal, the Fred Meyer position was focused on an ability to achieve rational management and find technical solutions to achieve the public good. It was this belief that rational scientific management could remedy development problems that resulted in a City Council decision in favour of the shopping complex's construction. Two of the council members voted against the amendment in the belief that the project would have a detrimental impact on the area's neighbourhoods and business district. Two voted in support of the amendment on the basis of its perceived economic benefits. Mayor Clark, who had been a neighbourhood activist himself, was the crucial vote. Citing the need for economic development and the potential for mitigation of any use-related problems, he voted in favour of the proposed plan amendment.

The land use case did not end with that decision. Members of the neighbourhood organizations and their legal representative, a well known figure in Oregon's land use law debates, appealed against the council's decision. During the early stages of the appeal process there was a certain confidence on the part of the appellants. The intent of the Comprehensive Plan, it was felt, was clearly at odds with the decision on behalf of Fred Meyer. The pressure that business might place on local government could not be felt by members of the State Land Use Board of Appeals or the courts. As land use lawyer Ed Sullivan explained, the effort was motivated by concern for both the proposed Fred Meyer development and the precedent set by the council's action (*Oregonian*, March 1986).

In March 1988, nearly three years after the case was introduced, the State Supreme Court upheld the City Council's decision approving the plan amendment. At each level of the appeal it was decided that no procedural error had been made and that the amendment was consistent with goals expressed in the plan. The official interpreters of the text gave approval to the council's alternative reading.

CONCLUSION

The Fred Meyer case became a paper war over interpretation of a planning document. Behind what appeared to be simply battling interests were the 'politics of reading'. Different textual communities found within the legal document alternative meanings and, consequently, multiple readings. Duelling technical consultants and issues of public participation and representation filled hours of public hearings. I would argue that these tensions and ambiguities are inherent within the discursive field of liberal ideology, with its attempt to manage society for the public good while supporting the rights of individuals. The difficulty of achieving this balance is highlighted by the varied interests within a community. Defining the public good becomes a matter of power.

In a poor business climate, confidence in Portland's ability to manage growth had been shaken and, consequently, a new interpretation of the city's goals gained political favour. The availability of jobs and the concern with economic health were more conspicuous in people's evaluation of the quality of life in the

1980s. The changing economic conditions provided a new context for evaluating issues of planning theory and practice. As Terry Eagleton describes, the act of reading provides a 'frame of reference within which to interpret what comes next' (1983: 87).

The Republican Secretary of State Norma Paulus expressed some concern that Oregonians had become 'brazen hussies, throwing ourselves on anyone with a shekel in his pocket' (Kern 1985: 20). This statement, with its anti-business rhetoric unfortunately expressing antisemitism as well, suggests the concern with which members of the textual communities continue to deal with issues of planning and growth. Future cases will determine the manner in which the document will serve as a text in the production and maintenance of the Portland landscape.

12

TEXTS, HERMENEUTICS AND PROPAGANDA MAPS

John Pickles

> From cities of brick to cities in books to cities on maps is a path of increasing conceptualization. A map seems the type of the conceptual object, yet the interesting thing is the grotesquely token foot it keeps in the world of the physical, having the unreality without the far-fetched appropriateness of the edibles in Communion, being a picture to the degree that the sacrament is a meal. For a feeling of thorough transcendence such unobvious relations between the model and the representation seem essential, and the flimsy connection between acres of soil and their image on the map makes reading one an erudite act.
>
> (Harbison 1977: 124)

INTRODUCTION

Geographers have long pondered their unique relationship to maps. Yet the theory of maps has received comparatively little attention amidst the burgeoning literature dealing with the new theoretically informed geography. The traditional theory of cartography remains bound by: representational notions of correspondence where images are mirrors of nature; technical issues dealing with the transformation of space; and psychological accounts of behavioural spatial learning. In the first view, the relationship of the map to the world remains unproblematic. In the second, the manipulation of spatial data is the main focus of concern rather than a theory of maps *per se*. And in the third the map is dissolved into the context of its production.

This chapter deals with theories of maps, and begins with a critical assessment of the traditional approach that is predicated upon notions of correspondence and representation. The chapter argues that maps have the character of being textual in that they have words associated with them, that they employ a system of symbols with their own syntax, that they function as a form of writing (inscription), and that they are discursively embedded within broader contexts of social action and power. The chapter then contrasts the traditional theory of maps as unproblematic mirrors of nature with a hermeneutical approach that recognizes the problematic nature of all texts. Treating the map as a text in this way has two tightly interconnected implications. First, maps are seen as socially produced, as discursive tools by which to persuade others. For this reason the geographer is not merely representing objects that correspond unproblematically to things in the world. This does not imply, however, that the geographer is cut loose from all responsibility for critique and evaluation. Not all images are of equal value, equally successful or equally useful. Second, and following from the previous point, discursive texts – above all, propaganda maps – must be situated within a critical theory of judgement and evaluation. Several options are available. Contextualist and pragmatic accounts of judgement have recently gained some popularity, but I will adopt another route, hermeneutics.

The chapter focuses on one form of discursive text – the propaganda map as a particularly useful way to illustrate the text metaphor. Specifically, it is the extreme case, where traditional theories of cartography most easily break down. Having shown that propaganda maps do not fit within the framework of traditional theories of maps, we can then turn to 'regular' maps to see the same problem. As such, the extreme case shows that the traditional theories of maps fail in dealing with all maps.

By considering propaganda maps I will illustrate some of the possibilities and problems involved in interpreting embedded texts. From these brief examples of *map interpretation* I will then draw out some principles of hermeneutic theory for reading maps. In concluding I will ask whether traditional theories of maps are adequate to the task of 'reading' such texts. I presuppose throughout this chapter that the propaganda map provides a partial surrogate and heuristic for the consideration of other texts (such as social action and landscapes). But this essay is primarily an exploration of the ways in which maps have been used to influence the social construction of reality, and the failure of technicist and representational views of cartography to come to grips with the essential qualities of maps as texts. I will suggest that the communication model or cognitive model of communication adopted by modern cartography is riddled with contradictions. If the map arises from a series of graphical codes, a theory of signs must be invoked: representationalism founded on naive correspondence between image and reality must be questioned. If cartography is a form of discourse, as I shall suggest, then the cartographer and the map are at the centre of debates over technocracy and power in the modern world, and must be brought within the compass of social criticism and assessed from the perspective of social theory.

Modern cartography and communication models of meaning

Present-day cartography is a product of the Cartesian world, in which the map is the scaled representation of the real. Under this view, maps are devices of information transmission involving the basic rules of graphic communication based on a one-to-one correspondence of the world and the message sent and received (Muehrcke 1972). Communication involves the mechanical transfer of messages from a sender (inputs) through a medium (transfer) to a receiver (outputs) (Monmonier 1975). As with any communication system the model requires that information from the sender be encoded and that the receiver decode the information (Fig. 12.1). Information is conveyed, and, in so far as the cartographer, map, and map reader all receive the same information, distortion is avoided (Robinson and Petchenik 1975). (Fig. 12.2.) The measure of communication efficiency in the mapping process is related to the amount and accuracy of information transmitted. The cartographer's task is to devise better approximations between raw data and the map image (Muehrcke 1972). Thus the map is an objective tool for transmitting information. In so far as the technical production does not distort the data collected from the real world the 'good cartographer' is successful. By contrast, if the cartographer deliberately selects information to support his or her argument, and seeks to produce a map that has visual impact, then he or she becomes the 'propaganda cartographer' (Ager 1977) (Fig. 12.3.).

In this model the sources of error are limited and can be specifically ascribed to either a domain of technical control or a domain of ideological/political control (Fig. 12.4). Error is introduced (1) as a result of technical error (the poor choice of symbols, inappropriate projection or incorrect scale); (2) as a result of malicious intent to deceive (as with Haushofer's (1928) cartography, where some ulterior motive lies behind the construction of the map and where technique is subsumed to the attempt to persuade); or (3) from untrained decoders (map readers) or perceptual or value differences among interpreters.

SENDER → encoding → MEDIUM → decoding → RECEIVER

Figure 12.1

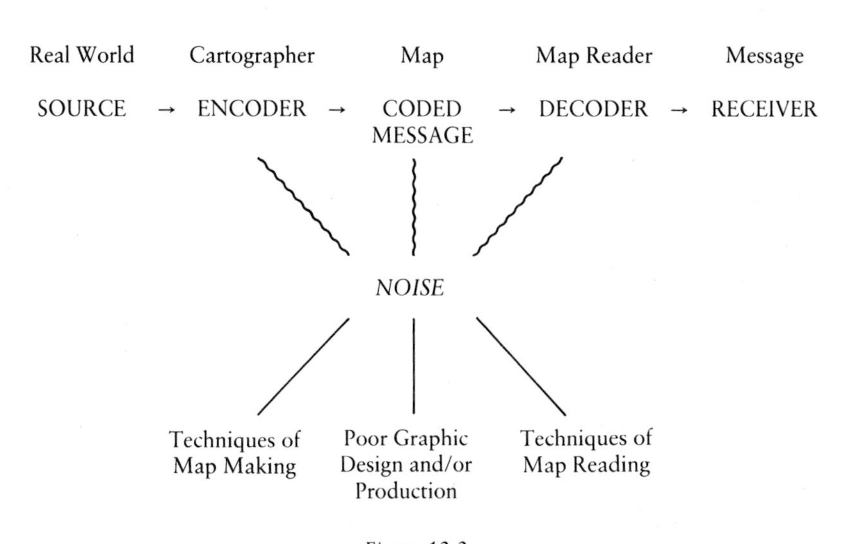

Figure 12.2

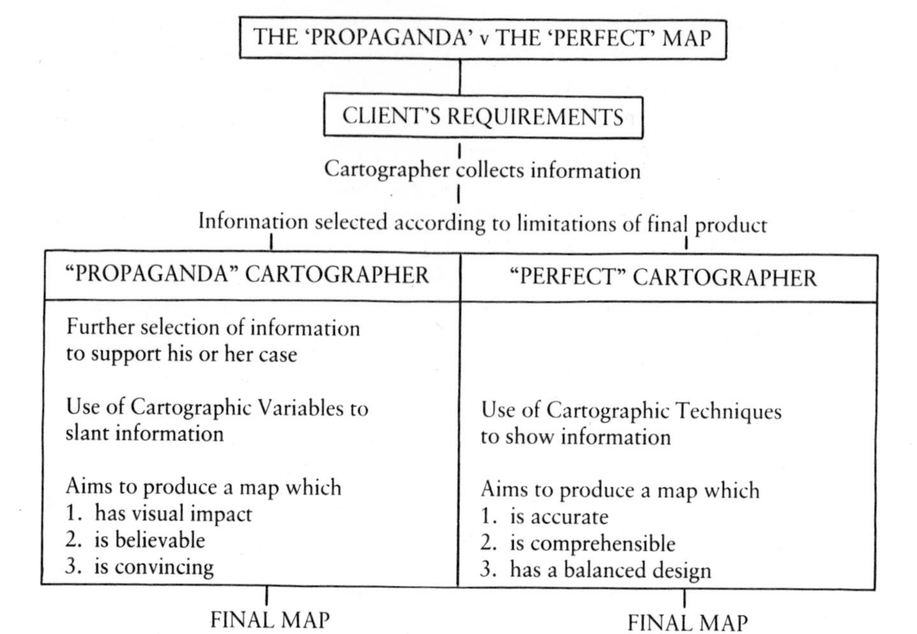

Figure 12.3 It should be pointed out that in reality there is no clear division between 'propaganda' cartographers and 'perfect' cartographers; the two types are at opposite ends of a spectrum within which fall all cartographers, their positions varying in accordance with the production of each map (from Ager 1977)

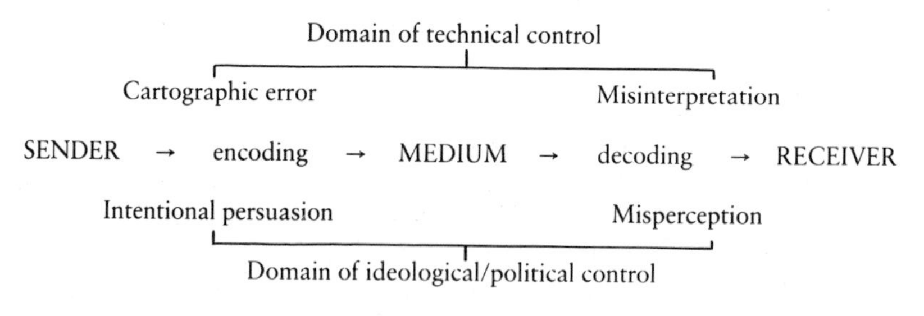

Figure 12.4

Such a neat division between the technical and the political produces, in turn, useful practical distinctions for the working cartographer: cartographic technique is seen as an on-going approximation to the real, presupposing a correspondence or representational theory of truth. The distinction between fact is thereby mirrored by the separation of the good cartographer and the propaganda cartographer (the latter being banished from the halls of science). In this sense, adapting correspondence rules between world, fact and map raises no fundamental and problematic questions of interpretation or of context.

Representationalism, correspondence rules, and the separation of fact and value, however, *do* present immense conceptual difficulties for accounts of distortion and error. The ideological is to be expunged but from a world that disavows its own ideology. The only remedies open in such a world are to retreat into technique, to reject interpretation as bias, and to accept an 'openness to alternatives' which permits claims of the sort that 'because interpretation is always a factor the cartographer must exercise care and judgement'. Not surprisingly, then, in the present century, as objective science has increasingly assumed that its methods give access to the true, the view of propaganda has altered radically. Science is seen not as a persuasive enterprise but as a claim to true knowledge. A good map is one in which the image received by the map user corresponds to that intended (inscribed) by the map maker and where the image inscribed (and received) is an accurate representation of the real world. Map making and map reading are seen to involve the straightforward transmittal of information in a philosophically and practically unproblematic manner. In particular, cartography does not seek to persuade, to convince or to argue; it does not select techniques of representation on the basis of their visual impact; and in the choice of subject matter, what is centred on the page, what is consigned to the edge of the map, and which scale and projection shall be used, the cartographer is guided by rules of scientific procedure and convention.

Propaganda maps

While the cartographer tries to present information accurately, comprehensively, with a balanced design, and without favouring one side of an issue, the propaganda cartographer seeks 'to produce a map which has visual impact and is not only believable, but goes a stage further – is convincing' (Ager 1977: 1). Propaganda maps are to be guarded against because the cartographer has used the wrong method and has 'failed to communicate correctly with the user' (Ager 1977: 14). The cartographer's colour choice, use of lines, orientation of north to the top of the page, and choice of material which will appear at the centre of the map, as opposed to at the edges, are all elements which are 'extraneous to the scientific purpose of the map' (Speier 1942: 313). The propagandist exploits these elements 'The propagandist's primary concern is never the truth of an idea but its successful communication to a public. Geography as a science and cartography as a technique become subservient to the demands of effective symbol manipulation' (Speier 1942: 313). (Plate 12.1.)

In what is probably the most systematic treatment in English of propaganda maps to date Tyner (1974: 2) suggests the name 'persuasive cartography' to distinguish propaganda, suggestive, advertising, journalistic and subjective cartography from other forms. Persuasive cartography is a 'type of cartography whose main object or effect is to change or in some way influence the reader's opinion, in contrast to most cartography which strives to be objective'. Persuasive cartography thus seeks to manipulate symbols in order to influence some

ONE SITUATION - TWO MAPS

ANGOLA ON NOVEMBER 15 1975

1. Stressing Encirclement.

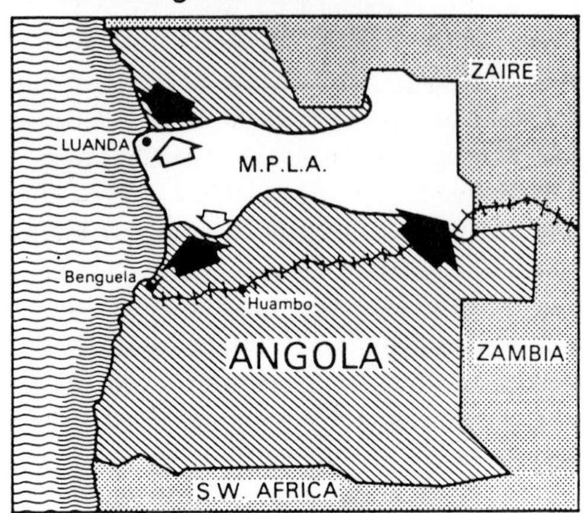

2. Underplaying Encirclement.

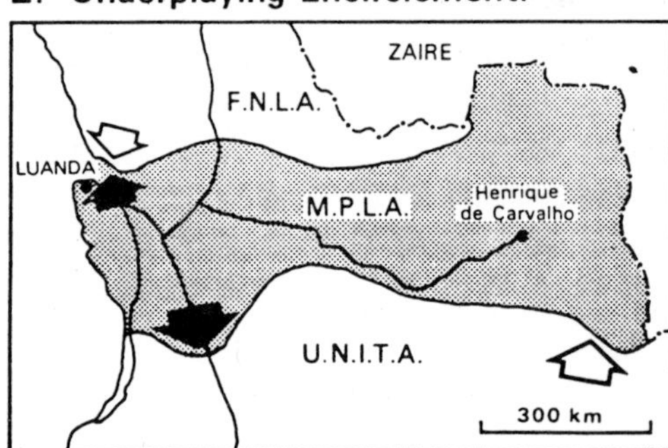

Plate 12.1 Shading, dynamic symbols such as bold arrows, and the careful selection of detail and coverage permit the cartographer to represent the same situation but with different messages (from Ager, 1977)

group about the value of some idea, opinion or action. Where some malicious intent lies behind this manipulation, propaganda is used. Thus:

> [T]he popular image of propaganda has undergone radical changes and the word has come to acquire overtones implying a process which is frequently sinister, lying, and based on the deliberate attempt on the part of an individual or group to manipulate, often by concealed or underhanded means, the minds of others for their own ulterior ends.

And:

> In part because maps appear to represent facts pertaining to mother earth herself, veracity and authority are frequently attributed to them, beyond their deserts. In what may be called 'cartohypnosis', or 'hypnotism by

> cartography', the map user or audience exhibits a high degree of suggestibility in respect to stimuli aroused by the map and its explanatory text.
> (Boggs 1947: 469)

As Ager (1977) points out:

> in reality there is not a clear division between 'Propaganda' cartographers and 'Perfect' cartographers, but both are at opposite ends of a spectrum in which all cartographers fall, and their positions vary in accordance with the production of each map.

Ager's conclusion is that in practice the distinctions between cartography and propaganda are difficult to draw:

> Regardless of the objectivity with which they were prepared, a great percentage of existing historical maps present some information which some individuals honestly consider 'propaganda'. For certain areas, the historical issues are so complicated and the record goes back so far that an unbiased map presentation becomes almost impossible.
> (Thomas 1949: 76)

It should be clear from this discussion that current theories of cartography, working with a distinction between objective cartography and propaganda maps, are founded on erroneous epistemological grounds. The distinction is articulated in terms of poles of objectivity and bias. When interrogated by cartographers themselves, however, the distinction breaks down, and leaves us with a series of difficult questions to answer. Are only those maps which use certain graphical techniques to create a distorted impression to be considered propaganda maps? Or are there certain categories of map *use* that constitute propaganda, while other uses do not? Are all maps propaganda maps?

CONTEXTUAL INTERPRETATION: THE SOCIOLOGY OF PROPAGANDA MAPS

Solving the problem of distortion and representation by adopting the view that all maps are propaganda maps does not deal with our problem. It merely sidesteps the issue. Since all maps are constructed images, and since all images are interpretations of a particular context, we gain little by merely repeating that maps and propaganda maps are both interpretations and distortions. We remain caught within the metaphysics of presence which presupposes some foundational object against which the distortions and interpretations can be measured: that some interpretation-free image could be produced that does not distort the world. Moreover, in arguing that all maps are propaganda maps we seem to deny the value of any theory which seeks to judge maps critically. Because all maps distort, what is important is the intention behind the construction of the map, and the use to which a map is put. However, against these views I will argue in this chapter that the problem is not how to dissolve the distinction between propaganda maps and other maps, but how to work out whether we can establish any criteria of judgement once we recognize the discursive nature of texts such as maps.

Propaganda maps and hegemony

If individuals are reciprocally constituted by and in turn themselves reproduce the apparatuses of the state, how is it that ruling groups maintain power for so long? Force alone cannot account for it. Some form of mass consent or acquiescence must be involved. But how is it obtained? In other words, how is hegemony established over a people?

Generally speaking, geographers have paid little attention to the ways in which power is exercised through social discourse, although this inattention is being redressed through works influenced in particular by Michel Foucault, Jacques Derrida and Jacques Ellul. Yet in spite of the growing influence of the work of these scholars, between the 1960s and the 1980s little attention was paid to an analysis of propaganda and even less to the topic of propaganda maps. 'One reason has been the frustration experienced in trying to define the term. There has never been a clear agreement on exactly what propaganda is' (Jowett 1987: 101). In its benign form propaganda has been seen as any association, systematic scheme or concerted movement for the propagation of a particular doctrine or practice; that is, the systematic propagation of any opinion, creed or practice. In this view to propagandize is to disseminate principles by organized effort. In time this term was applied less to the organization that set out to propagate a doctrine than to the doctrine itself, and 'by the nineteenth century it also came to mean the techniques employed to change opinions and spread doctrines, the modern-day use of the term' (Jowett 1987: 97).

The increasingly sophisticated techniques of propaganda have often been seen as problematical for a society which protects free speech. By the late nineteenth and early twentieth centuries such techniques were seen to pose a threat to democratic practices and institutions themselves (see, for example, Lasswell 1927). Liberal political thought saw propaganda as a potentially dangerous form of distortion of true claims and accurate positions, which in the hands of the unscrupulous and corrupt might pose a threat to liberal democracy itself:

> Given a greatly expanded franchise with its corollary of the need to base authority on the support of public opinion, political society invited the attention of the professional controller of public opinion. When to the demand for new methods of publicity there were added revolutionary advances in the technique of communication, and the latest discoveries in social psychology, mankind had to fear more than ever 'the cold-blooded manipulation of popular impulse and thought by professional politicians'.
>
> (Wallas 1921: 17, 51)

Gramsci (1981: 80, n. 49) saw the techniques of public persuasion (of which propaganda was one) as central elements in forging the relationship between consent and coercion in the establishment of hegemony:

> The 'normal' exercise of hegemony on the now classical terrain of the parliamentary regime is characterised by the combination of force and consent. Indeed, the attempt is always made to ensure that force will appear to be based on the consent of the majority, expressed by the so-called organs of public opinion – newspapers and associations – which, therefore, in certain situations, are artificially multiplied.

Propaganda aims at persuading large groups of people to believe something or act in a way that they would not, in the normal course of events. Propaganda techniques are, then, techniques of persuasion which fail to abide by established and accepted norms of accuracy and truth. They seek to manipulate relationships in order to persuade people about a particular claim to truth.

With such techniques institutions concerned with the process of establishing hegemony capture the discursive field and reconstitute the discourse of the age and the place. One means by which the state has attempted in the past to capture the discursive field is precisely not only through the appropriation of space (and the map) to its purposes, but by the symbolic constitution of mapped space as national space. In so doing, fledgeling national territories sought to establish a national identity abroad, and to create a national ideology at home, sometimes in the face of internal disunity or rebellion (Plate 12.2). The map has been used frequently in this way to express some sense of national identity. Here the link between map and symbol becomes clear. The expansion of the frontier is captured in the stretching wings of the bald eagle, or the unity of President and country is captured in the unusual 1912 Roosevelt map of the United States (Plate 12.3). Here multiple texts are evoked: the anthropomorphized map unifies the land and territory; the resultant unitary state is personified through a single figure; and the superimposition of both images evokes multiple other texts not *represented* here.

In another context, the fledgeling revolution of Soviet communism is captured in the breaking of the chains around the globe, breaking the bonds of past injustices, and literally opening up the world anew (Plate 12.4). Such icons may resonate from the political domain into others, such as the scientific, as the play on images deepens with each repetition (Plate 12.5). Interpretation requires that such images be placed in the context to which they respond: the symbol of the Russian revolution and its promise to release the oppressive bonds of the past; the promise of the free world under socialism; the centrality of the worker as the objective force breaking the chains; the symbolic hammer; the analogy with a radical geography seeking to break the bonds that bind a discipline to its bourgeois modes; hidden tribute to the pantheon of intellectual leaders for such a new geography; and the earth and the worker (and geographer?) united in the task of production and emancipation.

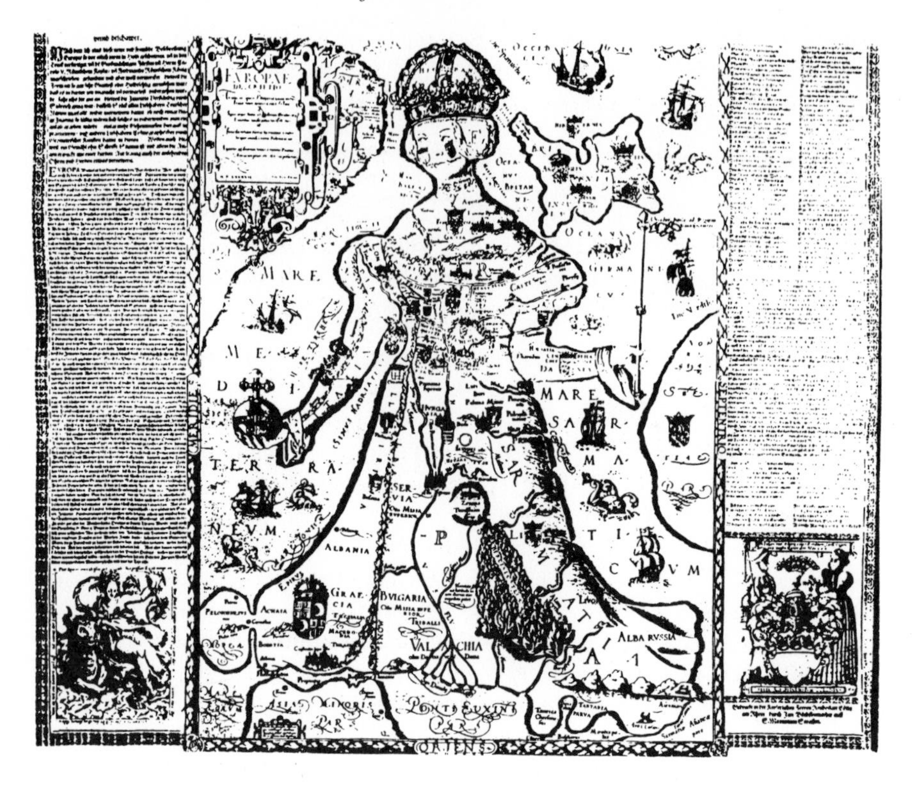

Plate 12.2 Europe as a queen (Library of Congress, Washington, D.C.)

Propaganda maps and war

As the previous example illustrates, the use of symbols and maps to establish a hegemonic project is not necessarily destructive or dangerous. It is perhaps the association between propaganda maps and the uses made of them by Nazi Germany that has diverted attention from the ubiquitous use of such maps in all sorts of hegemonic projects: commercial advertising, the everyday work of state institutions, and the construction of our own 'enemies'. Moreover, this association between Nazi Germany and 'propaganda maps' has led to them being seen as aberrant cartographic forms (often also *abhorrent* forms, if the writings of cartographers are to be believed). This notion of propaganda maps as aberrations has discouraged their consideration within a theory of maps – surprisingly, given their widespread and effective use in the selling of both products and wars (Plate 12.6). Lord Northcliffe's observation after the First World War that 'The bombardment of the German mind was almost as important as the bombardment by the cannon' was taken to heart by Hitler and his propagandists. In *Mein Kampf* Hitler argued that propaganda must be addressed to the emotions, not to

Plate 12.3 Theodore Roosevelt's face superimposed on a map of the United States, 1910–12; scale 1:40 million, c F22977, 25 September 1912 (Library of Congress, Washington, D.C.)

the intelligence. It must concentrate on a few simple themes. It should be presented in black and white. 'Propaganda consists in attracting the crowd, and not in educating those who are already educated.'

Goebbels openly admitted that propaganda had little to do with the truth: 'Historical truth may be discovered by a professor of history. We, however, are serving historical necessity. It is not its task, any more than it is the task of art, to be objectively true.' The sole aim of propaganda is success. To achieve such success the truth, perhaps surprisingly, remains important. Goebbels was very careful not to tell whole lies, since they are easily shown to be false and are therefore ineffective. Instead '[h]e hid the nucleus of truth with all the veils of interpretation. He always had a channel of escape when anyone questioned the truth of his statement.'[1]

Generally such propaganda is said to ignore the facts or distort them intentionally: 'Where favorable facts are wanting, the propagandist manipulates available facts and plausible fancies to serve his end' (Thomas 1949: 78). Thus Karl Haushofer transformed geopolitics into 'a dynamic *Weltanschauung* to further the expansive claims for *Lebensraum* of Germany' in part with the propaganda or suggestive map (Weigert 1941: 529). Such dynamic suggestive maps relied on the strength of the initial idea and the use of symbolism – the new cartography was to be visually violent – to accost the map maker and to present a clear message. Often such maps are either not maps in the strict sense or the map is

Plate 12.4 Soviet poster, source unknown

Plate 12.5 Cover of *Antipode*. The juxtaposition of the two images illustrates many of the issues involved in interpreting complex texts. The image of the worker breaking the chains that bind the earth is almost universal, and can be seen in many different contexts and countries. The design of the *Antipode* logo may have no connection whatsoever with the Soviet poster in Plate 12.4, but does that constrain our interpretation? Do we develop a forced interpretation if we connect the images and find cleverly provocative in the second the absence of the figures found in the first?

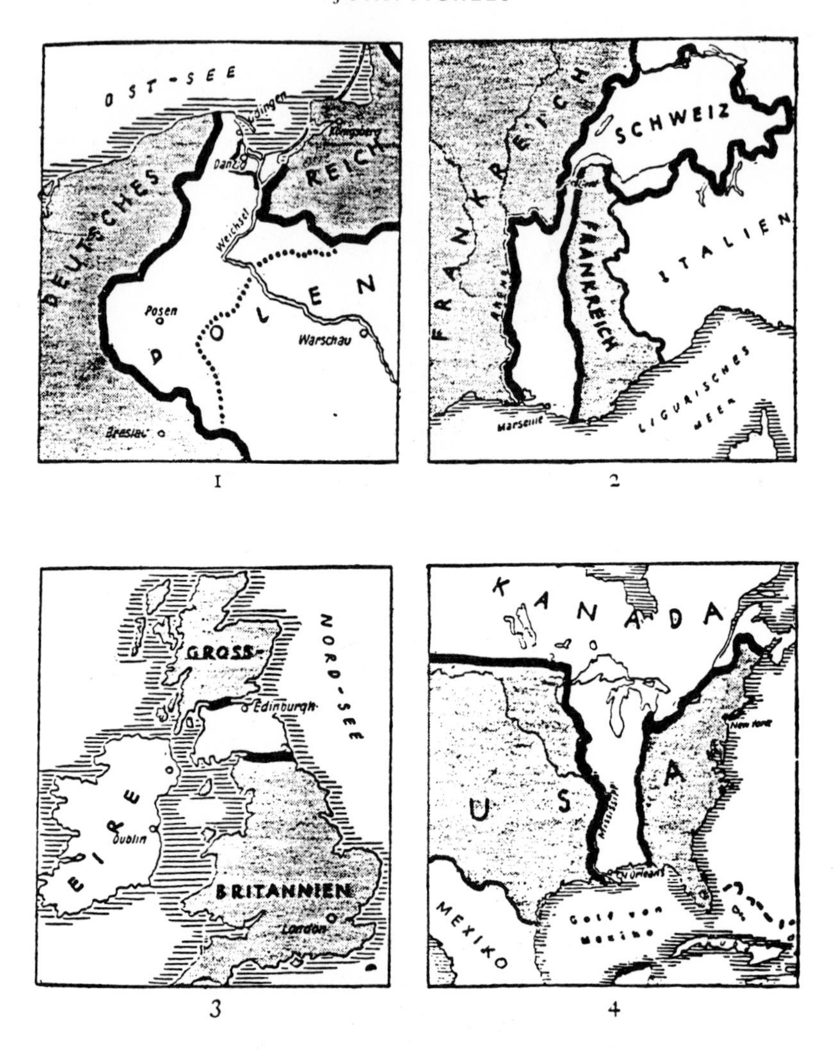

Plate 12.6 'Polish corridors elsewhere', poster from the *Schwarze Korps*, Berlin, reprinted as here in *The Living Age*, September 1939, p. 29. The following captions in German accompanied the map. '(1) One glance at this map makes discussion unnecessary; but since democratic statesmen regard this as a fair and reasonable state of affairs, they have only to make their choice. (2) Why should Switzerland lack rights which Poland possesses? The Swiss also need access to the sea; and that could be easy to obtain. M. Daladier need only say Yes, as he did in the case of the Sanjak! (3) The poor Irish – haven't they a claim on direct access to the North Sea? "Good old Chamberlain" should understand this better than anybody! (4) The United States is so huge that Canada should really be granted the boon of a coastline on the Gulf of Mexico. Think of your great predecessor Wilson, Mr Roosevelt! Come across!'

only part of a collage of images which wilfully exploit the inherent limitations of maps to distort and exaggerate (Quam 1943: 21). Such dynamic symbolic maps relied on the strength of the initial idea and the use of often visually violent symbolism to achieve its end – to accost the map reader and to present a strong message. A particularly clear example of such a map is one dropped on the Allies at Dunkirk during the Second World War (Plate 12.7). The map depicted the position of the troops as hopeless. They were shown to be completely

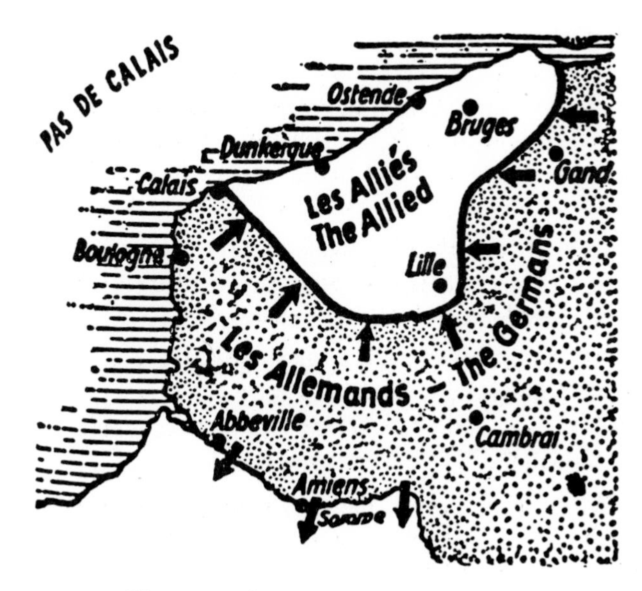

Camarades!

Telle est la situation!
En tout cas, la guerre est finie pour vous!
Vos chefs vont s'enfuir par avion.
A bas les armes!

British Soldiers!

Look at this map: it gives your true situation!
Your troops are entirely surrounded —
stop fighting!
Put down your arms!

Plate 12.7 German leaflet dropped on the allies at Dunkirk to encourage surrender (Speier 1942)

surrounded, with little hope of escape and nowhere to escape to: the Allies were surrounded, the Germans were on the move (indicated by the use of bold arrows throughout occupied territory). The technical manipulation of the visual field of the map made the call for men to lay down their arms appear reasonable in such an island of desperation. Hope was removed visually from the map by the failure to show the south-east coast of England thirty or so miles across the Channel.

The clearest statement of the map/propaganda map relation is provided by the very cartographers who were engaged in combating German attempts to use propaganda maps during the Second World War. For Weigert (1941: 530) the map is a double-edged weapon:

> in unskilled hands it easily becomes a subject of ruthless and stupid propaganda. But in the hands of the expert who knows the rules of the war of words as well as those of modern cartography, it is a good weapon.... it can bring hope to the suppressed nations and fright to their suppressors. And here too, the attack is the best defense.

Quam (1943: 32) claimed that map makers 'must strive to make their maps accurate and in harmony with the democratic ideals of our cause', and in 'War on the visual front' Heinz Soffner (1942: 465) argued that as the:

> global war progresses the harder it is for even the generally well informed and earnestly interested citizen to keep track of all its rapidly changing aspects and the more difficult grows the task incumbent upon the various media of information.

The propaganda maps used to illustrate parts of this chapter are, on the whole, rather unsophisticated (albeit effective) cartographic forms, and it is difficult to imagine any single image having a major effect in the general propagation of a particular truth claim. But not all graphical imagery is unsophisticated, and few images should be thought of as isolated from broader contexts. The failure to see graphic images as part of a series of wider texts and contexts often weakens the analysis of the map and results in a failure to see the full power of the image. Commercial designers have always seen the benefit of the temporal composition of an image through a series of incomplete visual forms, a technique that is currently being developed with immense sophistication in American television commericial and public-service advertising (Levi jeans, anti-smoking ads, and 'just say no' ads) (McDermott 1969). In these visual 'clips' (the commercial equivalent of the 'sound bite') narrative structure is generally absent but arises temporally as each clip builds on the previous one to form a coherent and often powerful composite impression. The images of Cold War geopolitics have long been fostered through similar partial, and in themselves often meaningless, visual 'clips'. The Russian bear is a form repeated in various guises, playing off earlier uses of the image and depending for its impact on them. A particularly distorted elaboration of this image was published by *Time* magazine in 1955 (Plates 12.8–9). The careful selection of shading and symbolism permitted the cartographer to

Plate 12.8 The cover of a military code book. The humorous image of the bear reflects a serious theme in Cold War geopolitics, and has been a central metaphor of much Cold War cartography

illustrate the red menace reaching round the Chinese mainland. The threat from the Soviet Union, North Korea and Vietnam is visually focused first on China, but then on the island of Okinawa, where stand the stars and stripes. Moreover, the gross exaggeration of the Himalayas closing in on the margins of China emphasizes its isolation from the West, whose surrogates India and Pakistan are shown in the recessive colours of light green. The message is strong but not obvious. The whole map is a study in suggestion, in which cartographic techniques are used to depict a particular situation in such a way that both the

Plate 12.9 *Time* magazine map of 'Red China', 1955. The metaphor of the bear and its conceptual foundation (the domino theory of geopolitics) is only slightly less obvious than in Plate 12.8

intrinsic meaning and the suggested meaning resonate with other texts and images beyond this single map.

Meaning thus arises from the merging of multiple horizons. In this way graphical images are rhetorical devices. They are part of a social discourse about the world, and as such must be approached as we would any rhetorical object. As Weigert (1941: 528) suggests:

> it is surprising to see that we are not all conscious of the important part which the map and the art of map-making plays in the process of creating a new conception of the world. We simply rely on maps as if they were facts in this transformation of thinking and seeing. The astounding observation that, in the discussion of the vital problems of the day, the maps as they are presented to us are being taken as stable and indisputable facts, as mere tools which do not themselves reflect aims and opinions of their creators – this naive confidence in the truthfulness of the map indicates that many of us are not aware that maps are weapons. Like the written and spoken word, like photographs and cartoons, the map has become a psychological weapon in a warring world where the souls of men are as strongly attacked as their lives.

And this is surely the point: mapping is an interpretive act, not a purely technical one, in which the product – the map – conveys not merely the facts but also and always the author's intention, and all the acknowledged and unacknowledged conditions and values any author (and his/her profession, time and culture) bring to a work. Thus, like all works, the map carries along with it so much more than the author intended. Also, like any text, the map takes on a life (and a context) of its own beyond the author's control. The map is a text, like any other in this regard, whose meaning and impact may go far beyond the limits of technique, *mens auctoris*, and the mere transmittal of information.

Propaganda maps and time

The perception of graphical images is not a purely psychological reception of information but a complex social play of images present and absent, in the context of other symbolic, ideological and material concerns. All cartography operates within and makes use of such unacknowledged pre-conditions and more or less accepted symbolic forms and mapping conventions. Yet the impact of these techniques and effects becomes particularly clear when we turn to obviously propagandistic texts. At a very basic level a particular iconography has been appropriated to the goals of political and commercial propaganda. The globe and the map both stand as icons, repeated as an unchallenged vocabularly of advertising, Cold Warism or national boosterism.

On occasion the globe and the map have become so successful as symbolic images that their 'shadows' can be presumed in images which contain no map form at all. To the globe motif has been added one especially rich evocative example: the motif of the spider/octopus. Here the historical repetition and reworking of the same image has permitted the spatial specification to be removed. No map or country location is given, but one is presumed. At this point the map and the cartoon fuse. In this case the map form is present as an *absence*, it is constitutive of the image, and must be incorporated in any interpretation of the poster. More important for our present purposes, our theory of maps must incorporate this domain of absent images as a concrete *presence* evoked by the skills and conventions of the propagandist and cartographer. (Plates 12.10–15.)

Temporal sequencing and contextual memory are vital to the work of all graphic designers and artists. In modern cartography they are made more powerful by new techniques of design and production. Principles and techniques including scale and projection, assimilation and contrast, negative and positive, grouping and line, symmetry, double symmetry and asymmetry, reversible images, and the perceptual influence of angles, provide the graphic designer with powerful optical devices to manipulate and create the required field of visual effects (Plates 12.16–17). Modern cartography linked with new technologies of image production and transmittal has many new and powerful techniques for the creation of new forms of dynamic imagery. More complex techniques of visual surface kinetics such as the moiré effect permit the graphic designer to set off a two-dimensional surface into an apparently three-dimensional pulsation, and thus give the designer the ability to construct spatially and temporally complex

Plate 12.10 'Bolshevism', poster for the Anti-bolshevik Exhibition at Karlsruhe in the 1930s

Plate 12.11 'His amputations continue systematically', wartime German poster for distribution in France

Plate 12.12 'Work and fight to free the Indies', a British poster of 1944 for the Dutch government (Pat Keely)

Plate 12.13 Uncle Sam gobbling chunks of Central America

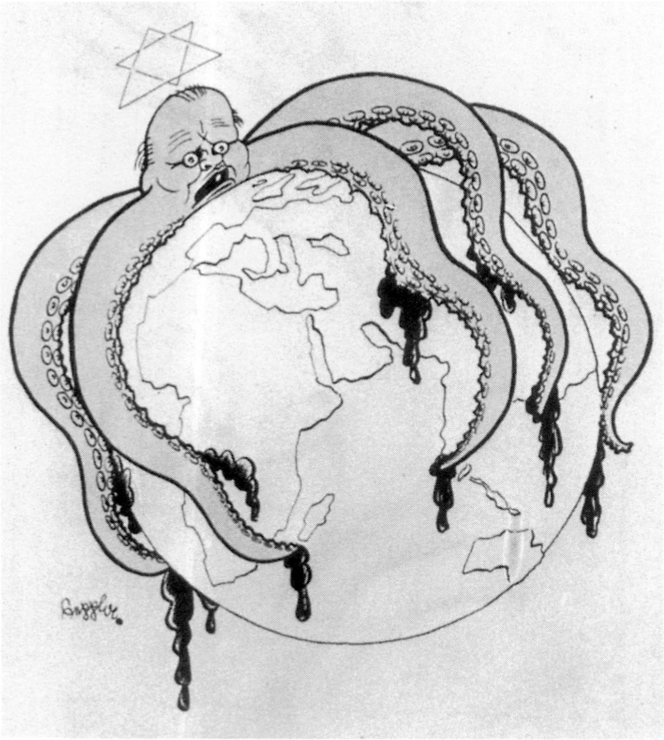

Plate 12.14 German poster showing Churchill as a greedy octopus, 1940 (Josef Plank)

Plate 12.15 German poster showing England threatened by a Jewish bolshevik conspiracy

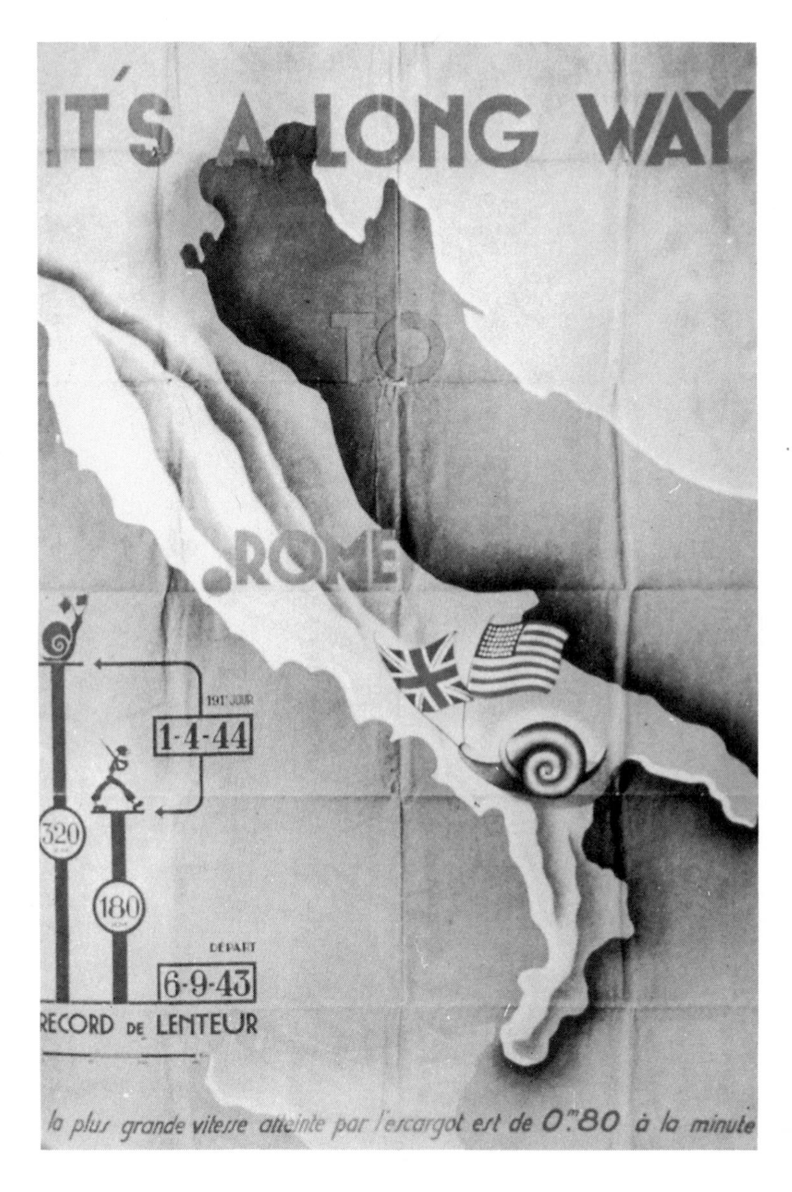

Plate 12.16 German poster of 1944, the allies advancing at a snail's pace

Plate 12.17 US leaflet warning the Japanese of the renewed might of the United States following the surrender of Germany, 1945

and dynamic images. These images are already incorporated into the new forms of computer cartography and urgently raise the question of the relationship between technique, content, control and meaning in complex visual media (Plate 12.18).

TEXTUAL INTERPRETATION OF THE PROPAGANDA MAP

Traditional theories of mapping and maps are of little use to us when we push them to incorporate propaganda maps. Without the foundation of an unproblematic theory of representation to fall back upon cartographers are forced to retreat to the position that all maps are distorting and hence all maps function as propaganda maps. But as we have already seen this is merely to sidestep the issue, and raises other serious questions about the sorts of claims we can make and the work we can do. Both approaches fail because they do not appreciate the textual qualities of maps. In this section I will begin to unpack this textuality in the form of a 'nested' account of maps as texts. This nested account will address three issues: the world and the text, the text in a text, and the analysis of the work itself.

We encounter ambiguity the moment we ask, what is the content of a graphic image? Clearly it is the real world, the real situation, the landscape, the scene. The map maker reduces this object-field according to established principles of objectification, abstraction, reduction and idealization to create the map. In this sense all maps are thematic abstractions involving reduction of one form or another. In a way different from the photographic image, however, this reduction is a *transformation.* In order to move from the real situation to the map it is necessary to divide up this reality into units and to constitute these units as signs substantially different from the object they communicate (as the opening quotation from Harbison suggests). The map is thus a coded message whose relationship to the object-world it portrays is complex.

What is the nature of the coding? The map is a message. As the previous discussion suggests, cartographers and geographers have traditionally taken this message to involve a source, a medium and a receiver. The source is the cartographer (and his or her body of received techniques and style), the medium is the map (and the often ignored immediate contexts within which the map is embedded) and the receiver is the map reader (as a public 'readership'). This view construes the map too narrowly, however. It ignores the other texts within which the map is itself embedded and with which it is co-determined. It ignores the context into which the map is projected. It one-sidedly places emphasis on the intended message and fails to consider possible unintended meanings. Finally, it has no way of accounting for the ability of graphic images to conjure up other texts (maps, photographs, books, etc.) and embed them in any reading of their own codes. By way of illustration let us ask, what is the medium of the map's message? In the communication models discussed above, the medium is the map.

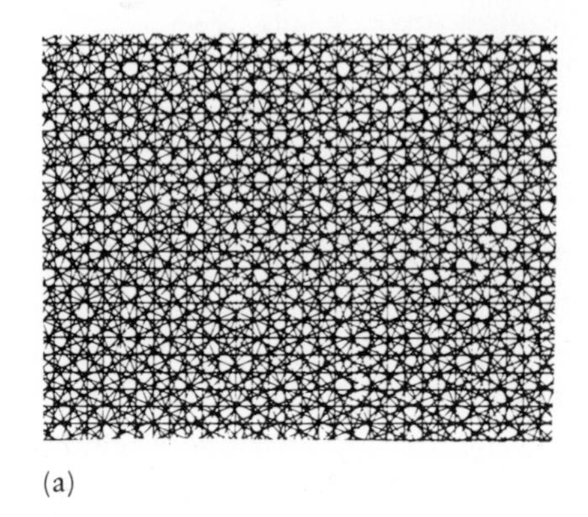

(a)

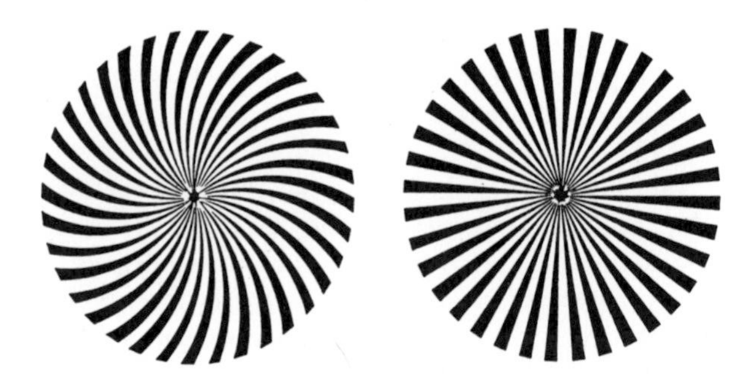

(b)

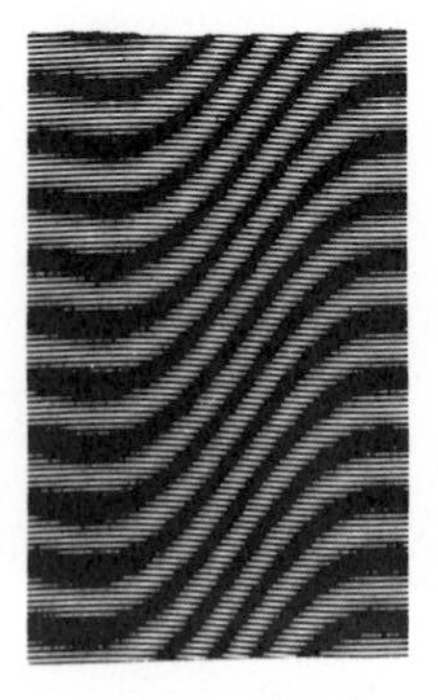

(c)

Plate 12.18 Selected examples of graphical illusions based on the effects of (a) randomly intersecting lines and (b, c) moiré effects

But how can this be? The medium is the report, the article, the book, the magazine, within which the map appears. More precisely, as Barthes (1987h: 15) says of the photograph, the medium is 'a complex of concurrent messages with the photograph as centre and surrounds constituted by the text, the title, the caption, the lay-out and, in a more abstract but no less "informative" way, by the very name of the publication'.

We are faced with layers of textuality: the map itself, the immediate context of the map (its caption, the chapter and the work of which it is a part) and the wider context of the map (the opus of the individual cartographer or school, the opus to which the text itself belongs, the socio-cultural context of the work). But although the map is an embedded figure, the map is also an object that has a structural autonomy independent of both its production and its use, and thus requires an analysis of the work itself. This will not be definitive, but will always have to be situated alongside a sociological analysis of text and context; of production and use. Even an analysis of the work itself cannot divorce the map entirely from its context, for the map is not an isolated object. It has a title, and fits within the body of a text, along with a set of other maps, or, if it is a single map, it is framed and displayed in some manner.

In terms of the internal construction of the map, the message of the map is carried by two different structures, one of which is graphical, the other of which is linguistic. Yet the consideration given to the linguistic components of the map has been mainly restricted to the design and effectiveness of the graphicality of lettering (size, print style, placement). The importance of the linguisticality of the map has been ignored. While in other graphic forms (photography, painting) the graphical and linguistic elements are complementary, in the map they operate almost uniquely as inseparable from each other. This inseparability is also typical of certain forms of advertising, poster art and modernist art forms such as Dadaism. Here the linguistic elements are embedded within the image, not incidentally but as intrinsic components of the whole picture (Plate 12.19).

In describing the linguisticality of the title of the photograph Barthes (1987h: 16) says:

> The two structures are co-operative but, since their units are heterogeneous, necessarily remain separate from one another: here [in the text] the substance of the message is made up of words; there [in the photograph] of lines, surface, shades. Moreover, the two structures of the message each occupy their own defined spaces, these being contiguous but not 'homogenized', as they are for example in the rebus which fuses words and images in a single line of reading. Hence, although a press photograph is never without a written commentary, the analysis must first of all bear on each separate structure; it is only when the study of each structure has been exhausted that it will be possible to understand the manner in which they complement one another.

But this position is untenable for the theory of maps. In the map the symbolic

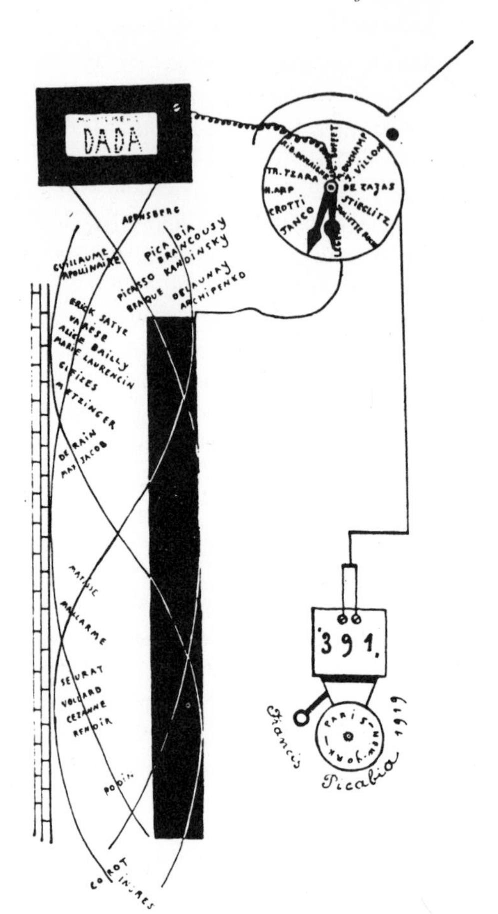

Plate 12.19 Francis Picabia, *The Dadaist Movement*, pen and ink, 1919 (Museum of Modern Art, New York)

graphic image is embedded in a written text (a paper, a book, an atlas) and rarely has an existence beyond the body of the text and the discursive aims of the research of which it is a part. Moreover the symbols and words in the map are interbedded: the names of places, features and other descriptors are integral to the visual image, and call for a special form of construction and present specific difficulties for analysis. In particular, such interbedded texts (maps, poster, commercial and Dadaist art) are correspondingly much closer to the tract: the commercial, political poster or art-work. It is in this intersection that much of what has been called propaganda mapping arises. The issue becomes clearer on closer analysis.

For Barthes all 'imitative' arts comprise two messages: a denotive message and a connotive message. The realist painter and photographer stake their reputations on their work being predominantly denotive, in the sense that the representation of the objects is a re-presentation in which the objects represented are objects from the world, without transformation. Traditional cartographic theory presents the map as a purely denotive message. But as we have already seen in the mapping process, objects to be represented are transformed and reconstituted as signs and symbols substantially different from the objects they communicate. That is to say, the map is a coded message.

There is another important sense in which the map differs from the photograph. In most photographs (except where the object of the photograph contains language) the caption constitutes a parasitic message which adds to and circumscribes the meaning of the photograph. In the map the issue is more complicated. The caption here is also a parasitic, albeit essential, part of the map. First, it merely illustrates the image, often through a repetition of the more obvious content of the map image itself. Second, 'the text loads the image, burdening it with a culture, a moral, an imagination' (Barthes 1987h: 26). That is, the caption also reinterprets the map and points us to specific or specified meanings; the caption circumscribes our reading of the map. Third, the map image itself is also linguistic. Here the interplay of codes and words constitutes a distinctive image form, in which the message is achieved largely in terms of the interplay and duality of graphic and linguistic meaning.

The transmission and reception of the map image are not the straightforward, linear process presumed in the communication model. The coded image (the map, linguistic and graphic) is also connotive. Through the fusion of horizons between the reader's world and the world of the map (and the map maker) the map *connotes* a variety of meanings. Thus the reading of the map is always historical and 'depends on the reader's "knowledge" just as though it were a matter of a real language, intelligible only if one has learned the signs' (Barthes 1987h: 28). The map is a purposive cultural object with reasons behind its construction and values associated with its reading. To suggest otherwise is to fail to see its status as *made* object. The map is always and necessarily an expression of an idea. In mediating the transformative processes of abstraction, reduction, thematization and idealization, the cartographer selects, sifts and emphasizes this or that aspect of the world under consideration, and articulates an image in the rebus linking graphic and linguistic codes.

If Barthes's distinction between denotive and connotive meaning allows us to prise open a first step in the analysis of the work, further reflection on the distinction forces us to abandon it. The experience of modern art forces us to rethink the very nature of this distinction. John Berger (1965: 55) explains this change in showing how the revolutionary vision of Cubism arose out of an inheritance passed to us from the nineteenth century:

> Nature in the picture is no longer something laid out in front of the spectator for him to examine. It now includes him and the evidence of his senses and his constantly changing relationships to what he is seeing. Before Cézanne, every painting was to some extent like a view seen through a window. Courbet had tried to open the window and climb out.

> Cézanne broke the glass. The room became part of the landscape, the viewer part of the view.

Thus the challenge of modern art and modern science is to work through the implications of accepting the inevitability of our participation. For Heisenberg (1959) this meant that 'Natural science does not simply describe and explain nature; it is part of the interplay between nature and ourselves; it describes nature as exposed to our method of questioning.' Failure to come to terms with this participation has serious consequences. It was the power of Cubist painters before 1914 that they were able to link Courbet's materialism with Cézanne's dialectical view of the image. But one without the other would have led and did lead to a sterile art. Materialism became literal and mechanical. An ungrounded dialectical view became disembodied and overly abstract. The danger for a theory of maps/texts is obvious. A representational view of the image divorced from an investigation of the role of the one who constructs the image becomes literal and mechanistic. Conversely, overemphasis on the viewer and the viewer's responses becomes idealistic and equally reductionistic.

These textual qualities are, of course, common to all maps. The question arises, then, where will we find a theory that can deal with maps as texts, without reducing all map forms to forms of propaganda maps? The answer might be found in a reconsideration of hermeneutic theory and critical hermeneutics.

HERMENEUTICS

In the nineteenth century the clock became the metaphor for mechanical approaches to the social sciences. In the twentieth century the electrical circuit became the metaphor for systems approaches in the social sciences. These and related metaphors have left a lasting impression on twentieth-century social science. Yet in the second half of the twentieth century a new metaphor – the text-metaphor – has arisen as the template for understanding and framing social life. In this period the text-metaphor has colonized certain domains of study – painting, film, landscape and most recently social life.

Extensive use has been made of the analogy of reading and the text-metaphor throughout the history of modern geography: Sauer's reading of the origins and development of past landscapes from the tracings and antecedents in the contemporary landscape; Lewis's axioms for reading the landscape; Samuels's biography of landscapes; Meinig's symbolic landscapes; Jackson's close interpretation of vernacular artefacts as symbols that reflect broader social changes; Sitwell's equation of elements of landscape with figures of speech; Duncan's studies of the language and semantics of cultural and symbolic inscription; and, of course, the map as an encoded artefact. As I have argued elsewhere (Pickles 1987), the adoption of the reading metaphor in these interpretive approaches has given rise to a methodology which is largely implicit, often derived from years of apprenticeship and practice, and informed by the keen eye of the fieldworker, but unguided by the detailed and explicit debates about hermeneutic methodology developed elsewhere in the social sciences and in philosophy. Interpretation theory has yet to be rigorously grounded in a detailed methodology. Only recently have geographers begun to explore the necessity for, and the character of, an explicit hermeneutic canon and its associated interpretive theories and approaches, an exploration given further impetus by the recent consideration given by geographers to narrative (Sayer 1989), rhetoric (Sugiura 1983) and judgement (Meyer 1984).

Central to each of these explorations is the question of whether or not we can develop rigorous interpretive methods. Without these methods and a corresponding theory of interpretation – *hermeneutics* – our ability to develop intersubjectively accepted criteria for assessing claims must be in question. However, hermeneutics has received a limited reading in geography, first by its appropriation to individualist philosophies of meaning, and second by its introduction through Habermas's triad of knowledge-constitutive interests – technical–empirical; hermeneutic–interpretative; critical–emancipatory – in which hermeneutic knowledge is merely one form among others and not itself the manner of all knowing. The consequence has been both an idealist and an instrumentalist conception of hermeneutic method. Hermeneutic method has been seen as a technique appropriate only to *some* aspects of social enquiry, but also as a hindrance to materialist critique (Walker 1989). In this context the radical and universal nature of hermeneutic understanding has yet to be fully explicated, and it is a sign of this absence that virtually all discussion about hermeneutics in geography of late has come not from hermeneutic traditions themselves but from attempts to incorporate the interpretive, rhetorical and narrative structure of social action into a realist perspective (Sayer 1989; Walker 1989). Even in these careful works, however, the task of constructing a rigorous methodology appropriate to interpretation, case studies, locality studies, social critique, narrative structure, or social action, has not been attempted.[2]

How, then, do we read maps, especially those in which problems of interpretation are compounded by distortion, error and lies? More generally, how do geographers read texts? Like the map the landscape is a particularly good example of a 'text' which has been presumed to require a straightforward literal reading, but which actually poses great problems of interpretation and requires a rigorous hermeneutic analysis. Indeed, as texts the map and the landscape present innumerable problems of determining authorship, establishing a syntax and structure by which to read (and knowing what not to read), and distinguishing and relating the various levels of determination that historically constituted any particular map or landscape. In the case of the propaganda map an additional problem is always apparent (although it may not always be absent from the landscape, the film or the novel). From its conception the propaganda map aims to be a convincing distortion. The theory of interpretation which deals with problematic texts – their origin, correct ascription, intended meaning, received meaning, etc. – is hermeneutics. The propaganda map is thus the archetypical problematical text requiring hermeneutic interpretation, and provides a potentially good starting point for elaborating the methods of interpretation; philology, hermeneutics and criticism. The previous sections of this chapter have attempted to begin the process of hermeneutic analysis. This section will abstract the lessons and principles of that analysis.

Philology places strong demands on the act of interpreting texts (be they poems, landscapes, maps or social actions). Is the text the one it is claimed to be: is the ascribed authorship correct, did the text fulfil the role it is claimed to have filled, etc.? Is it a coherent whole? What does the text say about its own world? What does the text now mean? What is the relationship between the meaning of a text and the intention of the author in creating it? Given that some of these texts may have been authored by people who are no longer known or who were anonymous at the time of production, that they may have originated in worlds about which we now know little or nothing, and that only fragments (like the Anaximander fragment of a single map) may now be extant, are we really able to retrieve the *mens auctoris* (the author's intention)? And if we are, then in what sense can we claim to have access to the *mens auctoris*? Does the work constitute something independent of and different from that intention? And, if we cut our interpretation loose from the author's intention, how do we avoid the danger of arbitrary and forced interpretations?

Strict concern for the *mens auctoris* would, of course, place us in an untenable position as social scientists. The antiquarian may be able to bracket his/her present world and become immersed 'fully' in the world of the other, of the past, of the author. This option is never open to the social scientist (nor, practically, even to the antiquarian). We ask questions always from the standpoint of the present, and in every case we carry out a retrieval of the author, his/her intentions, and the work, in order to make them meaningful in our present worlds (be they conceptually, temporally or geographically removed). We cannot derive the text's 'own meaning' in this way, but must recognize that in any interpretation the work has an autonomy of its own beyond the intensions of its author(s). In this process the author retains a claim on the surface details of the work: the site, the literal and symbolic content intended, the date of production, the materials and techniques used in production. But with regard to the meaning of the word *as* a work, the author's claim is more tenuous. The author's intention is not fully determinative of the meaning of the text. It is for this reason that all interpretations of works must also be *destructive retrievals*, i.e. they must situate the work in a context meaningful to the reader. Consequently, where the interpreter is able to make explicit what the author left implicit or did not recognize or understand, he or she may know more about a work than its author. The philological concern thus corresponds to a lower hermeneutic, which is concerned to establish a critical edition of a text, to verify that the text is the text it is claimed to be, that it has not been falsified, that it is (or is not) a coherent whole and is not (or is) a pastiche of several authors, that it is authentic and that it is complete. Higher hermeneutics takes as its task the proper understanding of the meaning of a text, how it related to its own world (and subsequent worlds in which it has had an existence) and how it is to be related to our present world.

These claims become clearer when we recognize that symbols may be of one of two types: univocal or equivocal. Univocal signs, like symbols in symbolic logic or mathematics, have one designated meaning. Cartographic symbols (church, castle, urban area) have often been seen to be univocal symbols in this manner. Their correspondence, however, is not of the same kind as that of those symbols of logic or mathematics where the equivalence is complete. In the case of the symbol 'church' the equivalence, as with mathematical notation, is purely formal. Except in this trivial formal equivalence the symbol is actually equivocal: the age, style, denomination and size of the church remain open to interpretation from the context of the whole map and the text within which it is embedded. Equivocal symbols may have several layers of meaning and are the true focus of hermeneutics (Ricoeur 1971).

For any rigorous interpretation several conditions must pertain: the integrity of the meaning of the text must be preserved in such a way that meaning is derived from, not projected into, the text (canon 1).[3] Consequently the interpreter has the responsibility to bring him/herself into a harmonious relationship or proper attunement with the text (canon 2). This is neither a call for a slavelike adherence to the text or the tradition to which it belongs, nor to argue that any reading can fully preserve the meaning of the text or bring the reader into full harmony with it. It is an argument that critique must be rooted in the claims, conventions and forms of the text. The interpreter must give an optimal reading of the text and of the meaning the text must have had for those for whom it was written *and* show what the text means for us today in the context of modern views, interests and prejudices (canon 3).

Of course, the difficulty for any interpretation is precisely where to begin and where to end in reconciling these three canons. This is the importance of the canon of the hermeneutic circle (canon 4): the whole must be understood from its parts, and all the parts must be understood from the whole. All texts have a certain anticipation of their parts from the whole, leading to a wider understanding of the whole and the parts. Part–whole relationships permeate all readings of texts at all levels of analysis and critique, and specifically include: (*a*) the relationship of the text to its own intrinsic parts; (*b*) the relationship of language–text–lauguage; (*c*) the relationship of cultural context–text–cultural context; (*d*) the relationship of author and his/her world-text as part of this world. Finally, since clear and unambiguous texts do not need interpretation (by definition), all other texts must be complemented with suitable assumptions in order for the interpreter to make things explicit which the author (and/or subsequent readings) left implicit (canon 5). In this way we can say that the interpreter understands the text better than the author.

All rigorous readings of texts must use (implicitly or explicitly) these hermeneutic canons. They are standard methods of textual exegesis, general philology, linguistics and literary criticism. With Dilthey and Heidegger they became standard methods for the study of human works, with Ricoeur they function as the system of interpretation used to reach the meaning in myths and symbols, and with Derrida and his followers hermeneutic methods (in amended form) become central to the deconstruction of texts to overcome the metaphysics of presence. It is the adoption of a theory of interpretation predicated on the notion that symbols are the mirror of the world that has caused contemporary theories of maps to be one-sidedly concerned with the *mens auctoris* and the objectivity of the map, marginalized the importance of propaganda maps (and distortion generally), and led to the current practical gulf between social theory and the theory of maps.

CONCLUSION: WRITING AND THEORY

It would be wrong to interpret the claims made here as in any way an attempt to establish propaganda maps as a separate category of text. The point is, first, that these maps are currently not part of the critical canon of the theory of maps. Second, that an effective critique of the distortive and ideological nature of propaganda maps must be based on a wider conception of what constitutes propaganda. That is, the ideological and propagandistic elements of contemporary 'scientific' maps must also be assessed at those points where the cartographer shares the ideology of his/her age, where accepted practices are founded on particular ideologies, and where unchallenged interests influence the form and content of the theory and practice of mapping. Examples of the ways in which cartography shares and reproduces the values of the age are numerous and some are well known: the continued public use of the Mercator and adapted Mercator projections, the ideological fixation on 'north at the top' maps, and the ridicule which greets, for example, 'the Australian's view of the world'. Other examples are less well recognized, and perhaps more significant: the focus in Western cartography on private property boundaries and lines and the failure to give equal form to public rights of access and usufruct; the focus of mapping convention on natural and built physical objects, rather than developing universal conventions dealing with symbol, affect or movement.

Interpreting the meaning of maps also requires that other issues be considered. Two symbolic systems are involved: graphical images and writing systems. Not only does the image exist in a reductive relationship to the world, but graphical systems always also exist as an interplay between images and linguistic texts and contexts, thus creating a multiplicity of cross-cutting structures. As a writing system, maps also contain within them the spoken and written in a relationship which is never exactly correspondent (i.e maps 'play' in two registers). For cartographers this complexity of meaning has generally been seen as a technical problem to be either dissolved by careful adherence to established mapping practices or explained in terms of the creative ability of the map maker. In seeing this complexity of meaning in terms resolvable by technical means or reducible to individual skills traditional theories of maps have failed to address the issue as one which is inherent to the form and always in need of unpacking. Thus Thomas (1949: 76) admits that:

> Regardless of the objectivity with which they were prepared, a great percentage of existing historical maps present some information which some individuals honestly consider 'propaganda'. For certain areas, the historical issues are so complicated and the record goes back so far that an unbiased map presentation becomes almost impossible.

For Thomas (and other commentators on propaganda maps) the issue is one of primary (intentional) propaganda, where opinion dominates in the use of technique for persuasive purposes, and secondary (inadvertent) propaganda, where opinion creeps in as a result of the great difficulty experienced in dealing with historical data or complex issues. The first is distortive, the second is attributed to the fallibility of technical objectivity.

We have seen how all attempts to ground interpretation in terms of some 'original' meaning (the author's intention) or correspondence between the image and the real world (representationalism) are characteristic of a foundationalism which is increasingly being challenged because no such firm ground is available to us. But in rejecting the foundationalist project must we also accept that we are without a vantage point? Are all interpretations of texts valid? Are they all equally valid? The critique of modernity and the turn to postmodern social theory juggles uneasily with this question, and for many the solution does lie in a groundless acceptance of all positions. For others, acceptance of the critique of representationalism and foundationalism does not leave us without a vantage point. In both views claims to truth are constructed rhetorically within particular contexts. But in the second view some basis for judgement must be established. That is, we need to think through the discursive nature of texts and to clarify the canons of interpretation necessary to the critique of an embedded text. The hermeneutic canons presented in this chapter are only the necessary first canons of any rigorous interpretation. Cartographers and geographers must clarify their own interpretive frameworks if we are to understand and control our production and use of maps, and we might begin by questioning our current understanding of texts, writing and 'reading'.

First, it would be foolish in the present age to attempt to construct a theory of maps without reference to the new technologies of production and transmittal. Modern cartography is currently bound to the many new forms of computer-assisted information storage and retrieval, graphic display and manipulation, and image production. We need a theory which incorporates some way of dealing with the possibilities and impact of the 'current upheavals in the forms of communication, the new structures emerging in all the formal practices, and also in the domains of the archive and the treatment of information' (Derrida 1981a: 13); specifically we need a means of dealing with the various forms of the 'communication media', especially, but not confined to, the electronic media. Map drawing is now also map construction, transformation and electronic production. That this has changed both the character of the map and the nature of the map makers' craft seems undeniable. We must understand what it means for a theory of maps.

Second, we need a theory of writing and reading which moves beyond naive empiricism and representationalism, and which does not trivialize the tracings and inscriptions of culture or literalize them, but which integrates and reforms the modes of discourse appropriate to reading. We need a broader conception of the nature of writing:

> And thus we say 'writing' for all that gives rise to an inscription in general, whether it is literal or not and even if what it distributes in space is alien to the order of the voice: cinematography, choreography, of course, but also pictorial, musical, sculptural 'writing'. One might also speak of athletic writing, and with even greater certainty about military or political writing in view of the techniques that govern those domains today. All this to describe not only the system of notation secondarily connected with these activities but the essence and the content of these activities themselves.
>
> (Ulmer 1985: 9).

In traditional theory inscription has no *intrinsic* value. In this tradition the inscription records a discourse that has already taken place or an idea already formed (either in speech, in the mind of the author, or in action) and is therefore testable in terms of accuracy or truth – as an accurate portrayal or resemblance of what is 'engraved on the psychic surface' (Derrida 1981b: 184–8). Here inscription is a representation or copy of the *mens auctoris*. This instrumentalist and technicist view of writing valorizes essence over the written form, and is to be overcome by focusing on the exteriority of the written work. In this view writing is not merely the external expression of speech, and writing and speech are not merely the external expression of thought (Ulmer 1985: 7).

Third, we need to address the problem of *technicity*, which, in its production of the modern sciences, has fragmented and hierarchically organized eye, hand, ear and touch. In the age of techno-political writing – the age of electronic media – while the modern sciences fragment our analytical frameworks, the techno-politicians unify their manipulation of the media, such that any single message or medium cannot provide the basis of criticism. Each is linked in a total onslaught – one which demands a different reading; one in which text and context take on very different meanings. The consolidation of power and the increasing ability to manipulate the media demand urgent response from a comprehensive theory of interpretation and criticism. Specifically we need a grammar which transcends, and opens up, the various specialized 'grammars' of the sciences – speaking, writing and mapping.

Fourth, propaganda maps are not merely one more medium or form to be interpreted, but are in many ways an archetypical form of the age of *technicity*. They are exemplars of the manipulation of symbols and writing. They cannot be read without a broader grammatology than the one provided by 'map-reading

Plate 12.20 'Arise!' Cartoon from the *Chicago Daily News*, 1943 (Shoemaker)

skills', or indeed any 'readings' currently extant within either geography or cartography. The existing theories of reading and the corresponding regional grammars (reading texts, reading landscapes, reading maps) remain flawed in their partiality. The building of a critical human geography has occurred quickly and gone far in recent years, and through it we see now the necessity of penetrating and reconstructing the discourses which foster the distorted and uneven social relations from which we seek to ARISE (Plate 12.20). Perhaps a critical social theory of maps, propaganda and distortion will be there to assist.

NOTES

1 For contemporary examples see Taylor (1985) on the tactics employed by the multinational tobacco corporations to fight anti-smoking legislation; Cockburn (1988) on the ways in which the Reagan and Thatcher governments have prepared the ground for the dismantling of the social welfare system at home and for military adventurism abroad; and Hall (1982) on the way in which the Thatcher government and the media sold a more Conservative image of 'the British nation' to the British and won consent for war in the Malvinas (consent obtained by capturing the discursive field around specific definitions of 'a people' and 'the nation').

2 The hermeneutical canon and its exegetical, interpretative and critical aspects have passed from the fields of classical hermeneutics in biblical exegesis, the study of legal documents and philosophy into literary theory (Eagleton 1983), history (White 1973), linguistics (Barthes 1987h) and social theory (Ricoeur 1971; Thompson 1981). But no such working out of a canon for geographical interpretation and critique has been developed yet.

3 It might be objected that the reader always brings with her all the baggage of the present world view. This may be true, but it does not constitute an argument against the prescriptive nature of the canon. The rule of the road that 'when driving an automobile drivers must remain alert at all times' is, of course, not invalidated by the fact that some drivers rarely remain alert or that all drivers are occasionally distracted. The canon remains prescriptive and important in the same way.

ACKNOWLEDGEMENTS

I am grateful to Trevor Barnes, Hubertus Bloemer, Ruth Rowles, Henry Ruf and Michael Watts for their invaluable comments on earlier drafts of this paper.

13

DECONSTRUCTING THE MAP

J.B. Harley

I can never romanticize language again
never deny its power for disguise
for mystification
but the same could be said for music
or any form created
painted ceilings beaten gold
worm-worn Pietàs reorganizing victimization
frescoes translating violence
into patterns so powerful and pure
we continually fail to ask are they true for us.
(Adrienne Rich, from 'The Images', in *A Wild Patience has Taken me this Far: Poems, 1978–81*)

This chapter has three points of departure. The first is a belief in the value of maps – both old and new – for the practice of human and historical geography. The second is a frustration with many of the academic cartographers of today, who operate in a tunnel created by their own technologies without reference to the social world. The third is a desire to intervene. Epistemic time has largely stood still in cartography. One effect of accelerated technological change – as manifest in digital cartography and geographical information systems – has been to strengthen its positivist assumptions and it has bred a new arrogance in geography about its supposed value as a mode of access to reality. If it is true that 'New fictions of factual representation' (White 1978) are daily being foisted upon us, then the case for inserting a social dimension into modern cartography is especially strong. Maps are too important to be left to cartographers alone.

One consequence of these developments is that it is increasingly difficult to conduct a dialogue between the practice of cartography as it is now constituted and the interpretive strategies that are embraced in this volume. While we can talk freely of our different discourses, cartographers seem unable to situate their maps within the discourse of cartography. Along with other aspects of critical theory, I suspect that the notion of discourse, as defined by Foucault and others, would be alien and bizarre to most cartographers. The orthodox words in their vocabulary are 'impartial', 'objective', 'scientific' and 'true', but we seldom catch the resonances of class, gender, race, ideology, power and knowledge, or myth and ritual. A conceptual vacuum lies between cartography and human geography.

What I am seeking to do in this and related historical essays (Harley 1988a, 1988b, 1989, forthcoming) is to show how cartography also belongs to the terrain of the social world in which it is produced. Maps are ineluctably a cultural system. Cartography has never been an autonomous and hermetic mode of knowledge, nor is it ever above the politics of knowledge. My key metaphor is that we should begin to deconstruct the map by challenging its assumed autonomy as a mode of representation. From the viewpoint of human geography, maps are perhaps better understood – and used – not so much as discrete or 'unique' images but as accents within a wider theory of representation. Rather than accepting what cartographers tell us maps are supposed to be, the thrust of my deconstruction is to subvert the apparent naturalness and innocence of the world shown in maps both past and present. In the sense of Barthes, I believe we should demystify both cartographic process and the resulting images we call maps.

The notion of deconstruction (Derrida 1976; Norris 1982; Eagleton 1983; Norris 1987) is, of course, a password for the postmodern enterprise. Deconstructionist strategies can now be found not only in philosophy but also in localized disciplines, especially in literature, and in other subjects such as architecture, planning and, more recently, geography (Gregory 1987; Dear 1988; Knox 1988). I shall specifically use a deconstructionist tactic to break the assumed link between reality and representation which has dominated cartographic thinking, has led it in the pathway of 'normal science' since the Enlightenment, and has also provided a ready-made and taken-for-granted epistemology for cultural studies of maps as geographical or historical records. The objective is to suggest that an alternative epistemology, rooted in social theory rather than in scientific positivism, is more appropriate if we are to reassimilate cartography within human geography. It will be shown that even 'scientific' maps are a product not only of 'the rules of the order of geometry and reason' but also of the 'norms and values of the order of social . . . tradition' (Marin 1988: 173). Our task is to search for the social forces that have structured cartography and to locate the presence of power – and its effects – in all map knowledge.

The ideas that follow owe most to writings by Foucault and Derrida. My approach is deliberately eclectic, although the theoretical positions of these two authors are sometimes incompatible. Foucault anchors texts in socio-political realities and constructs systems for organizing knowledge of the kind that Derrida loves to dismantle (Skinner 1985). But even so, by combining different ideas on a new terrain, it may be possible to devise a sketch of social theory with which we can begin to interrogate the hidden agendas of cartography. Such a scheme offers no 'solution' to a humanistic interpretation of the cartographic record, nor a precise method or set of techniques, but as a broad strategy it may help to locate some of the fundamental forces that have driven map making in both European and non-European societies. From Foucault's writings the key revelation has been the omnipresence of power in all knowledge, even though that power is invisible or implied, including the particular knowledge encoded in maps and atlases. While I do not accept Derrida's view that nothing lies outside the text – it clearly defeats the idea of a social history of cartography – his notion of the rhetoricity of all texts offers a provocative challenge. It demands a search for metaphor and rhetoric in maps where previously scholars had found only measurement and topography. Its central question is reminiscent of Korzybski's much older dictim, 'The map is not the territory' (1948), but deconstruction goes further to bring the issue of how the map represents place into much sharper focus.

Deconstruction urges us to read between the lines of the map – 'in the margins of the text' – and through its tropes to discover the silences and contradictions that challenge the apparent honesty of the image. We begin to learn that cartographic facts are facts only within a specific cultural perspective. We start to understand how maps, like art, far from being 'a transparent opening to the world', are but 'a particular human way ... of looking at the world' (Blocker 1979: 43).

In pursuing this strategy I shall develop three threads of argument. First, I will examine the discourse of cartography in the light of some of Foucault's ideas about the play of rules within discursive formations. Second, drawing on one of Derrida's central positions, I will examine the textuality of maps and, in particular, their rhetorical dimension. Third, returning to Foucault, I will consider how maps work in society as a form of power-knowledge.

THE RULES OF CARTOGRAPHY

One of Foucault's primary units of analysis is the discourse. A discourse has been defined as 'a system of possibility for knowledge' (Philip 1985: 69). Foucault's method was to ask, it has been said:

> what rules permit certain statements to be made; what rules order these statements; what rules permit us to identify some statements as true and others as false; what rules allow the construction of a map, model or classificatory system ... what rules are revealed when an object of discourse is modified or transformed ... Whenever sets of rules of these kinds can be identified, we are dealing with a discursive formation or discourse.
>
> (Philip 1985: 69)

The key question for us then becomes 'What type of rules govern the development of cartography?'

Cartography I define as a body of theoretical and practical knowledge that map makers employ to construct maps as a distinct mode of visual representation. The question is, of course, both culturally and historically specific: the rules of cartography vary in different societies. Here I refer particularly to two distinctive sets of rules that underlie and have dominated the history of Western cartography, first in Europe and later in its overseas colonial territories, since the seventeenth century. One set may be defined as governing the technical production of maps and are made explicit in cartographic treatises and writings. The history of these technical rules has been extensively written about in the history of cartography (Crone 1978), though not in terms of their social implications nor in Foucault's sense of discourse. The other set of rules relates to the cultural production of maps. These rules must be understood in a broader historical context than either scientific procedure or technique. They are, moreover, rules that are usually ignored by cartographers, so that they form a hidden aspect of their discourse.

The first set of rules for mapping can thus be defined in terms of a positivistic epistemology. From at least the seventeenth century onward there was an epistemic break in activities such as cartography and architecture (Peréz-Gomez 1983), and European map makers increasingly promoted what we would describe today as a standard scientific model of knowledge and cognition. The object of mapping is to produce a 'correct' relational model of the terrain. Its assumptions are that the objects in the world to be mapped are real and objective, and that they enjoy an existence independent of the cartographer; that their reality can be expressed in mathematical terms; that systematic observation and measurement offer the only route to cartographic truth; and that this truth can be independently verified. The procedures of both surveying and map construction came to share strategies similar to those in science in general. Cartography also documents a history of more precise instrumentation and measurement; increasingly complex classifications of its knowledge and a proliferation of signs for its representation; and, especially from the nineteenth century onward, the growth of institutions and a 'professional' literature designed to monitor the application and propagation of the rules (Wolter 1975). Moreover, although cartographers have continued to pay lip service to the 'art and science' of map making (Meynen 1973; Wallis and Robinson 1987), art has been progressively edged off the map. It has often been accorded a cosmetic rather than a central role in cartographic communication (Morris 1982). Even philosophers of visual communication – such as Arnheim, Eco, Gombrich and Goodman (Goodman 1968; Gombrich 1975; Eco 1976; Arnheim 1986) – have tended to categorize maps as a type of congruent diagram – as analogues, models or 'equivalents' creating a similitude of reality – and, in essence, different from art or painting. A 'scientific' cartography (so it was believed) would be untainted by social factors.

The acceptance of the map as 'a mirror of nature' (to employ Richard Rorty's phrase, 1979) also results in a number of other characteristics of cartographic discourse even where they are not made explicit. Most striking is the belief in linear progress: that, by the application of science, ever more precise representations of reality can be produced. The methods of cartography should be able to deliver a 'true, probable, progressive, or highly confirmed knowledge' (Laudan 1977: 2). This mimetic bondage has led to a tendency not only to look down on the maps of the past (with a dismissive scientific chauvinism) but also to regard the maps of other, non-Western or early cultures (where the rules of map making were different) as inferior to European maps (Harley 1987–8). Similarly, the primary effect of the scientific rules was to create a 'standard' – a successful version of 'normal science' – that enabled cartographers to build a wall around their citadel of the 'true' map. Its central bastions were measurement and standardization, and beyond there was a 'not cartography' land where lurked an army of inaccurate, heretical, subjective, valuative and ideologically distorted images. Cartographers developed a 'sense of the other' in relation to non-conforming maps. Even maps such as those produced by journalists, where different rules and modes of expressiveness might be appropriate, are evaluated by many cartographers according to standards of 'objectivity', 'accuracy' and 'truthfulness'. In this respect the underlying attitude of many cartographers is revealed in a recent book of essays, *Cartographie dans les médias* (Gauthier 1988). One of its reviewers has noted how many authors attempt to exorcise from:

> the realm of cartography any graphic representation that is not a simple planimetric image, and to then classify all other maps as 'decorative graphics masquerading as maps' where the 'bending of cartographic rules' has taken place ... most journalistic maps are flawed because they are inaccurate, misleading or biased.
>
> (Andrews 1989: 219)

Or in Britain, we are told, there was set up a 'Media Map Watch' in 1984. 'Several hundred interested members [of cartographic and geographical societies] submitted several thousand maps and diagrams for analysis that revealed [according to the rules] numerous common deficiencies, errors, and inaccuracies along with misleading standards' (Balchin 1988). These cartographic vigilantes were defending the 'ethic of accuracy' with some ideological fervour. The language of exclusion is that of a string of 'natural' opposites: 'true and false', 'objective and subjective', 'literal and symbolic', and so on. The best maps are those with an 'authoritative image of self-evident factuality' (Lupton 1986: 53).

In cases where the scientific rules are invisible in the map we can still trace their play in attempting to normalize the discourse. The cartographer's 'black box' has to be defended and its social origins suppressed. The hysteria among leading cartographers at the popularity of the Peters projection (Peters 1983; Loxton 1985a, 1985b; Robinson 1985; Porter and Voxland 1986; Snyder 1988; Vujakovic 1989), or the recent expressions of piety among Western European and North American map makers following the Russians' admission that they had falsified their topographic maps to confuse the enemy give us a glimpse of how the game is played according to these rules. What are we to make of the 3 September 1988 newspaper headlines such as 'Russians caught mapping' (*Ottawa Citizen*), 'Soviets admit map paranoia' (*Wisconsin State Journal*) or (in the *New York Times*) 'In West, map makers hail "Truth"' and '"The rascals finally realized the truth and were able to tell it," a geographer at the Defense

Department said'? The implication is that Western maps are value-free. According to the spokesman, our maps are not ideological documents, and the condemnation of Russian falsification is as much an echo of Cold War rhetoric as it is a credible cartographic criticism.

This timely example also serves to introduce my second contention, that the scientific rules of mapping are, in any case, influenced by a quite different set of rules, those governing the cultural production of the map. To discover these rules, we have to read between the lines of technical procedures or of the map's topographical content. They are related to values, such as those of ethnicity, politics, religion or social class, and they are also embedded in the map-producing society at large, and in its other forms of representation. Cartographic discourse operates a double silence towards this aspect of the possibilities of map knowledge. In the map itself social structures are often disguised beneath an abstract, instrumental space, or incarcerated in the co-ordinates of computer mapping. And in the technical literature of cartography social values are also ignored, notwithstanding the fact that they may be as important as surveying, compilation, or design in producing the statements that cartography makes about the world and its landscapes. Such an interplay of social and technical rules is a universal feature of cartographic knowledge. In maps it produces the 'order' of its features and the 'hierarchies of its practices' (Foucault 1973: xx). In Foucault's sense the rules may enable us to define an *episteme* and to trace an archaeology of that knowledge through time (Foucault 1973: xxii).

Two examples of how such rules are manifest in maps will illustrate their force in structuring cartographic representation. The first is the well known adherence to the 'rule of ethnocentricity' in the construction of world maps. This has led many historical and modern societies to place their own territory at the centre of their cosmography or world map. While it may be dangerous to assume universality, and there are exceptions, such a 'rule' is as evident in cosmic diagrams of pre-Columbian North American Indians as it is in the maps of ancient Babylonia, Greece or China, or in the medieval maps of the Islamic world or Christian Europe (Harley and Woodward 1987, forthcoming). Yet what is also significant in applying Foucault's critique of knowledge to cartography is that the history of the ethnocentric rule does not march in step with the 'scientific' history of map making. Thus while the scientific Renaissance in Europe gave modern cartography co-ordinate systems, Euclid, scale maps and accurate measurement, the epistemic break was only partial. Philosophers continued to believe that mathematical thought constituted a privileged channel of communication between human minds and the divine mind, while the new geometrical cartography served to reinforce a new myth of Europe's ideological centrality (Peters 1983). Throughout the history of cartography ideological 'Holy Lands' are frequently centred on maps. Such centricity, a kind of 'subliminal geometry' (Harley 1988a), adds geopolitical force and meaning to representation. Though the link between actual mapping, as the principal source of our world vision, and *mentalité* still has to be thoroughly explored, it is also likely that such maps have in turn helped to codify, to legitimate and to promote the Eurocentric world views that have been prevalent in so much of modern world history (Henrikson 1987; Saarinen 1988).

A second example is how the 'rules of the social order' appear to insert themselves into the smaller codes and spaces of cartographic transcription. The history of European cartography since the seventeenth century provides many examples of this tendency. Pick a printed or manuscript map from the drawer at random and what stands out is the unfailing way its text is as much a commentary on the social structure of a particular nation or place as it is on its topography. The map maker is often as busy recording the contours of feudalism, the shape of a religious hierarchy, or the steps in the tiers of social class (Harley 1988a, forthcoming), as the topography of the physical and human landscape.

Why maps can be so convincing in this respect is that the rules of society and the rules of measurement are mutually reinforcing in the same image. My example is historical but it could apply equally to many modern maps. Writing of the map of Paris, surveyed in 1652 by Jacques Gomboust, the king's engineer, Louis Marin (1988: 173) points to 'this sly strategy of simulation – dissimulation':

> The knowledge and science of representation, to demonstrate the truth that its subject declares plainly, flow none the less in a social and political hierarchy. The proofs of its 'theoretical' truth had to be given, they are the recognisable signs; but the economy of these signs in their disposition on the cartographic plane no longer obeys the rules of the order of geometry and reason but, rather, the norms and values of the order of social and religious tradition. Only the churches and important mansions benefit from natural signs and from the visible rapport they maintain with what they represent. Town houses and private homes, precisely because they are private and not public, will have the right only to the general and common representation of an arbitrary and institutional sign, the poorest, the most elementary (but maybe, by virtue of this, principal) of geometric elements: the point identically reproduced in bulk.

Once again, much like 'the rule of ethnocentrism', this hierarchicalization of space is not a conscious act of cartographic representation. Rather it is taken for granted in a society that the place of the king is more important than the place of a lesser baron, that a castle is more important than a peasant's house, that the town of an archbishop is more important than that of a minor prelate, or that the estate of a landed gentleman is more worthy of emphasis than that of a plain farmer. Cartography deploys its vocabulary accordingly so that it embodies a systematic social inequality. The map discriminates: the distinctions of class and power are engineered, reified and legitimated by means of cartographic signs. The rule seems to be 'the more powerful, the more prominent'. To those who have strength in the world shall be added strength in the map. Using all the tricks of the cartographic trade – size of symbol, thickness of line, height of lettering, hatching and shading, the addition of colour – we can trace this reinforcing tendency in innumerable European maps. We can begin to see how maps, like art, become a mechanism 'for defining social relationships, sustaining social rules, and strengthening social values' (Geertz 1983: 99).

In the case of both these examples of rules, the point I am making is that the rules operate both within and beyond the orderly structures of classification and measurement. They go beyond the stated purposes of cartography. Much of the power of the map, as a representation of social geography, is that it operates behind a mask of a seemingly neutral science. It hides and denies its social dimensions at the same time as it legitimates. Yet whichever way we look at it the rules of society will surface. They have ensured that maps are at least as much an image of the social order as a measurement of the phenomenal world of objects.

DECONSTRUCTION AND THE CARTOGRAPHIC TEXT

To move inward from the question of cartographic rules – the social context within which map knowledge is fashioned – we have to turn to the cartographic text itself. The word 'text' is deliberately chosen in an awareness that it offers no simple set of techniques for reading maps. Some cartographers have resisted the metaphor of map as language (Robinson and Petchenik 1976), but it is now generally accepted that the model of text can have a much wider application than to literary texts alone. To non-book texts such as landscapes, musical compositions and architectural structures we can confidently add the graphic texts we call maps (McKenzie 1986; Duncan and Duncan 1988). It is true that literally they have no grammar and lack the temporal sequence of a syntax but 'what constitutes a text is not the presence of linguistic elements but the act of construction' so that maps, as 'constructions employing a conventional sign system' (McKenzie 1986: 35), become texts. With Barthes we could say they 'presuppose a signifying consciousness' that it is our business to uncover (Barthes 1986b: 110). 'Text' is certainly a better metaphor for maps than the mirror of nature. Maps are a cultural text: not one code but a collection of codes, few of which are unique to cartography.

By accepting the textuality of maps we are able to embrace a number of different interpretive possibilities. Instead of just the transparency of clarity we can discover the pregnancy of the opaque. To fact we can add myth, and instead of innocence we may expect duplicity. Rather than working with a formal science of communication, or with a cognitive psychology saying nothing about the social world, or perhaps, worse still, with a sequence of loosely related technical processes, our concern is redirected to a history and anthropology of the image. We learn to recognize the narrative qualities of cartographic representation (Wood 1987) as well as its claim to provide a synchronous picture of the world. All this, moreover, is likely to lead to a rejection of the neutrality of maps as we come to define their intentions rather than the literal face of representation, and as we begin to accept the social consequences of cartographic practices.

What, therefore, does deconstruction have to offer? Most important is that it demands a closer and deeper reading of the cartographic text. Deconstruction does not solve the problem: but it aims at as many meanings as possible, even if some aspects of those meanings are undecidable (Hoy 1985). It may be regarded as a search for alternative meanings. 'To deconstruct,' it is argued:

> is to reinscribe and resituate meanings, events and objects within broader movements and structures; it is, so to speak, to reverse the imposing tapestry in order to expose in all its unglamorously dishevelled tangle the threads constituting the well heeled image it presents to the world.
> (Eagleton 1986: 80)

The published map also has a 'well heeled image' and our reading has to go beyond the assessment of geometric accuracy, beyond the fixing of location, and beyond the recognition of topographical patterns and geographies. Such interpretation begins from the premise that the map text may contain 'unperceived contradictions or duplicitous tensions' (Hoy 1985: 540) that undermine the surface layer of standard objectivity. Maps are slippery customers. In the words of W.J.T. Mitchell (1986: 8), writing of languages and images in general, we may need to regard them more as 'enigmas, problems to be explained, prison-houses which lock the understanding away from the world'. We should regard them 'as the sort of sign that presents a deceptive appearance of naturalness and transparence concealing an opaque, distorting, arbitrary mechanism of representation' (Mitchell 1986: 8). Throughout the history of modern cartography in the West, for example, there have been numerous instances of where maps have been falsified, of where they have been censored or kept secret, or of where they have surreptitiously contradicted the rules of their proclaimed scientific status (Harley 1988b).

Taking practices such as these, map deconstruction would focus on aspects of maps that many interpreters have glossed over. Writing of 'Derrida's most typical deconstructive moves', Christopher Norris (1987: 19) notes that:

> deconstruction is the vigilant seeking-out of those 'aporias', blind spots or moments of self-contradiction where a text involuntarily betrays the tension between rhetoric and logic, between what it manifestly *means to say* and what it is nonetheless *constrained to mean*. To 'deconstruct' a piece of writing is therefore to operate a kind of strategic reversal, seizing on precisely those unregarded details (casual metaphors, footnotes, incidental turns of argument) which are always, and necessarily, passed over by interpreters of a more orthodox persuasion. For it is here, in the margins of the text – the 'margins', that is, as defined by a powerful normative consensus – that deconstruction discovers those same unsettling forces at work.

A good example of how we could deconstruct an early map – by beginning with what have hitherto been regarded as its 'casual metaphors' and 'footnotes' – is provided by recent studies reinterpreting the status of decorative art on the European maps of the seventeenth and eighteenth centuries. Rather than being inconsequential marginalia, the emblems in cartouches and decorative title pages can be regarded as basic to the way such maps convey their cultural meaning, and they help to demolish the claim of cartography to produce an impartial graphic science (Harley 1984; Clarke 1988; Harley 1988a, forthcoming).

Such a strategy need not be limited to historic 'decorative' maps. A recent essay by Wood and Fels on the official state highway map of North Carolina (Wood and Fels 1986) indicates a much wider applicability for a deconstructive strategy by beginning in the 'margins' of the contemporary map. Wood and Fels also treat the map as a text and, drawing on the ideas of Roland Barthes (1986b) of myth as a semiological system, develop a forceful social critique of cartography which though structuralist in its approach is deconstructionist in its outcome. They begin, deliberately, with the margins of the map, or rather with the subject matter that is printed on its verso:

> One side is taken up by an inventory of North Carolina points of interest – illustrated with photos of, among other things, a scimitar horned oryx (resident in the state zoo), a Cherokee woman making beaded jewelry, a ski lift, a sand dune (but no cities) – a ferry schedule, a message of welcome from the then governor, and a motorist's prayer ('Our heavenly Father, we ask this day a particular blessing as we take the wheel of our car . . .'). On the other side, North Carolina, hemmed in by the margins of pale yellow South Carolinas and Virginias, Georgias and Tennessees, and washed by a pale blue Atlantic, is represented as a meshwork of red, black, blue, green and yellow lines on a white background, thickened at the intersections by roundels of black or blotches of pink. . . . To the left of . . . [the] title is a sketch of the fluttering state flag. To the right is a sketch of a cardinal (state bird) on a branch of flowering dogwood (state flower) surmounting a buzzing honey bee arrested in midflight (state insect).
>
> (Wood and Fels 1986: 54)

What is the meaning of these emblems? Like the contents of the Chinese encyclopedia in Borges's short story, referred to by Foucault in the preface to *The Order of Things* (1973), they are bizarre until we've cracked the code. Are they merely a pleasant ornament for the traveller or can they inform us about the social production of such state highway maps? A deconstructionist might claim that such meanings are undecidable, but it is also clear that the state highway map of North Carolina is making other dialogical assertions behind its mask of innocence and transparence. I am suggesting not that these elements hinder the traveller in getting from point A to point B, but that there is a second text within the map. No map is devoid of an intertextual dimension that involves an essentially plural and diffuse play of meanings across the boundaries of individual maps. It has been remarked that 'in the concept of "text" the boundaries which enclosed the "work" are dissolved; the text opens continually into other texts, the space of intertextuality' (Burgin 1988: 51). To read the map we have to dismantle first of all the frame that the cartographer has placed around it.

So it is with our state highway map. The discovery of intertextuality enables us to scan the image as more than a neutral picture of a road network (Bakhtin 1981). It shares the intertextuality of all discourse. The 'users' of the map are not only the ordinary motorists but also the State of North Carolina that has appropriated its publication (distributed in millions of copies) as a promotional device. The map has become an instrument of state policy and an instrument of sovereignty (Wood and Fels 1986). At the same time it is more than an affirmation of North Carolina's dominion over its territory. It also constructs a mythic geography, a landscape full of 'points of interest', with incantations of loyalty to state emblems and to the values of a Christian piety. The hierarchy of towns and the visually dominating highways that connect them have become the legitimate natural order of the world. The map finally insists 'that roads really *are* what North Carolina's all about' (Wood and Fels 1986: 60). The map idolizes our love affair with the automobile. The myth is believable.

Cartographers' stock response to this deconstructionist argument might well be to cry 'foul.' The argument would run like this: 'Well, after all, it's a state highway map. It's designed to be at once popular and useful. We expect it to exaggerate the road network and to show points of interest to motorists.' They might also invoke the favourite cartographic axiom that it is a derived rather than a basic map. Deconstruction, however, by making explicit the play of forces such as intention, myth, silence and power in maps, will tend to dissolve such an opposition for interpretive purposes except in the very practical sense that one map is often copied or derived from another. Or again it may be claimed that it is not a 'scientific map'. The appeal to the ultimate scientific map is always the cartographers' last line of defence when seeking to deny the social relations that permeate their technology.

It is at this point that Derrida's strategy can help us to extend such an interpretation to all maps, scientific or non-scientific, basic or derived. Just as in the deconstruction of philosophy Derrida was able to show 'how the supposedly literal level is intensively metaphorical' (Hoy 1985: 44) so too we can show how cartographic 'fact' is also symbol. In 'plain' scientific maps, science itself becomes the metaphor. Such maps contain a dimension of 'symbolic realism' which is no less a statement of political authority and control than a coat of arms or a portrait of a queen placed at the head of an earlier decorative map. The metaphor has changed. The map has attempted to purge itself of ambiguity and alternative possibility (Eagleton 1983). Accuracy and austerity of design are the new talismans of authority, culminating in our own age with computer mapping.

We can trace this process very clearly in the history of Enlightenment mapping in Europe. The topography as shown in maps, increasingly detailed and planimetrically accurate, has become a metaphor for a utilitarian philosophy and its will to power. Cartography inscribes this cultural model upon the paper and we can examine it in many scales and types of maps. Precision of instrument and technique merely serves to reinforce the image, with its encrustration of myth, as a selective perspective on the world. Thus maps of local estates under the *ancien regime*, though derived from instrumental survey, were a metaphor for a social structure based on landed property. County and regional maps, though founded on scientific triangulation, were an articulation of local values and rights. Maps of the European states, though constructed along arcs of the meridian, served still as a symbolic shorthand for a complex of nationalist ideas. And maps, though increasingly drawn on mathematically defined projections, nevertheless gave a spiralling twist to the manifest destiny of European overseas conquest and colonization (Harley 1988a). In each of these examples we can trace the contours of metaphor in a scientific map. This in turn enhances our understanding of how the text works as an instrument operating on social reality.

In deconstructionist theory the play of rhetoric is closely linked with that of metaphor. In concluding this section of the essay I will argue that notwithstanding 'scientific' cartography's efforts to convert culture into nature, and to 'naturalize' social reality (Eagleton 1983), it has remained an inherently rhetorical discourse. Another of the lessons of Derrida's criticism of philosophy is 'that modes of rhetorical analysis, hitherto applied mainly to literary texts, are in fact indispensable for reading *any* kind of discourse' (Norris 1982: 19). There is nothing revolutionary in the idea that cartography is an art of persuasive communication. It is now commonplace to write about the rhetoric of the human sciences in the classical sense of the word 'rhetoric'. Even cartographers – as well as their critics – are beginning to allude to the notion of a rhetorical cartography, but what is still lacking – with a few notable exceptions – is a rhetorical close reading of maps (Wood and Fels 1986; Goffart 1988).

The issue in contention is not whether some maps are rhetorical, or whether other maps are partly rhetorical, but the extent to which rhetoric is a universal aspect of all cartographic texts. Thus for some cartographers the notion of 'rhetoric' would remain a pejorative term. It would be an 'empty rhetoric' which was unsubstantiated in the scientific content of a map. 'Rhetoric' would be used to refer to the 'excesses' of propaganda mapping or advertising cartography or an attempt would be made to confine it to an 'artistic' or aesthetic element in maps as opposed to their scientific core. My position is to accept that rhetoric is part of the way all texts work and that all maps are rhetorical texts. Again we ought to dismantle the arbitrary dualism between 'propaganda' and 'true', and between modes of 'artistic' and 'scientific' representation as they are found in maps. All maps strive to frame their message in the context of an audience. All maps state an argument about the world, and they are propositional in nature. All maps employ the common devices of rhetoric such as invocations of authority. This is *especially* so in topographical maps, with their reliability diagrams, multiple referencing grids and magnetic error diagrams, or in thematic maps, with their 'trappings of F-scaled symbols and psychometrically divided greys' (Wood and Fels 1986: 99). Maps constantly appeal to their potential readership through the use of colour, decoration, typography, dedications or written justifications of their method (Marin 1988). Rhetoric may be concealed but it is always present, for there is no description without performance.

The steps in making a map – selection, omission, simplification, classification, the creation of hierarchies, and 'symbolization' – are all inherently rhetorical. In their intentions as much as in their applications they signify subjective human purposes rather than reciprocating the workings of some 'fundamental law of cartographic generalisation' (Robinson *et al.* 1984: 127). Indeed, the freedom of rhetorical manoeuvre in cartography is considerable: as with any other text, the map maker merely omits those features of the world that lie outside the purpose of the immediate discourse. There have been no limits to the varieties of maps that have been developed historically in response to different purposes of argument, aiming at different rhetorical goals, and embodying different assumptions about what is sound cartographic practice. The style of maps was not fixed in the past, nor is it today. It has been said that 'The rhetorical code appropriates to its map the style most advantageous to the myth it intends to propagate' (Wood and Fels 1986: 71). Instead of thinking in terms of rhetorical versus non-rhetorical maps it may be more helpful to think in terms of a theory of cartographic rhetoric which accommodated this fundamental aspect of representation in all types of cartographic text. Thus I am not concerned to privilege rhetoric over science, but to dissolve the illusory distinction between the two in reading the social purposes as well as the content of maps.

MAPS AND THE EXERCISE OF POWER

For the final stage in the argument I return to Foucault. In doing so I am mindful of Foucault's criticism of Derrida, that he attempted 'to restrict interpretation to a purely syntactic and textual level' (Hoy 1985: 60), a world where political realities no longer exist. Foucault, on the other hand, sought to uncover 'the social practices that the text itself both reflects and employs' and to 'reconstruct the technical and material framework in which it arose' (Hoy 1985: 60). Though deconstruction is useful in helping to change the epistemological climate, and in encouraging a rhetorical reading of cartography, my final concern is with its social and political dimensions, and with understanding how the map works in society as a form of power-knowledge. This closes the circle to a context-dependent form of cartographic understanding.

We have already seen how it is possible to view cartography as a discourse, a system which provides a set of rules for the representation of knowledge embodied in the images we define as maps and atlases. It is not difficult to find for maps – especially those produced and manipulated by the state – a niche in the 'power/knowledge matrix of the modern order' (Philip 1985: 76). Especially where maps are ordered by government (or are derived from such maps) it can be seen how they extend and reinforce the legal statutes, territorial imperatives and values stemming from the exercise of political power. Yet to understand how power works through cartographic discourse and the effects of that power in society further dissection is needed. A simple model of domination and subversion is inadequate, and I propose to draw a distinction between *external* and *internal* power in cartography. This ultimately derives from Foucault's ideas about power-knowledge, but this particular formulation is owed to Joseph Rouse's book *Knowledge and Power* (1987), where a theory of the internal power of science is in turn based on his reading of Foucault.

The most familiar sense of power in cartography is that of power *external* to maps and mapping. This serves to link maps with the centres of political power. Power is exerted *on* cartography. Behind most cartographers there is a patron; in innumerable instances the makers of cartographic texts were responding to external needs. Power is also exercised *with* cartography. Monarchs, ministers, state institutions, the Church, have all initiated programmes of mapping for their own ends. In modern Western society maps quickly became crucial to the maintenance of state power – to its boundaries, to its commerce, to its internal administration, to control of the population, and to its military strength. Mapping soon became the business of the state: cartography is early nationalized. The state guards its knowledge carefully: maps have been universally censored, kept secret and falsified. In all these cases maps are linked with what Foucault called the exercise of 'juridical power' (Foucault 1980: 88; Rouse 1987). The map becomes a 'juridical territory': it facilitates surveillance and control. A recent textbook on geographical information systems even boasts that it is 'a science of surveillance'. Maps are still used to control our lives in innumerable ways. A mapless society, though we may take the map for granted, would not be politically unimaginable. All this is power *with* the help of maps. It is an external power, often centralized and exercised bureaucratically, imposed from above, and manifest in particular acts or phases of deliberate policy.

I come now to the important distinction. What is also central to the effects of maps in society is what may be defined as the power *internal* to cartography. The focus of enquiry therefore shifts from the place of cartography in a juridical system of power to the political effects of what cartographers do when they make maps. Cartographers manufacture power: they create a spatial panopticon. It is a power embedded in the map text. We can talk about the power of the map just as we already talk about the power of the word or about the book as a force for change. In this sense, just as with other artefacts and technologies, maps do have politics (Winner 1980). It is a power that intersects and is embedded in knowledge. It is universal. Foucault (1978: 73) writes of:

> The omnipresence of power: not because it has the privilege of consolidating everything under its invincible unity, but because it is produced from one moment to the next, at every point, or rather in every relation from one point to another. Power is everywhere; not because it embraces everything, but because it comes from everywhere.

Power comes from the map and it traverses the way maps are made. Maps are a technology of power, and the key to this internal power is cartographic process. By this I mean the way maps are compiled and the categories of information selected; the way they are generalized, a set of rules for the abstraction of the landscape; the way the elements in the landscape are formed into hierarchies; and the way various rhetorical styles that also reproduce power are employed to represent the landscape. To catalogue the world is to appropriate it (Barthes 1980; Wood and Fels 1986), so that all these technical processes represent acts of control over its image which extend beyond the professed uses of cartography. The world is disciplined. The world is normalized. We are prisoners in its spatial matrix. For cartography as much as other forms of knowledge, 'All social action flows through boundaries determined by classification schemes' (Darnton 1984: 192–3). An analogy is what happens to data in the cartographer's workshop and what happens to people in the disciplinary institutions – prisons, schools, armies, factories – described by Foucault (Rouse 1987): in both cases a process of normalization occurs. Or similarly, just as in factories we standardize our manufactured goods, so in our cartographic workshops we standardize our images of the world. Just as in the laboratory we create formulaic understandings of the processes of the physical world, so too, in the map, nature is reduced to a graphic formula. Indeed, cartographers like to promote this metaphor of what they do. One textbook claims:

> Geography thrives on cartographic generalization. The map is to the geographer what the microscope is to the microbiologist, for the ability to shrink the earth and generalize about it . . . The microbiologist must choose a suitable objective lens, and the geographer must select a map scale appropriate to both the phenomenon in question and the 'regional laboratory' in which the geographer is studying it.
>
> (Monmonier and Schnell 1988: 15)

The power of the map maker was generally exercised not over individuals but over the knowledge of the world made available to people in general. Yet this is not consciously done, and it transcends the simple categories of 'intended' and 'unintended' altogether. I am not suggesting that power is deliberately or centrally exercised. It is a local knowledge which at the same time is universal. It usually passes unnoticed. The map is a silent arbiter of power.

What have been the effects of this 'logic of the map' upon human consciousness, if I may adapt Marshall McLuhan's phrase ('logic of print', McLuhan 1962)? Like him I believe we have to consider for maps the effects of abstraction, uniformity, repeatability and visuality in shaping mental structures, and in imparting a sense of the places of the world. It is the disjunction between those senses of place and many alternative visions of what the world is, or what it might be, that has raised questions about the effect of cartography in society. Thus Theodore Roszak (1972: 410) writes:

> The cartographers are talking about their maps and not landscapes. That is why what they say frequently becomes so paradoxical when translated into ordinary language. When they forget the difference between map and landscape – and when they permit or persuade us to forget that difference – all sorts of liabilities ensue.

One of these 'liabilities' is that maps, by articulating the world in mass-produced and stereotyped images, express an embedded social vision. Consider, for example, the fact that ordinary road atlases are among the best-selling paperback books in the United States (McNally 1987) and then try to gauge how this may have affected ordinary Americans' perception of their country. What sort of image of America do these atlases promote? On the one hand, there is a patina of gross simplicity. Once off the interstate highways the landscape dissolves into a generic world of bare essentials that invites no exploration. It avoids the irregularities of lived experience. Context is stripped away and place is no longer important. On the other hand, the maps reveal the ambivalence of all stereotypes. Their silences are also inscribed on the page: where, on the page, is the variety of nature, where is the history of the landscape, and where is the space–time of human experience in such anonymized maps? (Roszak 1972; Szegö 1987.) This criticism is reminiscent of Roland Barthes's essay on 'The *Blue Guide*', where he writes of the *Guide* as 'reducing geography to the description of an uninhabited world of monuments' (we substitute 'roads') (Barthes 1986c: 74–7). Or in a similar vein Roszak (1972: 408) writes:

> We forfeit the whole value of a map if we forget that it is *not* the landscape itself or anything remotely like an exhaustive depiction of it. If we do forget, we grow rigid as a robot obeying a computer program; we lose the intelligent plasticity and intuitive judgement that every wayfarer must preserve. We may then know the map in fine detail, but our knowledge will be purely academic, inexperienced, shallow.

Certainly, in all such maps, the geometries of the *Lebenswelt* are nowhere to be seen.

The question has now become: do such empty images have their consequences in the way we think about the world? Because all the world is designed to look the same, is it easier to act upon it without realizing the social effects? It is in the posing of such questions that the strategies of Derrida and Foucault appear to clash. For Derrida, if meaning is undecidable so is the force of the map as a discourse of symbolic action. In ending, I prefer to align myself with Foucault in seeing all knowledge – and hence cartography – as thoroughly enmeshed with the larger battles which constitute our world (Rabinow 1984). Maps are not external to these struggles to alter power relations. The history of map use shows how often maps embody specific forms of power and authority. Since the Renaissance they have changed the way in which power was exercised. In colonial North America, for example, it was easy for Europeans to draw lines across the territories of Indian nations without sensing the reality of their political identity (Harley 1988c). The map allowed them to say, 'This is mine; these are the boundaries' (Boelhower 1984: 47, quoting from Wahl 1980: 41). Similarly, in innumerable wars since the sixteenth century it has been equally easy for the generals to fight battles with coloured pins and dividers rather than sensing the slaughter of the battlefield (Muehrcke 1986). Or again, in our own society, it is still easy for bureaucrats, developers and 'planners' to operate on the bodies of unique places without measuring the social dislocations of 'progress'. While the map is never the reality, in such ways it helps to create a different reality. Once embedded in the published text the lines on the map acquire an authority that may be hard to dislodge. Maps are authoritarian images. Without our being aware of it maps can reinforce and legitimate the *status quo*. Sometimes agents of change, they can equally become conservative documents. But in either case the map is never neutral. Where it seems to be neutral it is the sly 'rhetoric of neutrality' (Kinross 1985) that is trying to persuade us.

CONCLUSION

The interpretive act of deconstructing the map can serve three functions in the way we view maps in geographical culture. First, it allows us to challenge the epistemological myth (created by cartographers) of the cumulative progress of an objective science always producing better delineations of reality. Second, deconstructionist argument allows us to redefine the social importance of maps. Rather than invalidating their study, it enhances it by adding different nuances to our understanding of the power of cartographic representation as a way of building order into our world. If we can accept intertextuality then we can start to read our maps for alternative and sometimes competing discourses and they can overflow into a new range of problems. Third, a deconstructive turn of mind may allow geographical cartography to take a fuller place in the interdisciplinary study of text and knowledge. Intellectual strategies such as those of discourse in the Foucauldian sense, the Derridian notion of metaphor and rhetoric as inherent to scientific discourse, and the pervading concept of power-knowledge are shared by many subjects. As ways of looking at maps they are equally enriching. They are neither inimical to hermeneutic enquiry nor anti-historical in their thrust. The possibility of discovering new meanings in maps is enlarged. Postmodernism offers a challenge to read maps in ways that could reciprocally enrich the reading of other texts.

ACKNOWLEDGEMENTS

These arguments were presented in earlier versions at 'The Power of Places' conference, Northwestern University, in January 1989 and as a 'Brown Bag' lecture in the Department of Geography, University of Wisconsin at Milwaukee, in March 1989. I am grateful for the suggestions received on those occasions and for other helpful comments received from Derek Gregory and Cordell Yee. I am also indebted to Howard Deller of the American Geographical Society Collection for a number of references and to Ellen Hanlon for editorial help in reworking this version of the paper for press. It is reprinted with modifications from the article of the same title in *Cartographica* 26 (1989): 1–20.

14

AFTERWORD

James S. Duncan and Trevor J. Barnes

The essays in this volume all participate in the on-going project of deconstructing traditional forms of representation in late twentieth-century human geography. This project, largely pursued under the banner of postmodernism, is one of the most vital intellectual currents not only in the field of geography but in the humanities and social sciences more generally. While deconstruction in the strict sense of the term is employed in a minority of the essays in this volume, its implied critical and self-critical spirit pervades them all.

We conclude, in the spirit of deconstructionism, by turning a critical eye on the postmodernist project itself, of which this volume is a part. To do so we critically situate this work within a broader intellectual context, noting the contradictions and absences embedded within it.

While these essays have not for the most part been self-consciously post-modern, the fact that they argue against modernist forms of representation place them within that rather nebulous (one is even tempted to say byzantine), discursive field of postmodernism. To take issue with modernist representation in late twentieth-century human geography is to adopt a radical stance on a number of grounds. First, it is to set oneself in opposition to the goals of the vast majority of researchers within the field. Second, and more important, it represents a critique of the traditional categories used to characterize the recent history of human geography. Thus the various classificatory schemes developed since World War II, for example, one dividing the history of the discipline into empiricist, positivist and social theoretical schools, are, under the postmodernist view, all collapsed into a unitary modernist discursive field; one whose central characteristic is providing true accounts of phenomena in the world.

In this sense R. J. Johnston's *Geography and Geographers since 1945* (1987) could have been subtitled 'Modernism and its Discontents'. For its narrative structure is compellingly organized around the contest for intellectual supremacy among the triad of approaches: empiricist, positivist and modernist social theory. Although from Johnston's perspective the social theorists win the day, and the others are relegated to the dustbin of geographical history, it is only modernism that really triumphed, because it is the only game in town.

All these contending positions seek to provide the one best method to explain geographical phenomena. None has seriously entertained the notion that there is no best method (unless it is to make the rather innocuous point that some methods are better suited to studying certain types of phenomena). Harder to accept is the idea that there is no great truth about the world to be revealed. For the positivists the world is in principle (if not in practice, alas) reducible to scientifically formulated regularities revealed through an almost cosmo-magical belief in the efficacy of numbers. While, to the various brands of modernist social theorists, it is their own particular theory that holds the key to explaining social and geographical reality. In fact there have been endless acrimonious debates among those different brands over which theory is more mimetic, or more capable of affecting social change. For empiricists, mimesis is not a theoretical problem, since they tend to ignore theoretical problems altogether. Rather the truth is mined with the sledgehammer of empiricism as one toils in the field or the archive.

To argue that the three contending positions that have dominated academic discourse in geography since the war can be viewed as a single modernist discursive field is not to deny that there are interesting differences among them. Rather, it is to argue that their differences are relatively minor when compared to their shared assumptions. It is perhaps too soon to know whether postmodernism represents an important taxonomic moment within our field comparable to the shift that took place as cosmo-magical explanation was slowly displaced by rational empiricist explanation. To date, the exploration of postmodernist epistemologies have been largely theoretical. Intellectual route maps are plotted, political positions noted and literature surveys are undertaken. But few of the would-be postmodernist explorers have yet arisen from their armchairs. Furthermore, as we will discuss shortly, postmodernist epistemology is already being recuperated (gutted of its most radical and, we believe, central claims) by geographers and others who wish to incorporate certain of its important insights into a reconstructed modernist geographic project.

Dear (1988: 272) rather overstates the centrality of postmodern work within the academy when he claims that if geographers do not embrace the perspective they will be left out of the mainstream. It is highly questionable, however, whether postmodernism represents the mainstream even in literary criticism, its strongest bastion. In anthropology, political science, sociology, art history, philosophy and economics, as in geography, it is distinctly marginal. The ex-centricity of postmodernism in geography and elsewhere has less to do with its intellectual pedigree than with the sociology of the academy (although it is not clear whether postmodernism's chameleon-like characteristics will help or hinder its survival, and possible spread to new environs).

It is to the definitions of the postmodern that we now wish to turn. It is common to distinguish between postmodernity and postmodernism, although some scholars elide the two. Dear (1986) makes this distinction when he argues that postmodernism can be conceived of as an epoch, or as a method, or, as in the case of architecture, a style.[1] Following Hutcheon (1989: 23) we will refer to the method as *postmodernism* and to the epoch as *postmodernity.* This distinction is important because it is possible (in fact not uncommon) to hold one position and reject the other.

Gregory (1989a, 1989b) and Dear (1986, 1988) adopt the view that postmodernism is a critique of the Enlightenment project, a revolt against the rationality of modernism. Postmodernism in this sense is an attack on modernist epistemology and its search for truth. It is anti-foundational, rejecting the totalizing ambitions of modernist social science. All metanarratives are to be deconstructed. Every position, including that of the postmodernist researcher her or himself, is to be continually undermined. This is an extraordinarily radical epistemological position, for if taken seriously and not recuperated to modernist social science, as is so often the temptation, there can be no *reconstruction*, only *deconstruction.* Postmodernism as an epistemology is *inescapably* both radically relativist and nihilistic.

Geographers such as Dear (1986, 1988) and Gregory (1989a, 1989b) are aware of these radical implications of postmodernist epistemology and hold back from wholeheartedly embracing it. Dear (1988: 267–72) in particular rejects anti-foundationalism and deconstruction, in our opinion central tenets of postmodernism, because he wishes to 'privilege provisionally' a particular view of geography. He realizes that postmodernism will not allow him to achieve his academic goals, which are to reconstitute human geography in a particular mould and to put forward a progressive political programme. Not only will postmodernism disallow even the 'provisional privileging' that Dear wishes, but it will attack progressive political positions as ruthlessly as it will the most regressive ones (to use thoroughly modernist terms). Dear therefore recuperates postmodernism to a reconstructed modernism and gives only qualified support to the notion of a postmodern epoch.

Gregory (1989a, 1989b) also seems ambivalent about postmodernism, although perhaps less so than Dear. Both he and Dear follow Jameson in highlighting attention to spatial difference as a component of postmodernism. This emphasis upon spatial difference is clearly one of the things that have drawn them to postmodernism. Gregory, however, also appears to be uneasy with the radical relativism of postmodernism as he seeks a reconstituted human geography that is politically progressive. Soja (1989a, 1989b) employs the language of postmodernism but also recuperates it into the strong metanarrative of Western Marxism. Like Dear, his project is to *reconstruct* knowledge with geography at its centre. A number of other geographers, such as David Harvey (1989), appear to be more comfortable with the concept of postmodernity than they do with the idea of adopting postmodernist writing practices and philosophical positions.

Dear (1986: 374), one of the first geographers to espouse the concept of postmodernity, very largely follows Jameson (1984, 1985) in his definition of postmodernity. Postmodernity is 'A periodizing concept whose function is to correlate the emergence of new formal features in culture with the emergence of a new type of social life and a new social order' (Jameson 1985: 113). Dear (1986: 374) also draws upon Jameson's conception of the renewed significance of space in postmodernity. 'According to Jameson (1984, pages 83–84), old systems of organization and perception have been destroyed and replaced by a postmodern *hyperspace*. Space and time have been stretched to accommodate the multinational global space of advanced capitalism.'

Soja (1989a, 1989b), like Jameson, conflates postmodernism with postmodernity. His perspective emerges most clearly in his critique of what he terms modernist Marxism (1989a: 342). Modernist Marxism in geography, he claims, is based upon 'conditions and dilemmas which may no longer exist' and furthermore is committed to 'a grand narrative unable to contend with the possibility that *today it is space and geography, much more than time and history, that hide things from us*' (emphasis added). Here the confusion between postmodernism and postmodernity emerges most clearly. Because we have entered a new epoch we need a new epistemology. We have entered the age of spatial difference when the discipline of geography shall reign supreme, for it (not history) shall reveal the truth that is hidden from us. This kind of millenarianism may have appeal to geographers (in the United States, at least, where geography is seen to occupy a marginal and largely inconsequential position within the academy) but it bears little relation to postmodernism more generally.

The conflation of postmodernity as an epoch and postmodernism as an epistemology ultimately rests on the assumption that 'old systems of organization and perception have been destroyed' (Jameson 1984: 83), and that such a destruction produces a 'crisis of representation'. But – a crisis for whom? We would argue with Moore (1988: 178) that it is highly debatable whether the crisis of representation extends beyond a tiny coterie of 'hyper'-educated intellectuals. Again, this is not to deny that no crisis exists; rather it is to situate it within its sociological context (the academy). If this is the case, and we believe it to be so, then the claims for a postmodern era are overdrawn in that they erroneously generalize from an intellectual elite to the population at large.

This privileging of elite perceptions and subsequent 'colonization' of non-elite perceptions should not surprise us, given the intellectual background of the leading lights in postmodernist theory. Philosophers such as Lyotard, literary critics such as Jameson, sociologists such as Baudrillard, in spite of their disagreements share in common an authoritarian urge to speak for others. The postmodern call for difference does not for these thinkers imply drawing upon different voices in their work. As Wayne Hudson (1989: 158) puts it:

> Much of the current discussion is an in-house dialogue between Anglo-American and Franco-German philosophical and literary cultures, a dialogue in which having read Derrida or Heidegger or Wittgenstein or Adorno is much more important than understanding the world economy or having experience of famine in Africa.

To listen to the voices of the intellectuals' other, the people outside the academy who are being spoken for, would imply a different research strategy; doing field-work, leaving the academy and talking to those who are spoken for, plumbing *their* consciousness to see if they share the intellectuals' crises.

Umberto Eco (1986: 127), admittedly no paragon of multivocality himself, raises an intriguing (and disturbing) question with regard to an alleged crisis of postmodernity:

> I still don't understand what the hell it [the crisis of postmodernity] means. I cross the street on a red light, the cop blows his whistle and fines *me* (not someone else). How can this happen if the idea of the subject is in a state of crisis, along with the sign and the reciprocal?

To the extent that the essays in this volume fit under the general rubric of post-modernism (and once again we hasten to add that few have openly embraced this label), they appear to be more comfortable with postmodernism as an epistemology than do Harvey, Dear, Soja or Gregory. A number explicitly accept anti-foundationalism, the radical relativism that troubles the aforementioned geographers. Jameson's amalgam of Western Marxism and postmodernism is less influential in these essays, and the deconstruction of Derrida, the poststructuralism of Barthes and the work of postmodern ethnographers such as Clifford (1988), Clifford and Marcus (1986) and Tyler (1987a and b) have correspondingly greater influence.

Another difference between the papers in this volume and the work of other postmodern geographers is that the concept of space is not as important here. Rather the focus is upon the interpretation of landscapes and places, and the ways in which they can be represented. The issue of postmodernist representation of difference is the focus. However, it is precisely here, on the issue of the representation of difference, that we in this volume as well as other geographers interested in postmodernism run into yet another dilemma.

The problem lies in our conceptualization of difference. Difference for geographers is other spaces, other places, other regions or other landscapes. To embrace difference is in Gregory's (1989a: 71) words to embrace 'areal differentiation'. It is instructive, however, to compare the postmodernist geographers' conception of difference to that of postmodernist ethnographers. For the latter, difference is other people and, to be more precise, other peoples' voices. Their plea is for an end to monovocality and authorial authority. Let us hear direct, they say, from those for whom we as academics have for so long spoken.

Multivocality (taking others' voices seriously) is not something that can be easily accomplished in geography, for we have no tradition of ethnography. Unless we begin to focus more of our energies on developing our techniques for listening to others our calls for difference are highly suspect, for we will continue systematically to silence difference. Those outside the academy have no voice in our work other than the one we choose to give them. As Hutcheon (1989: 4) points out, to speak for another is not a politically innocent act. We have appropriated their voice – colonized their perspective.

There is an irony here, however, for the pursuit of difference and the call for deconstruction sit uneasily together. Put slightly differently, the goal of the post-modern academic is to question and attack in order to undermine the cultural order. But non-academics tend to have other goals. The dilemma is: how do we allow for other voices while also acting as critics whose job it is to defamiliarize and denaturalize commonsense perspectives? Barthes (1977b: 47) points to this problem when he identifies the tendency for people to fail to question the cultural order, to treat it as 'doxa' or, in his colourful phrase, 'the voice of nature'. The resulting irony – to welcome new voices but also to undermine them – is, as Hutcheon (1989: 1) points out, unavoidable in postmodernism. In fact, for some (Rorty 1989), an ironic stance is our only hope of achieving solidarity, which, of course, is an irony in itself.

NOTES

1 Within geography Ley (1987) has been the foremost proponent of architectural definitions of the postmodern.

BIBLIOGRAPHY

Abbot, Carl (1983) *Portland: Planning, Politics and Growth in a Twentieth Century City*, Lincoln, Neb.: University of Nebraska Press.

Abbott, W. (1946) 'The world's greatest splash', *Collier's*, 8 June.

Adamson, J.E., ed. (1985) *Niagara: Two Centuries of Changing Attitudes, 1697–1901*, Washington: Corcoran Gallery of Art.

Adey, Lionel (1978) *C.S. Lewis' 'Great War' with Owen Barfield*, Victoria, B.C.: University of Victoria.

Ager, J. (1977) 'Maps and propaganda', *Society of University Cartographers Bulletin* 11: 1–15.

Alexander, Christopher (1964) *Notes on the Synthesis of Form*, Cambridge, Mass.: Harvard University Press.

Allen, J.L.V. (1974) 'Permission is "up in air" for Falls high wire walk', *Buffalo Currier-Express*, 13 August: 12.

Alonso, W. (1964) *Location and Land Use*, Cambridge, Mass.: Harvard University Press.

Althusser, L. (1971) 'Ideology and ideological state apparatuses (notes towards an investigation)', in *Lenin and Philosophy and other Essays*, trans. Ben Brewster, London: New Left Books.

Althusser, L., and Balibar, E. (1970) *Reading 'Capital'*, London: New Left Books.

Alves, M.H.M. (1985) *State and Opposition in Military Brazil*, Austin, Texas: University of Texas Press.

Amin, A., and Robins, K. (1990) 'The re-emergence of regional economies? The mythical geography of flexible accumulation', *Environment and Planning D, Society and Space* 8: 7–34.

Andrews, C. Bryn (1935–6) *The Torrington Diaries*, London: Eyre & Spottiswoode.

Andrews, Sona Karentz (1989) review of *Cartography in the Media* in *The American Cartographer* 16: 219–20.

Anon. (1879) 'Niagara', in H.W. Longfellow (ed.) *Poems of Places*, Boston, Mass.: James R. Osgood.

Appeton, T.G. (1872) 'Niagara', in *Faded Leaves*, Boston, Mass.: Roberts, 27–30.

Arib, M.A., and Hesse, M.B. (1986) *The Construction of Reality*, Cambridge: Cambridge University Press.

Aries, P. (1974) *Western Attitudes toward Death: from the Middle Ages to the Present*, trans., P.M. Ranum, Baltimore: Johns Hopkins University Press.

Arnheim, Rudolf (1986) 'The perception of maps', in R. Arnheim (ed.) *New Essays on the Psychology of Art*, Berkeley: University of California Press, 194–202.

Arnold, Matthew (1969) *Culture and Anarchy*, Cambridge: Cambridge University Press (original, 1869).

Asplund, Johan (1987) *Det sociala livets elementära former*, Göteborg: Korpen.

—— (1989) *Rivaler och syndabockar*, Göteborg: Korpen.

Atencio, J. (1965) *Que es geopolitica*, Buenos Aires: Editorial Pleamar.

Atkin, R.H. (1974) *Mathematical Structure in Human Affairs*, New York: Crane Russak.

—— (1981) *Multidimensional Man*, Harmondsworth: Penguin.

Austin, H. (1990) *The Independent*, 29 November.

Backheuser, E. (1926) 'Das politische conglomerat Brasilien', *Zeitschrift für Geopolitik* 3: 625–30.

Baker, T.J. (1975) 'Woman defied Falls in '01: leap for Lib?' *Buffalo Currier-Express*, 29 December: 12.

Bakhtin, M.M. (1981) *The Dialogic Imagination: Four Essays*, ed. Michael Holquist, trans. Caryl Emerson and Michael Holquist, Austin, Texas: University of Texas Press.

Balchin, W.G.V. (1988) 'The Media Map Watch in the United Kingdom', in M. Gautier (ed.) *Cartographie dans les médias*, Quebec: Presses de l'Université du Québec, 33–48.

Ball, Patricia (1971) *The Science of Aspects*, London: Athlone Press.

Ballester, H.P., García, J.L., Gazcón, C.M., and Rattenbach, A.B. (1983) 'Poder militar y poder civil', *Revista Cruz de Sur* 3: 9–21.

Bardis, P.D. (1981) *History of Thanatology*, Washington, D.C.: University Press of America.

Barfield, Owen, (1928) *Poetic Diction: a Study in Meaning*, London: Faber & Gwyer.

Barham, W. (1845) 'Recollections of a trip to the Falls of Niagara in September, 1845', in W. Barham (ed.) *Description of Niagara*, Gravesend: The Compiler, 19.

Barnes, Trevor J., and Curry, Michael R. (1983) 'Towards a contextualist approach to geographical knowledge', *Transactions, Institute of British Geographers*, n.s. 8: 476–82.

Barrell, J. (1983) *English Literature in History, 1730–1780: Pastoral and Politics*, London: Hutchinson.

Barthes, R. (1977a) 'La Lumière du sud-ouest', in *Incidents*, Paris: Editions du Seuil, 13–20.

—— (1977b) *Roland Barthes by Roland Barthes*, trans. R. Howard, New York: Hill & Wang.

—— 1980. 'The plates of the *Encyclopedia*', in *New Critical Essays* New York: Hill and Wang.

—— (1982, original 1977) 'Inaugural lecture, Collège de France', in *A Barthes Reader*, ed. S. Sontag, New York: Hill & Wang, 457–78.

—— (1984, original 1957) 'The Eiffel Tower', in *The Eiffel Tower and other Mythologies*, trans. R. Howard, New York: Hill & Wang, 3–18.

—— (1986a original 1957) 'Myth today', in *Mythologies*, trans. A. Lavers, New York: Hill & Wang, 109–59.

—— (1986b, original 1957) *Mythologies*, trans. A. Lavers, New York: Hill & Wang.

—— (1986c, original 1957) 'The *Blue Guide*', in *Mythologies*, trans. A Lavers, New York: Hill & Wang, 74–7.

—— (1986d, original 1970–71) 'Semiology and the urban', in *The City and the Sign: an Introduction to Urban Semiotics*, eds. M. Gottdiener and A. Lagopoulos, New York: Columbia University Press, 87–98.

—— (1986e, original 1972) 'Pleasure/writing/reading', in *The Grain of the Voice: Interviews, 1962–80*, trans. L. Coverdale, New York: Hill & Wang, 88–108.

—— (1986f, original 1970) 'L'Exprès talks with Roland Barthes', in *The Grain of the Voice: Interviews*, 1962–80, trans. L. Coverdale, New York: Hill & Wang, 88–108.

—— (1986g, original 1971) 'Digressions', in *The Grain of the Voice: Interviews, 1962–80*, trans. L. Coverdale, New York: Hill & Wang, 113–27.

—— (1986h, original 1975) 'Twenty key words for Roland Barthes', in *The Grain of the Voice: Interviews, 1962–80*, trans. L. Coverdale, New York: Hill & Wang, 205–32.

—— (1986i, original 1977) 'Of what use is an intellectual?', in *The Grain of the Voice: Interviews, 1962–80*, trans. L. Coverdale, New York: Hill & Wang, 258–81.
—— (1986j, original 1972) 'The fatality of culture, the limits of the counterculture', in *The Grain of the Voice: Interviews, 1962–80*, trans. L. Coverdale, New York: Hill & Wang, 150–6.
—— (1986k, original 1970) 'On *S/Z* and *Empire of Signs*', in *The Grain of the Voice: Interviews, 1962–80*, trans. L. Coverdale, New York: Hill & Wang, 68–87.
—— (1986l, original 1974) 'What would become of a society that ceased to reflect upon itself?', in *The Grain of the Voice: Interviews, 1962–80*, trans. L. Coverdale, New York: Hill & Wang, 96–7.
—— (1986m, original 1980) 'The crisis of desire', in *The Grain of the Voice: Interviews, 1962–80*, trans. L. Coverdale, New York: Hill & Wang, 361–5.
—— (1987a, original 1953) *Writing Degree Zero*, trans. A. Lavers and C. Smith, New York: Hill & Wang.
—— (1987b, original 1970) *S/Z: an Essay*, trans. R. Miller, New York: Hill & Wang.
—— (1987c, original 1970) *Empire of Signs*, trans. R. Howard, New York: Hill & Wang.
—— (1987d, original 1975) *The Pleasure of the Text*, trans. R. Miller. New York: Hill & Wang.
—— (1987e, original 1971). 'From work to text', in *Image, Music, Text*, trans. S. Heath. New York: Hill & Wang, 155–64.
—— (1987f) 'Writers, intellectuals, teachers', in *Image, Music, Text*, trans. S. Heath, New York: Hill & Wang, 190–215.
—— (1987g) 'The death of the author', in *Image, Music, Text*, trans. S. Heath, New York: Hill & Wang, 142–8.
—— (1987h) *Image, Music, Text*, New York: Hill & Wang.
Bataille, Georges (1985) *Visions of Excess: Selected Writings, 1927–39*, Minneapolis: University of Minnesota Press.
Bateson, Gregory (1972) *Steps to an Ecology of Mind*, New York: Chandler.
Baudet, H. (1988) *Paradise on Earth: Some Thoughts on European Images of non-European Man*, Middletown, Conn.: Wesleyan University Press.
Baudrillard, J. (1983) *Simulations*, New York: Semiotext(e).
—— (1987) *The Evil Demon of Images*, Sydney: Power Institute Publications.
—— (1988) *America*, trans. Chris Turner, New York: Verso.
Becker, E. (1973) *The Denial of Death*, New York: Free Press.
Beckett, Samuel (1979) *The Beckett Trilogy: Molloy, Malone Dies, The Unnamable*, London: Picador. (French originals, 1950–2).
Bender, G., Coleman, J., and Sklar, R. eds. (1985) *African Crisis Areas and U.S. Foreign Policy*, Berkeley: University of California Press.
Berdoulay, Vincent (1976) 'French possibilism as a form of neo-Kantian philosophy', *Proceedings of the Association of American Geographers* 8: 176–9.
—— (1978) 'The Vidal–Durkheim debate', in David Ley and Marwyn Samuels (eds.) *Humanistic Geography: Prospects and Problems*, Chicago: Maaroufa Press, 77–90.
—— (1982) 'La métaphore organiciste: contribution à l'étude du langage des géographes', *Annales de Géographie* 91: 573–86.
Beresford, M. (1980) 'The face of Leeds, 1780–1914', in D. Frazer (ed.) *A History of Modern Leeds*, Manchester: Manchester University Press, 72–112.
Berg, M. (1983) 'Political economy and the principles of manufacture', 1700–1800', in M. Berg, P. Hudson and M. Sonnenscher (eds) *Manufacture in Town and Country before the Factory*, Cambridge: Cambridge University Press, 33–58.
Berger, J. (1965) *The Success and Failure of Picasso*, New York: Pantheon.
Berlin, I. (1957) *The Hedgehog and the Fox*, New York: Mentor.
—— (1980) 'The divorce between the sciences and the humanities', in H. Hardy (ed.)

Against the Current: Essays in the History of Ideas, New York: Viking, 80–110.

Bermingham, Ann (1986) *Landscape and Ideology; the English Rustic Tradition, 1740–1860*, Berkeley: University of California Press.

Berry, Wendell (1970) 'The regional motive', in *A Continuous Harmony: Essays Cultural and Agricultural*, New York: Harcourt Brace Jovanovitch, 63–70.

Bevan, Edwyn (1938) *Symbolism and Belief*, London: Allen & Unwin.

Bicchieri, C. (1988) 'Should a scientist abstain from metaphor?' in A. Klamer, D.N. McCloskey and R.M. Solow (eds) *The Consequences of Economic Rhetoric*, Cambridge: Cambridge University Press, 100–14.

Billinge, Mark, Gregory, Derek, and Martin, Ron (1983) *Recollections of a Revolution: Geography as a Spatial Science*, New York: St Martin's Press.

Bird, I.L. (1856) *The English Woman in America*, London: Murray.

Bissell, R. (1979) 'Soviet activity in Africa: should we care?' *South Africa International* 10: 199–210.

—— (1980) 'How strategic is South Africa?' in C. Crocker and R. Bissell (eds) *South Africa into the 1980's*, Boulder: Westview Press.

Black, M. (1962) *Models and Metaphors: Studies in Language and Philosophy*, Ithaca, N.Y.: Cornell University Press.

Blake, H.T. (1903) *Niagara: Monarch Supreme in Nature's Glorious Realms*, Hartford: Connecticut Magazine Co.

Blanchot, Maurice (1986) *The Writing of the Disaster*, Lincoln: University of Nebraska Press. (French original, 1980).

—— (1987) *Michel Foucault as I Imagine him*, New York: Zone Books (French original, 1986).

Blocker, H.G. (1979) *Philosophy and Art*, New York: Scribner.

Bloor, D.C. (1982) 'Durkheim and Mauss revisited: classification and the sociology of knowledge', *Studies in the History and Philosophy of Science* 13: 267–97.

—— (1983) *Wittgenstein: a Social Theory of Knowledge*, New York: Columbia University Press.

Boelhower, William Q. (1984) *Through a Glass Darkly: Ethnic Semiosis in American Literature*, Venice: Edizioni Helvetia.

Boggs, S.W. (1947) 'Cartohypnosis', *Scientific Monthly* 64: 469–76.

Bonnycastle, R. (1849) *Canada and the Canadians*, London: Colburn.

Booth, W.C. (1979) 'Metaphor as rhetoric: the problem of evaluation', in S. Sacks (ed.) *On Metaphor*, Chicago: University of Chicago Press, 47–70.

Bowman, L. (1982) 'The strategic importance of South Africa to the United States: an appraisal and policy analysis', *African Affairs: Journal of the Royal African Society* 81: 159–92.

Bradley, A.C. (1949) *Shakespearean Tragedy: lectures on 'Hamlet', 'Othello', 'King Lear', 'Macbeth'*, New York: Macmillan.

Brooks, Van Wyck (1941) *The Opinions of Oliver Allston*, New York: Dutton.

Buckingham, J.S. (1845) 'Travels in the eastern and western states in 1837', in W. Barham, (ed.) *Descriptions of Niagara*, Gravesend: The Compiler.

Burgin, Victor (1988) 'Something about photography theory', in A.L. Rees and Frances Borzello (eds) *The New Art History*, Atlantic Highlands, N.J.: Humanities Press, 41–62, quoted on p. 51.

Burke, K. (1950) *A Rhetoric of Motives*, Berkeley: University of California Press.

Bury, Pol (1976) *Le Sexe de anges et celui des géomètres*, Paris: Galilée.

Butlin, Martin, and Joll, Evelyn (1984) *The Paintings of J.M.W. Turner*, revised edition, London and New Haven: Yale University Press.

Buttimer, A. (1971) *Society and Milieu in the French Geographic Tradition*, Chicago: Rand & McNally.

—— (1982) 'Musing on Helicon: root metaphors and geography', *Geografiska Annaler* 64 B: 89–96.

Cameron, I. (1983) 'Metaphor in science and society', *Bulletin of Science, Technology and Society* 3: 251–92.

Campbell, P.N. (1973) 'Scientific discourse', *Philosophy and Rhetoric* 6: 1.

Castle Dracula (n.d.) promotional brochure.

Chapman, C. (1875) *The Ocean Waves: Travels by Land and Sea*, London: Berridge.

Chateaubriand, F.A.R. (1801) *Atala, of the Amours of two Indians in the Wilds of America*, London: Lee.

Chester, G.J. (1869) *Transatlantic Sketches in the West Indies, South America, Canada, and the United States*, London: Smith Elder.

Chiavenatto, J.J. (1981) *Geopolítica: arma do fascismo*, São Paulo: Global Editora.

Child, J.C. (1979) 'Geopolitical thinking in Latin America', *Latin American Research Review* 14: 89–111.

—— (1985) *Geopolitics and Conflict in South America*, New York: Praeger.

City of Portland (1976) *Comprehensive Plan*, Portland, Oregon: City Planning Department.

Clark, Gordon, and Dear, Michael (1984) *State Apparatus: Structures and Language of Legitimacy*, Boston, Mass.: Allen & Unwin.

Clarke, C.N.G. (1988) 'Taking possession: The cartouche as cultural text in eighteenth-century American maps', *World and Image* 4: 455–74.

Clifford, J. (1986) 'Introduction: partial truths', in J. Clifford and G.E. Marcus (eds.) *Writing Culture: the Poetics and Politics of Ethnography*, Berkeley: University of California Press, 1–26.

—— (1988) *The Predicament of Culture: Twentieth Century Ethnography, Literature and Art*, Cambridge: Mass., Harvard University Press.

—— and Marcus G.E., eds. (1986) *Writing Culture: the Poetics and Politics of Ethnography*, Berkeley: University of California Press.

Cockburn, A. (1988) *Corruptions of Empire* London: Verso.

Cohen, G. (1978) *Karl Marx's Theory of History: a Defense*, Princeton: Princeton University Press.

—— (1982) 'Reply to Elster on "Marxism, functionalism, and game theory"', *Theory and Society* 11: 483–95.

Coker, C. (1986) *The United States and South Africa, 1968–85: Constructive Engagement and its Critics*, Durham, N.C.: Duke University Press.

Connell, E.J. (1975) 'Industrial Development in South Leeds, 1790–1914', unpublished Ph.D. dissertation, University of Leeds.

Continelli, L. (1988) 'Over and out', *Buffalo Magazine*, Buffalo News, 13 March: 6–15.

Cooper J.F. (1821) *The Spy*, reprinted 1960, New York: Hafner.

Cosgrove, Denis (1984) *Social Formations and Symbolic Landscapes*, London: Croom Helm.

—— and Daniels, S. eds. (1988) *The Iconography of Landscape: Essays on the Symbolic Representation, Design and Use of Past Environments*, Cambridge: Cambridge University Press.

Cottom, D. (1989) *Text and Culture: the Politics of Interpretation.* Minneapolis: University of Minnesota Press.

Couclelis, Helen (1983) 'On some problems of defining sets for Q-analysis', *Environment and Planning* B 10: 423–38.

Crocker, C. (1980) 'South Africa: strategy for change', *Foreign Affairs* 59: 323–51.

Crone, G.R. (1978) *Maps and their Makers; an Introduction to the History of Cartography*, fifth edition (first edition, 1953), Folkestone: Dawson; Hamden, Conn.: Archon Books.

Culler, J. (1982) *On Deconstruction: Theory and Criticism after Structuralism.* Ithaca, N.Y.: Cornell University Press.

—— (1983) *Barthes*, London: Fontana.

Curtis, G.W. (1852) *Lotus-Eating: a Summer Book*, New York: Harper.

Daniels, S. (1980) 'Moral Order and the Industrial Environment in the Woollen Textile Districts of West Yorkshire, 1780–1880', unpublished Ph.D. thesis, University College London.

—— (1981) 'Landscaping for a manufacturer: Humphry Repton's Commission for Benjamin Gott at Armley in 1809–10', *Journal of Historical Geography* 7: 379–96.

—— and Cosgrove, D. (1988) 'Introduction: iconography and landscape', in *The Iconography of Landscape: Essays on the Symbolic Representation, Design and Use of Past Environments*, Cambridge: Cambridge University Press.

Dante, A. (1948) *Inferno, The Divine Comedy*, trans. L.G. White, New York: Pantheon.

Darnton, Robert (1984) *The Great Cat Massacre and other episodes in French Cultural History*, New York: Basic Books.

Davidson, D. (1979)'What metaphors mean', in *On Metaphor*, ed. S. Sacks, Chicago: University of Chicago Press, 29–46.

Day Lewis, Cecil (1946) *The Poetic Image*, New York: Oxford University Press.

Dear, Michael (1986) 'Postmodernism and planning', *Environment and Planning D: Society and Space* 4: 367–84.

—— (1988) 'The postmodern challenge: reconstructing human geography', *Transactions, Institute British Geographers*, n.s. 13: 262–74.

De Certeau, Michel (1984) *The Practice of Everyday Life*, trans. Steven Rendell, Berkeley: University of California Press.

—— (1986) *Heterologies: Discourse on the Other*, Minneapolis: University of Minnesota Press.

Defoe, D. (1971, originally 1724–6) *A Tour through the whole Island of Great Britain*, P. Rogers, London: Allen Lane.

Deputy, The (1888) *The Niagara Falls Detective* or *Solving the Whirlpool Mystery*, New York: Munro's Publishing House.

Derrida, Jacques (1976) *Of Grammatology*, trans. Gayatri Chakratvorty Spivak, Baltimore: Johns Hopkins University Press.

—— (1979) *Spurs: Nietzsche's Styles / Éperons: Les styles de Nietzsche*, Chicago: University of Chicago Press.

—— (1981a) *Positions*, Chicago: University of Chicago Press.

—— (1981b) *Dissemination*, Chicago: University of Chicago Press.

—— (1985) 'Racism's last word', ('Le dernier mot du racism') trans. Peggy Kamuf, *Critical Inquiry* 12: 290–99.

—— (1986a) *Glas.* Lincoln, Neb.: University of Nebraska Press (French original, 1974).

—— (1986b) 'But beyond . . .' (open letter to Anne McClintock and Rob Nixon), trans. Peggy Kamuf, *Critical Inquiry* 13: 155–81.

de Saussure, F. (1966) *Course in General Linguistics*, New York: McGraw-Hill.

De Veaux, S. (1839) *The Falls of Niagara*, Buffalo: Hayden.

Dickens, C. (1843) *American Notes for General Circulation*, London: Chapman & Hall.

Dicy, Edward (1971) *Spectator of America*, ed. Herbert Mitgang, Chicago: Quadrangle Books.

Dipert, Randall R. (1986) 'Art, artifacts and regarded intentions', *American Philosophical Quarterly* 23: 401–8.

Dorfman, A. and Mattelart, A. (1975) *How to Read Donald Duck: Imperialist Ideology in the Disney Comic*, New York: International General.

Dowling, Rev. J. (1848) 'Sacred musings', *Table Rock Album*, Buffalo: Jewett Thomas, 31–2.

Dumouchel, Paul, ed. (1987) *Violence and Truth: On the Work of René Girard*, London: Athlone Press (French original, 1985).
Duncan, J.S. (1990) *The City as Text: the Politics of Landscape Interpretation in the Kandyan Kingdom*, Cambridge: Cambridge University Press.
—— and N. —— (1988) '(Re) reading the landscape', *Environment and Planning D, Society and Space* 6: 117–26.
—— and Ley, D.F. (1982) 'Structural Marxism in human geography: a critical assessment', *Annals, Association of American Geographers* 72: 30–59.
Dwight, Timothy (1969) *Travels in New England and New York* 4, Cambridge, Mass.: Harvard University Press.
Dyer, J. (1761) *Poems by John Dyer*, London: Dodsley.
Eagleton, T. (1983) *Literary Theory: an Introduction*, Minneapolis: University of Minnesota Press.
—— (1986) *Against the Grain*, London: Verso.
Eco, U. (1976) *A Theory of Semiotics*, Bloomington: Indiana University Press.
—— (1983) 'Travels in hyperreality', in *Travels in Hyperreality*, trans. William Weaver, New York: Harcourt Brace Jovanovich.
—— (1986) 'On the crisis of the crisis of reason', in *Travels in Hyperreality*, London: Picador, 125–132.
Eliade, Mircea (1959) *The Sacred and the Profane*, New York: Harcourt Brace Jovanovich.
El-Khawas, M.A., and Cohen, B. (1976) *The Kissinger Study of Southern Africa*, National Security Study memorandum 39, Westport, Conn.: Lawrence Hill.
Elster, J. (1979) *Ulysses and the Sirens: Studies in Rationality and Irrationality*, Cambridge: Cambridge University Press.
—— (1982) 'Marxism, functionalism and game theory', *Theory and Society* 11: 453–82.
Escola Superior de Guerra (1976) *Manual básico da Escola Superior de Guerra*, Rio de Janeiro: ESG.
Executive Committee to Promote Navy Island as Permanent Headquarters of the United Nations (1945) *Proposed United Nations headquarters: Navy Island, Niagara Falls.*
Fairgrieve, J. (1941) *Geography and World Power*, eighth edition, London: University of London Press.
Falk, B. (1938) *Turner the Painter*, London: Hutchinson.
Farinelli, Franco (1989) 'Intraduzione: dal bar di de Saussure alla balera di Girard' in Gunnar Olsson (ed.), *Linea senza ombre*, Rome: Edizione Theoria.
Fawcett, T. (1974) *The Rise of English Provincial Art: Artists and Patrons and Institutions outside London*, Oxford: Clarendon Press.
Fields, A. (1897) *Life and Letters of Harriet Beecher Stowe*, Boston: Houghton Mifflin.
Fischer C.S. (1972) 'Urbanism as a way of life: a review and an agenda', *Sociological Methods and Research* 1: 187–242.
Fish, S. (1980) *Is there a Text in this Class? The Authority of Interpretive Communities*, Cambridge, Mass.: Harvard University Press.
Fiske, J. (1989) *Reading the Popular*, London: Unwin Hyman.
Foucault, M. (1967) *Madness and Civilization: a History of Insanity in the Age of Reason*, trans. R. Howard, London: Tavistock.
—— (1972) *The Archaeology of Knowledge*, trans. A. Sheridan, London: Tavistock.
—— (1973) *The Order of Things: an Archaeology of the Human Sciences* (translation of *Les Mots et les choses*), New York: Vintage Books.
—— (1977) *Discipline and Punish: the Birth of the Prison*, New York: Pantheon (French original, 1975).
—— (1977) *Language, Counter-memory, Practice*, eds. D. Bouchard and S. Simon, Oxford: Blackwell.

—— (1978) *The History of Sexuality* I *An Introduction*, trans. Robert Hurley, New York: Random House.
—— (1980) *Power/Knowledge: Selected Interviews and other Writings, 1972–77*, ed. Colin Gordon, trans. Colin Gordon, Leo Marshall, John Mepham and Kate Sopher, New York: Pantheon.
—— (1987) *Maurice Blanchot: the Thought from Outside*, New York: Zone Books (French original, 1987).
Foucher, M. (1986) *L'Invention des frontières*, Paris: Fondation pour les Études de Défense Nationale.
Francis, R.W. (1895) *The Great Mystery of Niagara Falls*, or, *Detective Payrre's Greatest Case*, New York: Munro's Publishing House.
Frazer, D. (1980) 'Politics and society in the nineteenth century', in D. Frazer (ed.) *A History of Modern Leeds*, Manchester: Manchester University Press, 270–300.
Fuller, A.B., ed. (1856) *Summer on the Lakes*, reprinted 1970, New York: Haskell House.
Fuller, Peter (1988) 'The geography of Mother Nature', in Denis Cosgrove and Stephen Daniels (eds) *The Iconography of Landscape: Essays on the Symbolic Representation, Design and Use of Past Environments*, Cambridge: Cambridge University Press, 11–31.
Fulton, L. (1901) *Nadia: the Maid of the Mist*, Buffalo: White-Evans Penfold.
Gans, Herbert (1968) 'Urbanism and suburbanism as ways of life', in Raymond Pahl (ed.) *Readings in Urban Sociology*, Oxford: Pergamon, 95–118.
Garey, D., and Hott, L.R. (1985) *Niagara Falls: the Changing Nature of a New World Symbol*, Los Angeles: Direct Cinema Limited.
Garrison, W. (1978) *Strange Facts about Death*, Nashville, Tenn.: Arbingden.
Gasché, Rodolphe (1986) *The Tain of the Mirror: Derrida and the Philosophy of Reflection*, Cambridge, Mass.: Harvard University Press.
Gatrell, A. (1983) *Distance and Space: A Geographical Perspective*, New York: Oxford University Press.
Gauthier, M., ed. (1988) *Cartographie dans les médias*, Quebec: Presses de l'Université du Québec.
Geddes, Patrick (1915) *Cities in Evolution*, London: Oxford University Press.
Geertz, C. (1973) *The Interpretation of Cultures*, New York: Basic Books.
—— (1983) *Local Knowledge: Further Essays in Interpretive Anthropology*, New York: Basic Books.
—— (1988) *Works and Lives: the Anthropologist as Author*, Cambridge: Polity Press.
Georgescu-Roegen, N. (1971) *The Entropy Law and the Economic Process*, Cambridge, Mass.: Harvard University Press.
Gillette, K.C. (1894) *The Human Drift*, Boston, Mass.: New Era.
Girard, René (1965) *Deceit, Desire and the Novel: Self and other in Literary Structure*, Baltimore: Johns Hopkins University Press (French original, 1961).
—— (1977) *Violence and the Sacred*, Baltimore: Johns Hopkins University Press (French original, 1972).
—— (1986) *The Scapegoat*, Baltimore: Johns Hopkins University Press (French original, 1982).
—— (1987a) *Things Hidden since the Foundation of the World*, Stanford: Stanford University Press (French original, 1978).
—— (1987b) *Job: the Victim of his People*, Stanford: Stanford University Press (French original, 1985).
Goffart, Walter (1988) 'The map of the barbarian invasions: a preliminary report', *Nottingham Medieval Studies* 32: 49–64.
Golbery do Couto e Silva (1955) *Planejamento estratégico*, Rio de Janeiro: Biblioteca do Exército.

—— (1967) *Geopolitica do Brazil*, Rio de Janeiro: Jose Olympio.
—— (1981) *Conjunctura politica nacional o poder executivo & Geopolitica do Brazil*, Rio de Janeiro: Jose Olympio.
Gombrich, E. (1975) 'Mirror and map: theories of pictorial representation', *Philosophical Transactions of the Royal Society of London*, Series B, 270, Biological Sciences: 119–49.
Goodman, Nelson (1968) *Languages of Art: an Approach to a Theory of Symbols*, Indianapolis and New York: Bobbs-Merrill.
Gorer, G. (1970) 'The pornography of death', in I. Buchen (ed.), *The Perverse Imagination: Sexuality and Literary Culture*, New York: New York University Press, 75–81.
Gould, P. (1981) 'Letting the data speak for themselves', *Annals, Association of American Geographers* 71: 166–76.
—— (1983) 'Reflective distanciation through metamethodological perspective', *Environment and Planning* B, 10: 381–92.
—— and White, R. (1974) *Mental Maps*, Harmondsworth: Penguin.
Goyret, J.T. (1980) *Geopolítica y subversión*, Buenos Aires: Ediciones Depalma.
Graham, L. (1949) *Niagara Country*, New York: Duell Sloan & Pearce.
Gramsci, A. (1981) *Selections from the Prison Notebooks*, London: Lawrence & Wishart.
Gregory, Derek (1987) 'Postmodernism and the politics of social theory', *Environment and Planning* D, *Society and Space* 5: 245–48.
—— (1989a) 'Areal differentiation and postmodern human geography', in D. Gregory and R. Walford (eds) *New Horizons in Human Geography*, London: Macmillan, 67–96.
—— (1989b) 'The crisis of modernity? Human geography and critical social theory', in N.J. Thrift and R. Peet (eds) *New Models in Human Geography*, London: Unwin Hyman, 348–85.
—— and Walford, R. (1989c) 'Introduction: making geography', in D. Gregory and R. Walford (eds) *Horizons in Human Geography*, London: Macmillan, 1–7.
Grinfield, T. (1845) 'Hyman on Niagara', in W. Barham, (ed.) *Descriptions of Niagara*, Gravesend: The Compiler.
Gudeman, S. (1986) *Economics as Culture*, London: Routledge.
Guillet, E.C. (1938) *The Lives and Times of the Patriots*, Toronto: Nelson.
Hall, S. (1982) 'The battle for socialist ideas in the 1980s', in M. Eve and D. Musson (eds) *The Socialist Register 1982*, London: Merlin Press.
Haller, Rudolf (1988) *Questions on Wittgenstein*, Lincoln, Neb.: University of Nebraska Press.
Handy, Myrtle M., and McKelvey, Blake (1940) 'British travelers in the Genesee country', *Rochester Historical Society Publication* 18: 1–73.
Hanks, R. (1980) *The Unnoticed Challenge: Soviet Maritime Strategy and the Global Choke Points*, Cambridge, Mass., and Washington, D.C.: Institute for Foreign Policy Analysis.
—— (1981) *The Cape Route: Imperiled Western Lifeline*, Cambridge, Mass., and Washington, D.C.: Institute for Foreign Policy Analysis.
—— (1983) *Southern Africa and Western Security*, Cambridge, Mass., and Washington, D.C.: Institute for Foreign Policy Analysis.
Harbison, R. (1977) *Eccentric Spaces*, New York: Avon Books.
Harley, J.B. (1984) 'Meaning and ambiguity in Tudor cartography', in Sarah Tyacke (ed.) *English Map-making, 1500–1650: Historical Essays*, London: British Library Reference Division Publications, 22–45.
—— (1987–8) 'L'Histoire de la cartographie comme discours', *Préfaces* 5: 70–5.
—— (1988a) 'Maps, knowledge and power', in Denis Cosgrove and Stephen Daniels (eds.) *The Iconography of Landscape*, Cambridge: Cambridge University Press, 277–312.

—— (1988b) 'Secrecy and silences: the hidden agenda of cartography in early modern Europe', *Imago Mundi* 40: 111–30.

—— (1988c) 'Victims of a Map: New England Cartography and the Native Americans', paper read at the Land of Norumbega Conference, Portland, Maine.

—— (1989) 'Historical geography and the cartographic illusion', *Journal of Historical Geography* 15: 80–91.

—— (forthcoming) 'Power and legitimation in the English geographical atlases of the eighteenth century', in John A. Wolter and John Y. Cole (eds) *Images of the World: the Atlas Through History*, Washington, D.C.: Center for the Book and Geography and Map Division, Library of Congress.

—— and Woodward, David, eds. (1987) *The History of Cartography* 1, *Cartography in Prehistoric, Ancient, and Medieval Europe and the Mediterranean*, Chicago: University of Chicago Press.

—— eds. (forthcoming) *The History of Cartography* 2, Book 1, *Cartography in the Traditional Islamic and South Asian Societies*, and Book 2, *Cartography in the Traditional East Asian Societies*, Chicago: University of Chicago Press.

Harvey, D. (1967) 'Models of the evolution of spatial patterns in human geography', in R.J. Chorley and P. Haggett (eds) *Integrated Models in Geography*, London: Methuen, 549–608.

—— (1982) *The Limits to Capital*, Chicago: University of Chicago Press.

—— (1985a) *The Urbanization of Capital*, Baltimore: Johns Hopkins University Press.

—— (1985b) *Consciousness and the Urban Experience*, Baltimore: Johns Hopkins University Press.

—— (1989) *The Condition of Postmodernity: an Enquiry into the Origins of Cultural Change*, Oxford: Blackwell.

Hathaway, H. (1953) *Niagara*, produced by C. Brackett, written by C. Brackett, W. Reisch and R. Breen, released by Twentieth Century–Fox.

Haunted House and the Funhouse (n.d.) promotional brochure.

Haushofer, K. (1928) 'Die Suggestive Karte', *Bausteine zur Geopolitik*, Berlin: 342–48.

Hecht, S.B., and Cockburn, A. (1989) *The Fate of the Forest: Developers, Destroyers and Defenders of the Amazon*, London: Verso.

Hegel, G.W.F. (1977) *Phenomology of Spirit*, Oxford: Clarendon Press (German original, 1807).

Heidegger, Martin (1962) *Being and Time*, New York: Harper & Row.

Heisenberg, W. (1959) *Physics and Philosophy*, London: Allen & Unwin.

Henrikson, Alan K. (1987) 'Frameworks for the world', preface to Ralph E. Ehrenberg, *Scholars' Guide to Washington D.C. for Cartography and Remote Sensing Imagery*, Washington D.C.: Smithsonian Institution Press.

Hepple, L.W. (1986a) 'Geopolitics, generals and the state in Brazil', *Political Geography Quarterly* 5 (supplement): S79–S90.

—— (1986b) 'The revival of geopolitics', *Political Geography Quarterly* 5 (supplement): S21–S36.

Hesse, M. (1963) *Models and Analogies in Science*, London: Sheed & Ward.

—— (1980a) *Revolutions and Reconstructions in the Philosophy of Science*, Brighton: Harvester Press.

—— (1980b) 'The explanatory function of metaphor', in *Revolutions and Reconstructions in the Philosophy of Science*, Brighton: Harvester, 111–24.

High, Dallas (1972) 'Belief, falsification, and Wittgenstein', *International Journal for the Philosophy of Religion* 3: 240–50.

Hill, D. (1984) *Turner's Footsteps*, London: Murray.

Hobbes, T. (1962, originally 1652) *Leviathan*, London: Macmillan.

Holley, G.W. (1883) *The Falls of Niagara*, New York: Armstrong.

Holt-Jensen, A. (1982) *Geography: its History and Concepts*, Totowa, N.J.: Rowman & Allanheld.

Horgan, Paul (1980) *The Return of the Weed*, Flagstaff, Az.: Northland.

Houghton. G. (1882) 'Niagara', *Niagara and other Poems*, Boston, Mass.: Houghton Mifflin.

Howard, R.W. (1968) *Thundergate: the Forts of Niagara*, Englewood Cliffs, N.J.: Prentice-Hall.

Howells, W.D. (1888) *Their Wedding Journey*, Boston, Mass.: Houghton Mifflin.

'How to see Niagara', (1888), no publisher listed.

Hoy, David (1985) 'Jacques Derrida', in Quentin Skinner (ed.) *The Return of Grand Theory in the Human Sciences*, Cambridge: Cambridge University Press, 41–64.

Hudson, W. (1989) 'Postmodernity and contemporary social thought', in P. Lassman (ed.) *Politics and Social Theory*, New York: Routledge, 138–59.

Hutcheon, L. (1989) *The Politics of Postmodernism*, London: Routledge.

Irigaray, Luce (1985) *This Sex which is not one*, Ithaca, N.Y.: Cornell University Press (French original, 1977).

—— (1984) *Éthique de la différence sexuelle*, Paris: Minuit.

Jakobson, Roman and Marris, Halle (1956) *Fundamentals of Language*, The Hague: Mouton.

James, H. (1884) 'Niagara', in *Portraits of Places*, Boston, Mass.: Osgood.

James, William (1902) *The Varieties of Religious Experience*, New York: Doubleday.

Jameson, A.B.M. (1838) *Winter Studies and Summer Rambles in Canada*, London: Saunders & Otley.

Jameson, F. (1984) 'Postmodernism, or the cultural logic of late capitalism', *New Left Review* 146: 53–92.

—— (1985) 'Postmodernism and consumer society', in H. Foster (ed.) *Postmodern Culture*, London: Pluto Press.

Janik, Allan, and Toulmin, Stephen (1973) *Wittgenstein's Vienna*, New York: Simon & Schuster.

Jeans, D.N. (1979) 'Some literary examples of humanistic description of place', *Australian Geographer* 14: 207–14.

Johnson, J. (1981) 'Some structures and notation of Q-analysis', *Environment and Planning* B 8: 73–86.

Johnson, R.L. (1898) 'Apostrophe to Niagara', in *Nigara: its History, Incident and Poetry*, Washington, D.C.: Neale, 35–40.

Johnston, R.J. (1987) *Geography and Geographers: Anglo-American Human Geography since 1945*, third edition, London: Edward Arnold.

Jowett, G.S. (1987) 'Propaganda and communication: the re-emergence of a research tradition', *Journal of Communication*, winter: 97–114.

Kant, Immanuel (1965) *The Critique of Pure Reason*, trans. Norman Kemp Smith, New York: St Martin's Press (original 1781, 1787).

Keightley, A. (1976) *Wittgenstein, Grammar and God*, London: Epworth.

Kelly, P. (1984) 'Geopolitical themes in the writings of General Carlos de Meira Mattos of Brazil', *Journal of Latin American Studies* 16: 439–61.

—— and Child, J.C., eds. (1988) *Geopolitics of the Southern Cone and Antarctica*, Boulder: Lynn Rienner

Kemp, G. (1977) 'The new strategic map', *Survival* 19: 50–9.

—— (1978) 'U.S. strategic interests and military options in sub-Saharan Africa', in J. Whitaker (ed.) *Africa and the United States: Vital Interests*, New York: New York University Press, 120–52.

Kennan, G. (Mr. X) (1947) 'The sources of Soviet conduct', *Foreign Affairs* 25: 566–82.

Kern, J. (1985) 'Where is Oregone?' *Sales and Marketing Management* 10, April: 3–8.

Kinross, Robin (1985) 'The rhetoric of neutrality', *Design Issues* II, 2: 18–30.
Kjellen, R. (1917) *Der Staat als Lebensform*, Leipzig: Hirzel.
Kline, M. (1980) *Mathematics: the Loss of Certainty*, New York: Oxford University Press.
Knox, Paul L., ed. (1988) *The Design Professions and the Built Environment*, London: Croom Helm.
Koestler, A. (1976) 'Whereof one cannot speak . . . ?' in A. Toynbee (ed.) *Life after Death*, New York: McGraw-Hill.
Korzybski, Alfred (1948) *Science and Sanity: an Introduction to non-Aristotelian Systems and General Semantics*, third edition, Lakeville, Conn.: International Non-Aristotelian Library.
Kouwenhoven, John (1982) *Half a Truth is Better than None: Some Unsystematic Conjectures about Art, Disorder and the American Experience*, Chicago: University of Chicago Press.
Kristof, L.K.D. (1960) 'The origin and evolution of geopolitics', *Journal of Conflict Resolution* 4: 15–51.
Kuhn, T. (1970) *The Structure of Scientific Revolutions*, second edition, Chicago: University of Chicago Press.
Lacan, Jacques (1977) *Écrits: a Selection*, London: Tavistock (French original, 1966).
—— (1978) *The Four Fundamental Concepts of Psychoanalysis*, New York: Norton (French original, 1964).
LaCapra, D. (1983) *Rethinking Intellectual History: Text, Context, Language*, Ithaca, N.Y.: Cornell University Press.
Lacoste, Y. (1985) *La Géographie, ça sert, d'abord, à faire la guerre*, second edition, Paris: Editions La Découverte.
Lasswell, H. (1927) *Propaganda Techniques in the World War*, New York: Knopf.
Laudan, Larry (1977) *Progress and its Problems: toward a Theory of Scientific Growth*, Berkeley: University of California Press.
Lelyveld, J. (1985) *Move your Shadow: South Africa, Black and White*, New York: Penguin.
Lemaire, Anika (1977) *Jacques Lacan*, London: Routledge (French original, 1970).
Levin, H. (1970) *The Power of Blackness*, New York: Knopf.
Lewis, C.S. (1988) 'On stories', and 'An experiment in criticism', Lyell W. Dorsett (ed.) *The Essential C.S. Lewis*, New York: Collier.
—— (1947) *The Abolition of Man*, New York: Macmillan.
Lewis, Pierce (1985) 'Beyond description', *Annals of the Association of American Geographers* 75: 465–78.
Lewis, W. (1976) 'How a defense planner looks at Africa', in H. Kitchen (ed.) *Africa: from Mystery to Maze* Lexington, Mass.: Lexington Books, 277–304.
Ley, D. (1980) 'Liberal ideology and the postindustrial city,' *Annals of the Association of American Geographers* 70: 238–58.
—— (1987) '"Styles of the times": liberal and neo-conservative landscapes in inner Vancouver, 1968–86', *Journal of Historical Geography* 13: 40–56.
Lindsay, J. (1966) *J.M.W. Turner: his Life and Work*, London: Cory Adams & Mackay.
Liston, J.K. (1843) *Niagara Falls: a Poem in Three Cantos*, Toronto: Lawrence.
Lord, J.C. (1869) 'The genius of Niagara', in *Occasional Poems*, Buffalo: Breed & Tent, 19–22.
Lowenthal, David (1968) 'The American scene', *Geographical Review* 58: 61–88.
Loxton, John (1985a) 'The Peters phenomenon', *Cartographic Journal* 22: 106–8.
—— (1985b) 'The so-called Peters projection', *Cartographic Journal* 22: 108–10.
Ludlow, F.H. (1857) *The Hasheesh Eater*, New York: Harper; reprinted 1970, Upper Saddle River, N.J.: Literature House.
Lukermann, Fred (1961) 'The concept of location in classical geography', *Annals of the*

Association of American Geographers 51: 194–210.
—— (1965) 'The *calcul des probabilités* and the École Française de Géographie', *Canadian Geographer* 9: 128–37.
Lupton, Ellen (1986) 'Reading isotype', *Design Issues* 3, 2: 47–58.
Lyman, Horace (1903) *History of Oregon: the Growth of an American State* 4, New York: North Pacific.
Lynch, J. (1876) Pastoral letter, reprinted 1914 in F.H. Severance (ed.) *Peace Episodes on the Niagara*, Buffalo Historical Society Publications XVIII, Buffalo: Buffalo Historical Society, 104–5.
Lyotard, Jean-François (1984) *The Postmodern Condition: a Report on Knowledge*, trans. Geoff Bennington and Brian Massumi, Minneapolis: University of Minnesota Press.
McCall, Tom, with Steve Neal (1977) *Tom McCall: Maverick*, Portland: Binfort & Mort.
McCloskey, D.N. (1985) *The Rhetoric of Economics*, Madison: University of Wisconsin Press.
MacColl, E. Kimbark (1979) *The Growth of a City: Power and Politics in Portland, Oregon*, Portland: Georgian Press.
McDermott, P.D. (1969) 'Cartography in advertising', *Canadian Cartographer* 149–55.
McGreevy, P. (1985) 'Niagara as Jerusalem', *Landscape* 28: 26–32.
—— (1987) 'Imagining the future at Niagara Falls', *Annals of the Association of American Geographers* 77: 48–62.
MacIntyre, Alasdair C. (1984) *After Virtue: a Study in Moral Theory*, second edition, Notre Dame, Ind.: University of Notre Dame Press.
McKenzie, D.F. (1986) *Bibliography and the Sociology of Texts*, London: British Library.
McKinsey, E. (1985) *Niagara falls: Icon of the American Sublime*, London: Cambridge University Press.
McLuhan, Marshall (1962) *The Gutenberg Galaxy: the Making of Typographic Man*, Toronto: University of Toronto Press.
McNally, Andrew (1987) '"You can't get there from here," with today's approach to geography', *Professional Geographer* 39: 389–92.
Mallarmé, Stéphane (1977) *The Poems*, Harmondsworth: Penguin (translation of 'Un coup de dès', 1897).
Marcum, J. (1989) 'Africa: a continent adrift', *Foreign Affairs* 68: 159–79.
Marcus, G.E., and Fischer, M.M.J. eds. (1986) *Anthropology as Cultural Critique: an Experimental Moment in the Human Sciences*, Chicago: University of Chicago Press.
Marden, J. (1932) *Historical Niagara Falls*, Niagara Falls, Ontario: Lindsay Press.
Marin, Louis (1988) *Portrait of the King*, trans. Martha M. Houle, Minneapolis: University of Minnesota Press.
Markoff, J., and Baretta, S.R.D. (1985) 'Professional ideology and military activism in Brazil: critique of a thesis of Alfred Stepan', *Comparative Politics* 17: 175–191.
Mason, H.E. (1978) 'On the multiplicity of language games', Elisabeth Leinfellner *et al.* (eds) in *Wittgenstein and his Impact on Contemporary Thought: Proceedings of the Second International Wittgenstein Symposium*, Kirchberg/Wechsel, Austria, 29 August – 4 September 1977, Vienna: Hölder-Pichler-Tempsky, 332–5.
Mau, C. (1944) *The Development of Central and Western New York*, Rochester: Du Bois Press.
May, J.G. (1968) 'Report on a journey to England', in W.O. Henderson (ed.) *Industrial Britain under the Regency: the Diaries of Escher, Bodmer, May and de Gallois*, London: Cass, 139–41.
Meira Mattos, C de (1975) *Brasil: geopolítica e destino*, Rio de Janeiro: José Olympio.
Merleau Ponty, Maurice, (1964) *The Primacy of Perception*, ed. James M. Edie, Evanston, Ill.: Northwestern University Press.
Meyer, J.T. (1984) 'The Charm of Central Place Theory: Beauty and Knowledge in

Human Geography', unpublished doctoral dissertation, Pennsylvania State University.
Meynen, E., ed. (1973) *Multilingual Dictionary of Technical Terms in Cartography*, Wiesbaden: Steiner.
Minter, W. (1986) *King Solomon's Mines Revisited: Western Interests and the Burdened History of southern Africa*, New York: Basic Books.
Mirowski, P. (1984a) 'Physics and the "marginalist revolution"', *Cambridge Journal of Economics* 8: 361–79.
—— (1984b) 'The role of conservation principles in twentieth-century economic theory', *Philosophy of the Social Sciences* 14: 461–73.
—— (1988) *Against Mechanism: Protecting Economics from Science*, Totowa, N.J.: Rowman & Littlefield.
Mitchell, W.J.T. (1986) *Iconology: Image, Text, Ideology*, Chicago: University of Chicago Press.
Monmonier, Mark S. (1975) *Maps, Distortion, and Meaning*, Association of American Geographers Resource Paper 75, Washington D.C.: AAG.
—— and Schnell, George A. (1988) *Map Appreciation*, Englewood Cliffs, N.J.: Prentice-Hall.
Moore, S. (1988) 'Getting a bit of the other – the pimps of postmodernism', in R. Chapman and J. Rutherford (eds) *Male Order; Unwrapping Masculinity*, London: Lawrence & Wishart, 165–92.
Morris, J. (1982) 'The Magic of Maps: the Art of Cartography', unpublished M.A. dissertation, University of Hawaii.
Muehrcke, P. (1972) *Thematic Cartography*, Association of American Geographers Resource Paper 19, Washington D.C.: AAG.
—— (1986) *Map Use: Reading, Analysis, and Interpretation*, second edition, Madison: JP Publications.
Munz, Peter (1987) 'Bloor's Wittgenstein, or, The fly in the bottle', *Philosophy of the Social Sciences* 17: 67–96.
Nares, G. (1954) 'Farnley Hall, Yorkshire', *Country Life* 20: 1620.
Neil, J. Meredith (1975) *Toward a National Taste: America's Quest for Aesthetic Independence*, Honolulu: University of Hawaii Press.
New York World (1896) 'A huge structure: Leonard Henkle's plan to bridge the Niagara river', 9 February.
Niagara Falls Canada (n.d.) promotional brochure.
Niagara Falls Museum News (n.d., *c.* 1985) promotional brochure.
Nicholson, J. (1825) *Airedale in Ancient Times*, London: Seeley.
Nietzsche, Friedrich (1968) *The Will to Power*, New York: Vintage (German original, 1883–8).
Noer, T.J. (1985) *Black liberation: the United States and White Rule in Africa, 1948–68*, Columbia, Mo.: University of Missouri Press.
Norris, Christopher (1982) *Deconstruction: Theory and Practice*, London: Methuen.
—— (1987) *Derrida*, Cambridge, Mass.: Harvard University Press.
Nunn, F.M. (1976) *The Military in Chilean History: Essays on Civil–Military Relations, 1810–1973*, Albuquerque, N.M.: University of New Mexico Press.
Nyiri, J.C. (1982) 'Wittgenstein's later work in relation to conservatism', in Brian F. McGuinness (ed.) *Wittgenstein and his Times*, Chicago: University of Chicago Press, 44–68.
Oliver, Gordon (1985) 'Fred Meyer opens store in East County', *Oregonian*, 23 November.
Olsson, Gunnar (1980) *Birds in Egg/Eggs in Bird*, London: Pion.
—— (1982) '–/–', in Peter Gould and Gunnar Olsson (eds) *A Search for Common Ground*, London: Pion 223–31.

—— (1987) 'The social space of silence', *Environment and Planning D: Society and Space* 5: 74–87.

—— (1988) 'The eye and the index finger: bodily means to cultural meaning', in Reginald Golledge *et al.* (eds) *A Ground for Common Search*, Goleta, CA: Santa Barbara, Geographical Press.

—— (1990) *Antipasti*, Göteborg: Korpen.

Oregon: Pictures to Remember her by (1985) New York: Crescent Books.

Ortona, E. (1980) 'The threat to vital Western resources', *Atlantic Community Quarterly* 18: 201–17.

Pahl, Raymond (1968) 'The rural–urban continuum', in *Readings in Urban Sociology*, Oxford: Pergamon, 263–305.

Palmer F.B. (1901) 'Apostrophe to Niagara', in P.A. Porter *Official Guide: Niagara Falls*, Buffalo: Matthews Northrup, 289–90.

Parliamentary Papers (1806) *Report and Minutes of Evidence on the State of Woollen Manufacture* III, testimony by Robert Cookson.

Peréz-Gomez, Alberto (1983) *Architecture and the Crisis of Modern Science*, Cambridge, Mass., and London: MIT Press.

Peters, Arno (1983) *The New Cartography*, New York: Friendship Press.

Pevsner, Nicholas (1975) *Pioneers of Modern Design*, New York: Penguin.

Philip, Mark (1985) 'Michel Foucault', in Quentin Skinner (ed.) *The Return of Grand Theory in the Human Sciences*, Cambridge: Cambridge University Press, 65–82.

Pickles, John (1985) Review of A. Gatrell, *Distance and Space: a Geographical Perspective* in *Annals of the Association of American Geographers* 75: 443–6.

—— (1987) *Geography and Humanism*, Concepts and Techniques in Modern Geography 44, Norwich: Geo Books/Elsevier.

Pinochet Ugarte A. (1968) *Geopolitica. Diferentes etapas para el estudio geopoliticos de los estados*, Santiago, Chile: Biblioteca del oficial.

—— (1981) *Introduction to Geopolitics* (English-language edn), Santiago, Chile: Andres Bello.

Pion-Berlin, D. (1989) 'Latin American national security doctrine: hard and soft-line themes', *Armed Forces and Society* 15: 411–29.

Pittman, H. (1981) 'Geopolitics in the ABC Countries: a Comparison', unpublished Ph.D. thesis, American University, Ann Arbor, Mich.: University Microfilms 82–04557.

Porter, Phil, and Voxland, Phil (1986) 'Distortion in maps: the Peters projection and other devilments', *Focus* 36: 22–30.

Portland City Council (1986) hearing transcript, 29 January, Portland, Oregon.

Pred, A. (1967) *Behavior and Location* 1, Lund: Gleerup.

—— (1988) 'Lost words as reflections of lost worlds', in Reginald G. Golledge, Helen Couclelis and Peter Gould (eds) *A Ground for Common Search*, Goleta, CA: Santa Barbara Geographical Press, 138–47.

Preston, T.R. (1845) 'Remarks on Niagara', in W. Barham (ed.) *Descriptions of Niagara*, Gravesend: The Compiler.

Price, R. (1978) *U.S. Foreign Policy in Sub-Saharan Africa: National Interests and Global Strategy*, Berkeley: Institute of International Studies.

—— (1981) 'Can Africa afford not to sell minerals?' *New York Times*, 18 August.

—— (1982) 'U.S. policy towards southern Africa: interests, choices and constraints', in G. Carter and P.O'Meara (eds) *International Politics in Southern Africa*, Bloomington: Indiana University Press, 45–88.

Pringle, P. (1986) 'View from Washington', in E.P. Thompson and N. Kaldor (eds) *Mad Dogs: the U.S. Raids on Libya*, London: Pluto Press, 54–64.

Public testimony before the Portland Planning Hearings Officer: Fred Meyer Inc. Comprehensive Plan Amendment Request, November 1985, Portland, Oregon.

Putnam, Hilary (1977) 'Is semantics possible?' in Stephen Schwartz (ed.) *Naming Necessity, and Natural Kinds*, Ithaca, N.Y.: Cornell University Press, 102–18.

Quam, L.Q. (1943) 'The use of maps in propaganda', *Journal of Geography* 42, 1: 21–32.

Rabinow, Paul, ed. (1984) *The Foucault Reader*, New York: Pantheon.

—— (1986) 'Representations are social facts', in James Clifford and George Marcus (eds) *Writing Culture: the Poetics and Politics of Ethnography*, Berkeley: University of California Press.

Ratzel, F. (1896) 'Die Gesetze des räumlichen Wachstums der Staaten: ein Beitrag zur wissenschaftlichen politischen Geographie', *Petermanns Mitteilungen* 42: 97–107.

—— (1901) *Der Lebensraum. Eine Biogeographische*, Tübingen.

Reboratti, C.E. (1983) 'El encanto de la oscuridad: notas acerca de la geopolítica en la Argentina', *Desarrollo Económico* 23: 137–144.

Relph, Edward (1976) *Place and Placelessness*, London: Pion.

Richards, I.A. (1936) *The Philosophy of Rhetoric*, London: Oxford University Press.

Ricoeur, P. (1971) 'The model of the text: meaningful action considered as a text', *Social Research* 38: 529–62.

Rimmer, W.G. (1960) *Marshall's of Leeds, Flax Spinners, 1788–1886*, Cambridge: Cambridge University Press.

Ripley's 'Believe it or Not!' Museum (n.d.), promotional brochure.

Robinson, A.H. (1985) 'Arno Peters and his new cartography', *American Cartographer* 12: 103–11.

—— and Petchenik, B.B. (1975) 'The map as a communication system', *Cartographic Journal* 12, 11: 7–15.

—— (1976) *The Nature of Maps: Essays toward Understanding Maps and Mapping*, Chicago: University of Chicago Press.

Robinson, Arthur H., Sale, Randall, D., Morrison, Joel L., and Muehrcke, Phillip C. (1984) *Elements of Cartography*, fifth edition, New York: Wiley.

Roemer, J. (1982) *A General Theory of Exploitation and Class*, Cambridge, Mass.: Harvard University Press.

Rolph, T. (1836) *A brief account, together with observations during a visit in the West Indies, and a tour through the United States of America, in parts of the years 1832–3*, Dundas, Upper Canada: Hackstaff.

Rorty, Richard (1979) *Philosophy and the Mirror of Nature*, Princeton, N.J.: Princeton University Press.

—— (1982) *The Consequences of Pragmatism: Essays, 1972–80*, Minneapolis: University of Minnesota Press.

—— (1985) 'Texts and lumps', *New Literary History* 17: 1–16.

—— (1989) *Contingency, Irony, and Solidarity*, Cambridge: Cambridge University Press.

Rosenberg, A. (1979) 'Can economic science explain everything?' *Philosophy of the Social Sciences* 9: 509–29.

—— (1985) *The Structure of Biological Science*, Cambridge: Cambridge University Press.

Roszak, Theodore (1972) *Where the Wasteland Ends: Politics and Transcendence in Post-industrial Society*, New York: Doubleday.

Rouse, Joseph (1987) *Knowledge and Power: toward a Political Philosophy of Science*, Ithaca, N.Y.: Cornell University Press.

Rowell, G. (1974) *Hell and the Victorians*, Oxford: Clarendon Press.

Roy, W.T. (1980) 'South Africa and the Indian Ocean', *South Africa International* II: 191–8.

Rozen, M. (1985) 'Maximizing behavior: reconciling neoclassical and x-efficiency approaches', *Journal of Economic Issues* 19: 661–85.

Rybczynski, Wytold (1986) *Home: a Short History of an Idea*, New York: Penguin.
Rylance, R., ed. (1987) *Debating Texts: a Reader in Twentieth-century Literary Theory and Method*, Milton Keynes: Open University Press.
Saarinen, Thomas F. (1988) 'Centering of mental maps of the world', *National Geographic Research* 4, 1: 112–27.
Said, E. (1979) *Orientalism*, New York: Vintage.
Santayana, George (1955, original 1896) *The Sense of Beauty: Being the Outline of Aesthetic Theory*, New York: Dover Publications.
Sapir, Edward (1924) 'Culture, genuine and spurious', *American Journal of Sociology* 29: 401–29; reprinted in E. Sapir, *Culture, Language, and Personality*, Berkeley: University of California Press, n.d., 78–119.
Saul, J., and Gelb, S. (1986) *The Crisis in South Africa*, revised edition, New York: Monthly Review Press.
Sayer, A. (1976) 'A critique of urban modelling: from regional science to urban and regional political economy', *Progress in Planning* 6: 187–254.
—— (1989) 'The "new" regional geography and problems of narrative', *Environment and Planning* D: *Society and Space* 7, 3: 253–76.
Seibel, G., ed. (1967) *Niagara Falls, Canada: a History*, Niagara Falls: Kiwanis Club of Stamford, Ontario.
Severance, F.H. (1914) *Peace Episodes on the Niagara*, Publications of the Buffalo Historical Society XVIII, Buffalo, N.Y.: Buffalo Historical Society.
—— (1917) *An old Frontier of France*, Publications of the Buffalo Society XXI, Buffalo: Buffalo Historical Society.
Shanes, E. (1979) *Turner's Picturesque Views in England and Wales, 1825–38*, London: Chatto & Windus.
—— (1981) *Turner's Rivers, Harbours and Coasts*, London: Chatto & Windus.
Sharrock, W.W., and Anderson, R.J. (1985) 'Criticizing forms of life', *Philosophy* 60: 394–400.
Silverman, K. (1983) *The Subject of Semiotics*, New York: Oxford University Press.
Simmel, Georg (1971) 'The metropolis and mental life', in David Levine (ed.) *Georg Simmel on Individuality and Social Forms*, Chicago: University of Chicago Press, 324–39 (German original, 1903).
Skinner, Quentin, ed. (1985) *The Return of Grand Theory in the Human Sciences*, Cambridge: Cambridge University Press.
Smith, N. (1984) *Uneven Development*, Oxford: Blackwell.
Snyder, John P. (1988) 'Social consciousness and world maps', *Christian Century*, 24 February: 190–2.
Soffner, H. (1942) 'War on the visual front', *American Scholar* 11: 465–76.
Soja, E. (1989a) 'Modern geography, Western Marxism and the restructuring of critical social theory', in N.J. Thrift and R. Peet (eds) *New Models in Human Geography*, London: Unwin Hyman, 318–47.
—— (1989b) *Postmodern Geographies: the Reassertion of Space in Critical Social Theory*, London: Verso.
Sontag, S. (1978) *Illness as Metaphor*, New York: Farrar Straus & Giroux.
Sorre, Max (1962) 'The concept of "genre de vie"', in Philip K. Wagner and Marvin Mikesell (eds) *Readings in Cultural Geography*, Chicago: University of Chicago Press, 339–415 (original, 1948).
Speier, H. (1942) 'Magic geography', *Social Research* 8: 310–30.
Stack, J.C. (1985) 'Filosofía y biología: fundamentos de la geopolítica contemporánea', *Revista Chilena de Geopolítica* 3: 7–13.
Steiner, George (1971) *In Bluebeard's Castle: some Notes towards the Redefinition of Culture*, New Haven: Yale University Press.

—— (1978) *On Difficulty, and other Essays*, New York: Oxford University Press.

Stepan, A. (1971) *The Military in Politics: Comparative Patterns in Brazil*, Princeton: Princeton University Press.

Stevenson, W. (1812) *General View of the Agriculture of Dorset*, London: Board of Agriculture.

Stock, Brian (1986) 'Texts, readers and narratives', *Visible Language* XX, 3: 89–95.

—— (1990) *Listening for the Text: on the Uses of the Past*, Baltimore: Johns Hopkins University Press.

Stoddart, D.R. (1967) 'Organism and ecosystem as geographical models', in R.J. Chorley and P. Haggett (eds), *Models in Geography*, London: Edward Arnold, 511–48.

Sugiura, N. (1983) 'Rhetoric and Geographers' Worlds: the Case of Spatial Analysis in Human Geography', unpublished doctoral dissertation, Pennsylvania State University.

Szegö, Janos (1987) *Human Cartography: Mapping the World of Man*, trans. Tom Miller, Stockholm: Swedish Council for Building Research.

Taylor, Charles (1980) 'Theories of meaning', *Man and World* 13: 281–302.

Taylor, P. (1985) *The Smoke Ring: Tobacco, Money and Multinational Politics*, London and Sydney: Sphere Books.

Theweleit, K. (1987) *Male Fantasies* 1 *Women, Floods, Bodies, Histories*, Minneapolis: University of Minnesota Press.

Thomas, L.B. (1949) 'Maps as instruments of propaganda', *Surveying and Mapping* 9, 2: 75–81.

Thompson, G.R., ed. (1974) *The Gothic Imagination: Essays in Dark Romanticism*, Pullman, Washington: Washington State University Press.

Thompson, J.B. (1981) *Critical Hermeneutics: a Study in the Thought of Paul Ricoeur and Jurgen Habermas*, Cambridge: Cambridge University Press.

Time Magazine (1965) 'Resorts: let's go again to Niagara', 18 June.

Toynbee, A. (1976) 'Man's concern with life after death', in A. Toynbee (ed.) *Life after Death*, New York: McGraw-Hill.

Trollope, A. (1862) *North America*, London: Chapman & Hall.

Tudor, H. (1834) *Narrative of a Tour in North America*, London: Duncan.

Tyler, S. (1987a) *The Unspeakable: Discourse, Dialogue, and Rhetoric in the Postmodern World*, Madison: University of Wisconsin Press.

—— (1987b) 'Ethnography, intertextuality, and the end of description', in *The Unspeakable: Discourse, Dialogue and Rhetoric in the Postmodern World*, Madison: University of Wisconsin Press.

Tyner, J.A. (1974) 'Persuasive Cartography: an Examination of the Map as a Subjective Tool of Communication', unpublished Ph.D. dissertation, Department of Geography, University of California, Los Angeles.

Ulmer, G.L. (1985) *Applied Grammatology: Post(e)-Pedagogy from Jacques Derrida to Joseph Beuys*, Baltimore: Johns Hopkins University Press.

United Nations (1945) *Charter of the United Nations*, New York: Office of Public Information.

United States Department of State (1969–88) *Bulletin of the United States Department of State*, Washington, D.C.: U.S. Government Printing Office.

United States Senate (1985) *U.S. Policy towards South Africa*, Hearings before the Committee on Foreign Relations, United States Senate, 99th Congress, 1st Session, 24 April, 2 and 22 May 1985, Washington D.C.: US Government Printing Office.

Vidal de la Blache, Paul (1911) 'Les genres de vie dans la géographie humaine', *Annales de Géographie* 20: 193–212, 289–304.

—— (1928) *The Personality of France*, trans. H.C. Brentnall, London: Knopf (original, 1911).

Villeneuve, Paul (1984) 'Pour une géographie des genres de vie urbains', presented at the

annual meeting, Canadian Association of Geographers, Quebec City.
Volkmer, Walter (1969) *The Liberal Tradition in American Thought*, New York: Putnam.
Von Chrismar, J. (1968) *Geopolíticia: leyes que se deducen del estudio de la expansion de las estados*, Santiago: Biblioteca del oficial.
Von Wright, Georg Henrik (1968) 'An essay in deontic logic and the general theory of action', *Acta Philosophica Fennica* 21: 137–51.
Vujakovic, Peter (1989) 'Arno Peters' cult of the "new cartography": from concept to world atlas', *Bulletin of the Society of University Cartographers* 22, 2: 1–6.
Wahl, François (1980) 'Le désir d'espace', in *Cartes et figures de la terre*, Paris: Centre George Pompidou.
Walker, G. (1813 [1814]) *The Costume of Yorkshire.* Leeds: Robinson & Holdsworth [London: Longman].
Walker, R. (1989) 'What's left to do? Some principles to live by', *Antipode*, forthcoming.
—— and Greenburg, Douglas (1982) 'Post-industrialism and political reform in the city: critique', *Antipode* 14, 1: 17–36,
Walker, W. (1980) *The next Domino?*, Sandton, South Africa: Valient House.
Wallas, G. (1921) *Human Nature in Politics*, New York: Knopf.
Wallin, Erik (1980) *Vardagslivets generativa grammatik: vid gränsen mellan natur och kultur*, Lund: Liber.
Wallis, Helen M., and Robinson, Arthur H., eds. (1987) *Cartographical Innovations: an International Handbook of Mapping Terms to 1900*, Tring: Map Collector Publications and International Cartographic Association.
Ward, J.W. (1886) 'To Niagara', in *Niagara River and Falls*, Buffalo: Fryer.
Warde, A. (1985) 'Spatial change, politics and the division of labour', in D. Gregory and J. Urry (eds) *Social Relations and Spatial Structures*, London: Macmillan, 190–212.
Warner, C.D. (1897) *Their Pilgrimage*, New York: Harper.
Warren, M. (1987) 'The Marx–Darwin question: implications for the critical aspects of Marx's social theory', *International Sociology* 2: 251–69.
Watson, John Richardson (1970) *The Picturesque Landscape in English Romantic Poetry*, London: Hutchinson.
Weber, Max (1949) *Max Weber on the Methodology of the Social Sciences*, Glencoe, Ill.: Free Press.
Weigert, H.W. (1941) 'Maps are weapons', *Survey Graphic*, October: 528–30.
—— (1942) *Generals and Geographers: the Twilight of Geopolitics*, New York: Oxford University Press.
Wellsteed, J. (1849) 'The Falls of Niagara', *Western Literacy Messinger*, July.
Whitaker, T.D. (1816) *Loidis and Elmete*, Leeds: Robinson & Holdsworth.
White, H. (1973) *Metahistory: the Historical Imagination in Nineteenth-century Europe*, Baltimore: Johns Hopkins University Press.
—— (1978) *Tropics of Discourse: Essays in Cultural Criticism*, Baltimore: Johns Hopkins University Press.
Wilbor, W.C. (1907) *Ode to Niagara*, Buffalo: Brinkworth.
Williams, D. (1976) 'The hills of Niagara', *Buffalo Currier-Express*, 28 November.
Wilson, R.G. (1971) *Gentlemen Merchants: the Merchant Community in Leeds*, Manchester: Manchester University Press.
Winner, Langdon (1980) 'Do artifacts have politics?', *Daedalus* 109, 1: 121–36.
Winningstad, C.N. (1985) Correspondence with Oran Robertson, chairman of Fred Meyer Inc., 4 November 1985.
Wirth, Louis (1938) 'Urbanism as a way of life', *American Journal of Sociology* 44: 1–24.
—— (1969, original 1956) 'Rural–urban differences', in Richard Sennett (ed.) *Classic Essays in the Culture of Cities*, New York: Appleton-Century-Crofts, 165–9.
Wittgenstein, Ludwig (1961) *Tractatus Logico-philosophicus*, trans. D.F. Pears and Brian

F. McGuinness, London: Routledge (original, 1921).
—— (1967) *Zettel*, ed. G.E.M. Anscombe and G.H. Von Wright, Oxford: Blackwell.
—— (1968) *Philosophical Investigations*, third edition, trans. G.E.M. Anscombe, New York: Macmillan (German original, 1945, 1947–9; first English translation, 1953).
—— (1971) *'Remarks on Frazer's Golden Bough'*, in *The Human World*, trans. A.C. Miles, (original, 1931).
—— (1983) *Remarks on the Foundations of Mathematics*, revised edition, ed. Georg Henrik Von Wright, Rush Rhees and G.E.M. Anscombe, Cambridge, Mass.: MIT Press (original, 1937–44).
Wolch, J., and Dear, M., eds. (1989) *The Power of Geography: how Territory shapes Social Life*, Boston, Mass.: Unwin Hyman.
Wolfe, Bryan J. (1982) *Romantic Revision*, Chicago: University of Chicago Press.
Wolfe, Tom (1981) *From Bauhaus to our House*, New York: Farrar Straus & Giroux.
Wolter, John A. (1975) 'The Emerging Discipline of Cartography', unpublished Ph.D. dissertation, University of Minnesota.
Wood, Denis (1987) 'Pleasure in the idea/The atlas as narrative form', in R.J.B. Carswell, G.J.A. de Leeuw and N.M. Waters (eds) *Atlases for Schools: Design Principles and Curriculum Perspectives, Cartographica* 24, 1: 24–45 (Monograph 36).
—— and Fels, John (1986) 'Designs on signs/Myth and meaning in maps', *Cartographica* 23, 3: 54–103.
Woods, N.A. (1861) *The Prince of Wales in Canada and the United States*, London: Bradbury & Evans.
Younghusband, Sir Francis (1920) 'Natural beauty and geographical science', *Geographical Journal* 61: 1–13.
Zaring, J. (1977) 'The romantic face of Wales', *Annals of the Association of American Geographers* 67: 397–418.
Zumbach, Clark (1984) 'Artistic function and the intentional fallacy', *American Philosophical Quarterly* 21: 147–56.

INDEX